Water Quality Management

The global attention in recent years has focused primarily on water quantity and allocation issues. Water quality has received significantly less attention than water quantity. Commendable progress has been made by the developed world to control point sources of pollution, but commensurate progress in reducing non-point sources has not been made. In the third world countries both point and non-point sources of pollution are becoming increasingly a serious concern. Already, nearly all water bodies in such countries near and around urban centres have been severely polluted, with very high health and environmental costs.

The book assesses the current status of water quality management in both developed and developing worlds, as well as analysing the effectiveness of economic instruments and legal and institutional frameworks to control water contamination. It outlines the importance of building up social and political awareness to reverse the trend of continuing water quality deterioration, which is likely to be a most challenging task in the coming years.

This book was published as a special issue of *International Journal of Water Resources Development*.

Asit K. Biswas is the founder and Chief Executive of the Third World Centre for Water Management, Mexico. He is a Distinguished Visiting Professor at the Lee Kuan Yew School of Public Policy, Singapore, and Indian Institute of Technology, Bhubaneswar.

Cecilia Tortajada is the Scientific Director of the International Centre of Water and Environment, Zaragoza, Spain, and Past President of the International Water Resources Association.

Rafael Izquierdo is the Director of the Aragon Water Institute, Zaragoza, Spain.

Water Quality Management
Present Situations, Challenges and Future Perspectives

Edited by
Asit K. Biswas, Cecilia Tortajada and Rafael Izquierdo

LONDON AND NEW YORK

First published 2012
by Routledge
2 Park Square, Milton Park, Abingdon, Oxfordshire OX14 4RN

Simultaneously published in the USA and Canada
by Routledge
711 Third Avenue, New York, NY 10017

First issued in paperback 2014

Routledge is an imprint of the Taylor & Francis Group, an informa business

© 2012 Taylor & Francis

This book is a reproduction of the *International Journal of Water Resources Development*, vol. 27, issue 1. The Publisher requests to those authors who may be citing this book to state, also, the bibliographical details of the special issue on which the book was based.

All rights reserved. No part of this book may be reprinted or reproduced or utilised in any form or by any electronic, mechanical, or other means, now known or hereafter invented, including photocopying and recording, or in any information storage or retrieval system, without permission in writing from the publishers.

Trademark notice: Product or corporate names may be trademarks or registered trademarks, and are used only for identification and explanation without intent to infringe.

British Library Cataloguing in Publication Data
A catalogue record for this book is available from the British Library

ISBN 978-0-415-68885-7 (hbk)
ISBN 978-1-138-79807-6 (pbk)

Typeset in Times New Roman
by Taylor & Francis Books

Disclaimer
The publisher would like to make readers aware that the chapters in this book are referred to as articles as they had been in the special issue. The publisher accepts responsibility for any inconsistencies that may have arisen in the course of preparing this volume for print.

Contents

Foreword: Water Quality, the Challenge of the Future
Alfredo Boné   1

1. Water Quality Management: An Introductory Framework
Asit K. Biswas & Cecilia Tortajada   5

2. Water—A Reflection of Land Use: Understanding of Water Pathways and Quality Genesis
Malin Falkenmark   13

3. Impact of Agriculture on Water Pollution in OECD Countries: Recent Trends and Future Prospects
Kevin Parris   33

4. Regulating Nonpoint Source Water Pollution in a Federal Government: Four Case Studies
Susan Graham, Adam Schempp & Jessica Troell   53

5. Introduction to Environmental and Economic Consequences of Hypoxia
Robert J. Díaz & Rutger Rosenberg   71

6. Financing Water Quality Management
Céline Kauffmann   83

7. Water Governance in Aragon
Rafael Izquierdo   101

8. Water Management in the Ebro River Basin: An Approach to the 2010–15 Hydrological Plan
Manuel Omedas-Margelí   119

9. Water Quality in Zaragoza
Javier Celma   149

10. Water Quality Management in China: The Case of the Huai River Basin
Jun Xia, Yong-Yong Zhang, Chesheng Zhan & Ai Zhong Ye   167

11. Water Quality Management in Egypt
Safwat Abdel-Dayem   181

CONTENTS

12. A New Mindset for Integrated Water Quality Management for South Africa
    L. Boyd & R. Tompkins — 203

13. Water Quality and Health in Poor Urban Areas of Latin America
    Maria Onestini — 219

14. Conceptual Framework for Protecting Groundwater Quality
    C. Martínez-Navarrete, A. Jiménez-Madrid, I. Sánchez-Navarro, F. Carrasco-Cantos & L. Moreno-Merino — 227

15. Evolution of Water Management in Mexico
    Felipe I. Arreguín Cortés & Enrique Mejía Maravilla — 245

16. Agriculture and Water Pollution: Farmers' Perceptions in Central Mexico
    Rosario Perez-Espejo, Alonso Aguilar Ibarra & Jose Luis Escobedo-Sagaz — 263

Index — 275

FOREWORD

Water Quality, the Challenge of the Future

In November 2009, the International Centre for Water and Environment (CIAMA, in its Spanish acronym) of the Ministry of Environment of the Government of Aragon in Spain once again became the centre of the world for the analysis of experiences on the management of water resources worldwide with the 3rd International Meeting of Experts. The meeting brought together about 30 of the most recognized water-management scientists from all continents of the world to discuss a topic as exciting as it is necessary nowadays: "Management of water quality: challenges and expectations". This meeting, the result of which appears in this issue, made it clear that future challenges regarding water availability around the world are not so much related to the quantity of water available, but to its quality.

Water is a vitally important resource linked to the future development of the world: society's well-being depends completely on it being managed appropriately. We all know that without water there is no life.

We live in a society which, since the second half of the 20th century, has adopted a model of growth based on a notion of progress which often involves the unsustainable development of resources. This has led to great social inequality and considerable deterioration of the environment. We must be very clear that unlimited growth is not compatible with sustainable development.

Water management is one of the 21st century's most critical challenges on a world scale, fulfilling an essential role in reducing poverty and promoting economic development. Water management is one of the objectives of this millennium: guaranteeing the sustainability of the environment and reducing by half the number of people with no access to improved drinking water. This is why integrated, sustainable and equitable management of water is absolutely essential.

I would like to highlight some aspects discussed during the meeting, and which Asit K. Biswas, President of the Third World Centre for Water Management and winner of the 2006 Stockholm Water Prize, explained in the closing session. One of them is that the characteristics of each hydrographical basin and each country around the world are different and, therefore, the instruments developed and implemented to improve water quality must consider the characteristics of each specific location if they are to have a positive impact. We thus face the challenge of developing the tools needed, and improving the means for the exchange of solutions and experiences in terms of water management.

On the other hand, I believe it is important to emphasize the fact that the different attempts made by the United Nations to achieve progress in terms of drinking water supply and its treatment around the world have, for the most part, been in vain. This makes it even more necessary for us, as regions and states, to consider all the experiences gained from this and other similar meetings, and turn them into specific and relevant actions to disseminate solutions and generate a genuine impact on the ground. Water management is a global problem that can only be addressed from national standpoints and with solutions implemented at a local level.

As far as the challenge we have before us of improving the tools for exchange of experiences and solutions in water management is concerned, I can say that in Aragon we have used, and have exported in recent years, the innovative experiences implemented in the region to improve the quality of the waters. Addressing the comprehensive treatment of wastewater in a territory such as our Autonomous Region, which is extensive, geographically disperse and has complicated terrain, was a complex undertaking; it is, however, becoming a reality thanks to the major wastewater management plans implemented. These programmes have not only generated interest from numerous countries and regions, but have also been recognized by institutions such as the Organisation for Economic Co-operation and Development (OECD), which has disseminated them to country members.

From the moment I took on my responsibility as Minister of the Environment of the Government of Aragon in 2002, improving water quality has been more than a priority objective: it is my main goal. This is why we have implemented ambitious and complex global water-treatment plans, and developed innovative infrastructures for the proper management of animal solid waste as an attempt to reduce the contaminating load characterized by a large number of highly intensive pig-farming operations. In summary, there are hundreds of projects under way which, I can state with great satisfaction, have already become a reality and are bearing positive fruit in the quality of water of our rivers.

Our commitment has been translated into actions that have included not only the development of projects, but also raising awareness and disseminating information. From our modest position as a region in the world, in Aragon it has always been our objective to be a spokesman of the water needs of the different regions. Firstly, because water is a need which is intrinsic to this region, something we can legitimately claim leads to future development. And secondly, because we firmly believe that Zaragoza and Aragon should continue to be a forum for world debate on water and sustainable development, we continue with the scientific and intellectual legacy bequeathed by the Zaragoza International Exhibition of 2008.

This International Meeting of Experts, the first two sessions of which were held in 2006 ("Water Management Beyond the Year 2020") and in 2007 ("30 Years from Mar del Plata: Achievements and Expectations"), represents a further effort by the Ministry of the Environment, through CIAMA, to perpetuate this legacy and continue to promote debates and stimulate the dissemination of best practices in the area of water. It is also an effort to continue working with national and international institutions and associations around the world.

I would like to take this opportunity to offer my sincere thanks for the commitment and work of the experts and organizations who have made it possible for us to move forward in our challenge: to the experts attending this and all the other events organized through CIAMA for their participation and for the valuable ideas and experiences with which they have contributed; to the International Water Resources Association (IWRA) and the Third World Centre for Water Management in Mexico as organizers of the event and examples of commitment and professionalism in disseminating knowledge and fighting to reduce inequalities around the world; to the staff and experts at the Ministry of the Environment; and last, but not least, special thanks to Asit K. Biswas and Cecilia Tortajada for their knowledge, wisdom and contributions to the Aragon government's International Centre for Water and Environment, and for their work, which has resulted in the scientific achievements of CIAMA.

Alfredo Boné
Minister of Environment of the Government of Aragon, Spain

Water Quality Management: An Introductory Framework

ASIT K. BISWAS* & CECILIA TORTAJADA**
*Third World Centre for Water Management, Estado de México, México; **International Centre for Water and Environment, Zaragoza, Spain

ABSTRACT *Much has been written and discussed in recent years on the water crisis and the belief that the world may run out of water in the foreseeable future. The main issue, however, is not physical scarcity of water but poor management. It is primarily a crisis due to mismanagement. An important result of such poor management practices has been the continual deterioration of water quality on a global basis. The main emphasis in the past and present has been on water quantity management, including allocation. Managing water quality is still not receiving adequate attention, because it is significantly more complex, difficult and expensive compared with water quantity management.*

Much has been written and said about how the world is facing a water crisis because of the physical scarcity of this resource, and how the present situation can only get worse in the coming years. In fact, numerous books and papers have been published during the past decade about this so-called water crisis due to physical scarcity. The media regularly carry frightening stories as to what the world will do when "it runs out of water", as if water was a non-renewable resource like oil, and "water wars". Without question, the water crisis has captured the attention of the media and a good section of the general public, as well as numerous water professionals. However, scientific and rational analyses of all the available data indicate that the world is not facing a crisis because of physical scarcities of water, but that it is indeed facing a crisis arising from continued poor management of its water resources.

From the United States to the United Arab Emirates, and throughout the continents of Asia, Africa and Latin America, past and existing water management practices and processes have left, and continue to leave much to be desired, and in all countries there is significant room for improvement. If this could be achieved, the world's water needs for all activities can be successfully met on a long-term basis. We already possess sufficient knowledge, technology, management expertise and capacity to ensure that the current water situation can be significantly improved. However, continuation of the current business-as-usual management practices will undoubtedly give rise to serious water problems in different parts of the world in the foreseeable future. Implementation of new, innovative and business-unusual practices, which are already known and are being used in some areas, can ensure that the world will not face a water crisis in the foreseeable future.

While much global attention in the past and more recently has been focused primarily on water quantity and allocation issues, poor water management has created a serious problem in terms of quality in many parts of the world. While developed countries have made significant and commendable progress in controlling point sources of pollution during the past three to four decades, commensurate progress has simply not occurred in developing countries. Consequently, most water bodies within and around the urban centres of Asian, African and Latin American developing counties are already heavily contaminated due to poor water management and widespread neglect of water quality considerations, both politically and socially. Unfortunately, there are no visible signs that the governments and people of these developing countries are aware of the seriousness of the problem or the dangers it poses to human health and ecosystems now and in the future. Thus, if a future water crisis develops, it will be due not to the physical scarcity of the resource, but because water quality is steadily deteriorating in most developing countries. Anecdotal evidence indicates that the health and environmental costs of such water quality deterioration are already in billions of dollars each year. These costs will continue to rise significantly at least during the 2011–2020 period, if not for much longer.

Worldwide, the situation is significantly worse with regard to nonpoint sources of pollution. Currently one would indeed be hard pressed to find a single developing country with a serious and implementable plan for managing nonpoint sources, let alone one with any record of making a serious and sustained effort for their control. Consequently, nitrogen (N), phosphorous (P), and potassium (K) contaminations of rivers, lakes and aquifers are everywhere increasing, albeit at a much higher rate in developing countries than in industrialized nations. Accordingly, and not surprisingly, most water bodies in and around areas where agriculture and animal husbandry are practised at intensive scales in developing countries already show serious signs of contamination and eutrophication.

Even in developed countries, although limited progress has been made in controlling NPK contaminations, without effective control of nonpoint sources it is impossible to ensure good ambient water quality conditions. In general, developed countries have not regulated nonpoint sources with the same zeal and vigor as for point sources. There are two main reasons for this difference. First, conceptually and technically, it is far more complex and difficult to regulate nonpoint sources compared to point sources. For example, at present there is simply no technology available which can measure pollutant discharges from a single farm unit with any degree of confidence. Accordingly, concepts such as effluent charges or tradable discharge permits, which may work quite well with point sources, can at best be of limited value for managing nonpoint sources. Equally, defining upper limits of contamination per unit of time has not been of much help. Usable and cost-effective policy instruments to regulate nonpoint sources simply do not exist at present. Thus, not surprisingly, the main causes of contamination of surface water in the developed world at present are from nonpoint sources.

Even though developing countries have given more attention to managing point sources of water contamination, compared with small or at most limited effort for controlling nonpoint sources, unfortunately the current situation even in terms of point sources is often very grim. For example, assessments done by the Third World Centre for Water Management indicate that at present no more than about 12–14 per cent of point sources of wastewater in Latin America are collected, adequately treated and then discharged to the environment in a safe manner. In other words, even in 2010, more than 85% of wastewater in Latin America is being inadequately treated (Biswas *et al.*, 2006). Such an independent and objective assessment simply does not exist for Africa and Asia at present, but it can be said with some degree of confidence that the situation in Asian developing countries is probably similar to Latin America, and it is probably somewhat worse for Africa. Even for a developed Asian country such as Japan, according to the Japan Sewage Works Association (2010), the percentage of the population of Japan with access to sewerage services increased from 8.3% 3 to 72.7% in 2008. This means that more than 27% of the Japanese population did not have access to wastewater treatment by 2008 (Figure 1).

The current global water quality situation raises some serious questions regarding the use and relevance of the Millennium Development Goals (MDGs) for water and wastewater management in the developing world. It should be noted that when the MDGs were proclaimed, no goal was proposed for wastewater treatment. This goal was added retrospectively in 2002 during the Johannesburg Summit.

There are some fundamental problems with the current wastewater goal, especially in terms of how it is being defined at present and also how it is being monitored and interpreted, both nationally and internationally. During the run up to the United Nations Water Conference held in Mar del Plata, Argentina, in 1977, which was instrumental in proposing the 1980s as the International Water Supply and Sanitation Decade, the word "sanitation" meant that wastewater would be collected from households, taken to a wastewater treatment plant where it would be properly treated, and thereafter discharged to the environment in a safe manner. However, this simple idea was subsequently corrupted very significantly so that in its current interpretation it bypasses any serious water quality management considerations.

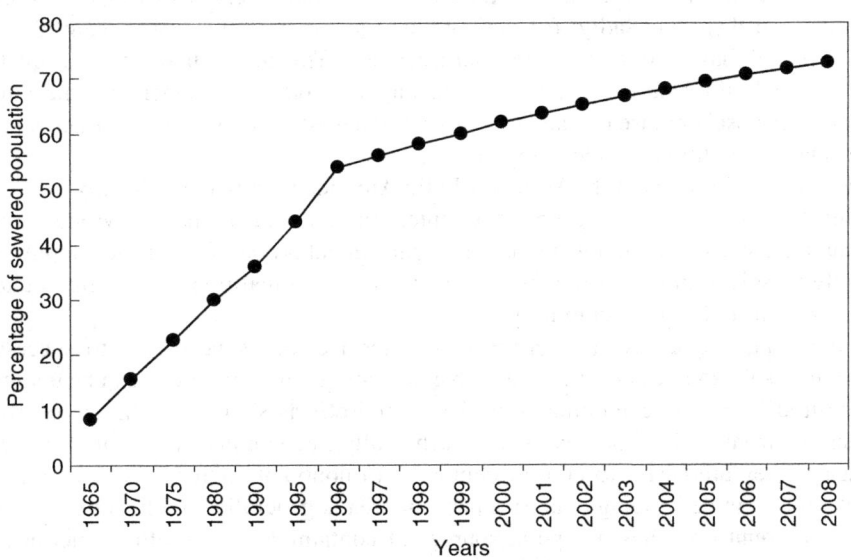

Figure 1. Sewered Population in Japan, 1965–2008. *Source*: Japan Sewage Works Association, 2010, Sewage Works in Japan, Tokyo.

At present, under the MDG interpretations, it is assumed that people have access to sanitation as long as wastewater is collected from their houses. Thus, inhabitants of cities such as Delhi or Mexico are supposed to have near total sanitation since the wastewater is taken away from their houses. Yet Delhi currently discharges almost all its wastewater into the river Yamuna virtually without any treatment, and Mexico City Metropolitan Area pumps its untreated wastewater down to Mezquital Valley. In spite of these unacceptable practices, which significantly damage the quality of surface and groundwater bodies, both international and national organizations claim that the two cities have almost full access to sanitation! This unfortunate situation and interpretation are widespread in the developing world, and condoned by various international organizations. Continuous discharge of untreated or partially treated wastewater has contributed to serious contamination of quality in nearly all water bodies in or around the urban centres of all developing countries because of the continuation of such unacceptable and unsafe practices. Yet the inhabitants are claimed by the various international bodies to have full access to sanitation. It is truly a sad and unacceptable situation which simply transfers the problem to outside the urban centres, and thus severely contaminates the water bodies that the downstream urban areas are forced to use.

According to the philosophy underpinning the Millennium Development Goals, the number of people who do not have access to sanitation should be reduced by half between 1990 and 2015. If the existing situation with wastewater management is correctly and objectively interpreted with regard to Latin America, it is unacceptable. In 1990, less than 10% of the Latin American population had access to proper wastewater treatment and disposal. By 2010, the overall situation had improved only incrementally to around 12–14%. On the basis of the current trends, significantly less than 20% of the Latin American population would have access to proper sanitation by 2015, when the Decade will end. In 2000, when the MDGs were proclaimed, the goal would have been that by 2015, 44% of the people in Latin America should have access to proper wastewater treatment and disposal. Sadly, for the continent as a whole, far fewer than 20% of its population will have access to wastewater treatment. This means that the water quality in most parts of Latin America will not show any appreciable improvement. On the contrary, continued disposal of untreated and partially treated wastewater can only further aggravate water quality conditions of receiving bodies.

The situation in terms of the MDG for Latin America is shown graphically in Figure 2. Continuation of the prevailing erroneous interpretation and the use of wrong statistical information mean that at best water quality management problems in the developing world can only be solved in an abstract and academic sense. In reality, the problem is likely to worsen significantly in the coming years.

In developed countries, a determined long-term effort is necessary to successfully control nonpoint sources in order that ambient water quality conditions can be maintained at a desired level. In developing countries, initial efforts should emphasize controlling discharges from point sources, since technically, economically, institutionally and politically they are easier to control compared to nonpoint sources. In addition, use of agrochemicals in developing countries per unit area is generally significantly less than in developed countries. Thus, nonpoint sources of contamination are often much less of a problem at present compared to industrialized countries.

No one single solution could address effectively and efficiently the different water quality concerns of all counties. Equally, different countries give different priorities to

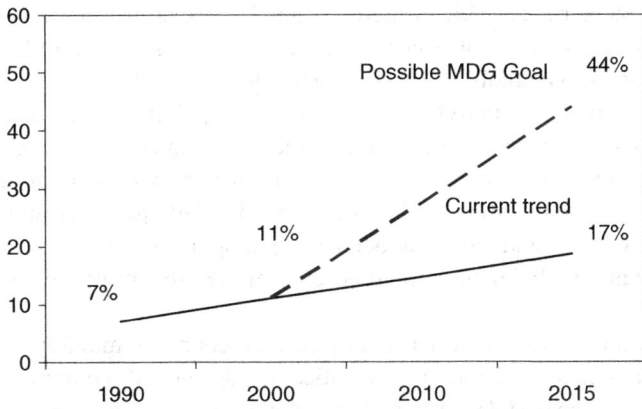

Figure 2. Appropriate MDG GOAL: Percentage of population in Latin America with wastewater treatment.

water quality management. Even for a single country, there is often no one single magic bullet that could address and solve the various water quality problems faced. Countries, and the various regions within a country, are often heterogeneous in terms of population densities, levels of industrial, agricultural and human activities, as well as physical climatic, economic, institutional and environmental conditions. Management and technical capacities may often vary significantly from one region to another, even within one specific country. Use of approaches such as command and control, public education and involvement, and use of economic and legal instruments are likely to be helpful under specific conditions. Equally, other requirements such as steps that could be taken to build up social and political awareness of the problems arising from continuing water quality deterioration, and pressure to take countermeasures to reduce the problem, could be helpful. A groundswell of public opinion that water quality is an important socio-political issue is necessary in order that country- and region-specific water quality management policies and programmes can be formulated and implemented.

What is often not realized at present is that water quality management is significantly more complex than water quantity management. In most countries, even now, institutionally and academically, water quality considerations often have significantly less weight than the importance accorded to water quantity issues. Technical and management capacities to control water pollution are often at a much lower level than for water quantity. Reversing these trends is likely to be difficult and challenging (Biswas, 2008).

The increasing complexity of water quality management can be demonstrated by a number of issues. Only one will be discussed here as an illustrative example: data requirements. For water quantity assessments only a few simple data need to be collected, such as cross-sections of channels and flow velocities. The total number of parameters for which data need to be collected are few, and these parameters do not change with time. In contrast, for water quality management, the type of data required varies with time, geographical locations, nature of pollutants being discharged, or likely to be discharged, and their potential impacts on human health and the environment, as well as a variety of other factors.

As people, especially in some select developed countries, have become more aware of the health and the environmental impacts of water quality deterioration and the proliferation of industrial activities which continually discharge old and new chemicals to the environment, the number of water quality parameters that need to be monitored has increased almost exponentially in recent years. The knowledge base has also increased significantly in terms of interrelationships between pollutants and human health. In addition, instrument capacities to detect extremely low doses of contamination in water bodies have improved tremendously, for example, it is now possible to detect very low doses of contamination in parts per billion compared to only parts per million a few years ago. Because of these and other associated developments, more and more new and emerging pollutants are being monitored because of their real or perceived adverse health impacts.

Accordingly, the number of water quality parameters being monitored in the drinking water supply systems of some major cities in developed countries has increased significantly during the past decades. Figure 3 shows one such case for the city of Ottawa, where the number of water quality parameters monitored between 1970 and 2000 increased nearly 15 times.

Such a significant increase in the monitoring of over 350 water quality parameters is an expensive process because of the sheer volume of data that has to be collected, analysed in sophisticated laboratories with state-of-the-art instruments, and the technical, management and administrative personnel required for different stages of collection, analysis of interpretation of data collected and their effective management. While major urban centres of developed countries can afford to establish such elaborate data collection, analysis, interpretation and management systems, most developing counties are finding it difficult to establish a successful and usable system which can regularly handle even around 25 water quality parameters. This is just one example of the complexities associated with a good water quality management system compared with maintaining only a water quantity management system (Biswas, 2007).

Because of the unsatisfactory status of water quality management in the world as a whole, the Department of Environment, Government of Aragon, Zaragoza, Spain, the

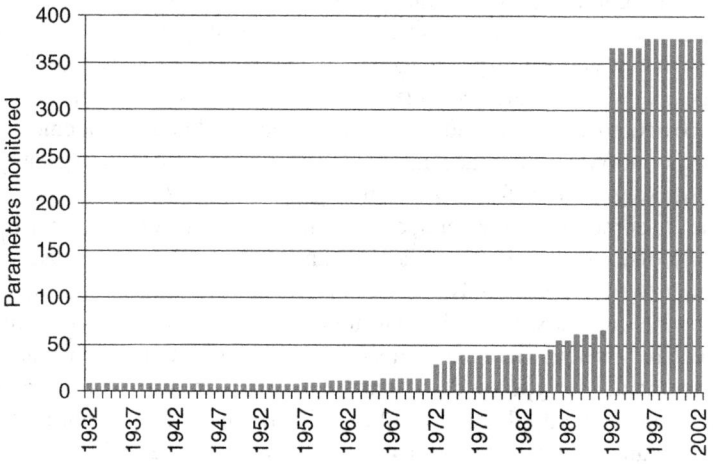

Figure 3. Number of water quality parameters monitored, Ottawa, 1932–2004. *Source*: Biswas (2007).

Third World Centre for Water Management, Mexico, and the International Water Resources Association organized an International Workshop on Water Quality Management in Zaragoza, Spain, in November 2009. Participation at this meeting was by invitation only, and leading experts from different areas of water quality management were invited from all over the world. Most of the current publication consists of the best papers prepared for this workshop including some excellent case studies from China by Jun Xia *et al.*; Egypt by Safwat Abdel-Dayem; Ebro River Basin by Manuel Omedas-Margelí; Zaragoza by Javier Celma; nonpoint sources of pollution by Kevin Parris and by Susan Graham *et al.*; financing requirements by Céline Kauffmann; environmental and economic consequences of hypoxia by Robert Díaz and Rutger Rosenberg; and an overall analysis of the water-land nexus by Malin Falkenmark. The workshop papers are supplemented by a few additional papers on the subject which very specifically complements them. We hope that the present publication will add to the current effort to improve water quality management on a global basis.

References

Biswas, A. K. (2007) Water as a human right in the MENA region: challenges and opportunities, *International Journal of Water Resources Development*, 23(2), pp. 209–225.
Biswas, A. K. (2008) Asian water development outlook: achieving water security for Asia, *International Journal of Water Resources Development*, 24(1), pp. 145–176.
Biswas, A. K., Tortajada, C., Braga, B. P. F & Rodríguez, D. (2006) *Water Quality Management in the Americas* (Berlin: Springer).
Japan Sewage Works Association (2010) *Sewage Works in Japan, 2010* (Tokyo: Japan Sewage Works Association).

Water—A Reflection of Land Use: Understanding of Water Pathways and Quality Genesis

MALIN FALKENMARK
SIWI and Stockholm Resilience Center, Stockholm, Sweden

ABSTRACT *The paper aims at a scientifically based synthesis of water quality genesis and pollution problems arising from human interventions in the landscape, physical as well as chemical. First, water quality genesis is explained in terms of sources, water pathways and some time scales involved. It goes on to look closer at chemical reactions along water pathways down a landscape catena, using the simple perception of a stream tube. The river quality outcome is explained in terms of a mix of water fractions with different hydrochemical signatures. Water quality is finally looked at in a 4000-year perspective, explaining some regional similarities and differences in the past. In looking towards the future, a potential further intensification and expansion in scale is seen as probable in response to driving forces at work, poor mitigation capabilities and the long response times involved.*

Introduction

Water is a unique solvent, chemically active and continuously moving in the water cycle. The implication of these phenomena is that every soluble chemical component that water encounters along its pathways through the landscape can be caught and carried by the moving water. The implications show up as water quality characteristics.

The landscape is a composite system where ecohydrological phenomena form important components. It can be thought of as a geological matrix, represented by its topography, its geomorphology, its mineral composition as well as the permeability characteristics of its various layers or components. When climatic forces such as rain and evaporation act on this matrix, the water flows are partitioned along pathways on the surface or in the ground. This has systematic impacts on hydrochemical and ecosystem characteristics. Land use affects determinants of water flow and can alter hydrochemical behaviour, for example by introducing pollutants along water pathways. For that reason, a land-use decision is also a water decision. Basically, water quality changes may result from altered water pathways, water flows, chemical determinants and introduced waste products.

In the past the scale and number of water quality and environmental changes have expanded rapidly. Environmental research initially focused on the biological consequences of disturbances in the landscape and the atmosphere. It was largely problem-driven and effect-oriented. Thus, impacts received more attention than the understanding of causes. Scientists were pushed by immediate reality more than pulled by their desire to understand the entire Earth system. The lack of interdisciplinary approaches has retarded both the necessary scientific and public understanding of water quality changes. Major obstacles to effective integrated system management have been the compartmentalization of problems and a concentration, one at a time, on narrowly focused problems. However, even when the water quality problem is well understood and the countermeasures well known, there is often an inability to act. The inaction is caused by such blocking effects as strong stakeholder interests, or—where environmental legislation is already in place—a lack of enforcement mechanisms, and a fragmented administration or bureaucracy. The public's expectations on today's politicians are tremendous. They have to balance between the need for curative action, on the one hand, by finding remedies to environmental problems; and preventive action, on the other hand, by trying to avoid letting the environmental problems become even worse.

This paper aims at a scientifically based synthesis of water quality genesis and pollution problems arising from human interventions in the landscape, physical as well as chemical. It builds on and extends from earlier papers in which the author has been lead author, treating water quality genesis and related issues (Falkenmark & Allard, 1989; Falkenmark & Mikulski, 1994; Falkenmark *et al.*, 1999; Falkenmark, 2000, 2005). It discusses water quality in terms of sources, water pathways and time scales involved. It looks closer at chemical reactions along water pathways down a landscape catena, using the simple perception of a stream tube. The river outcome is explained as a mix of water fractions with different hydrochemical signatures. Water quality is finally looked at in a 4000-year perspective with outlook to future water pollution-abatement challenges.

Multicause Water Pollution as Seen from an Overall Water Cycle Perspective

Pollutants Caught and Carried by the Cycling Water

Figure 1 provides an overall conceptual framework for a discussion of points of entry of multicause pollutants into the water cycle. It indicates the cycling of water from the sea to the atmosphere, to the continental land, to the water bodies and back to the sea. Basically

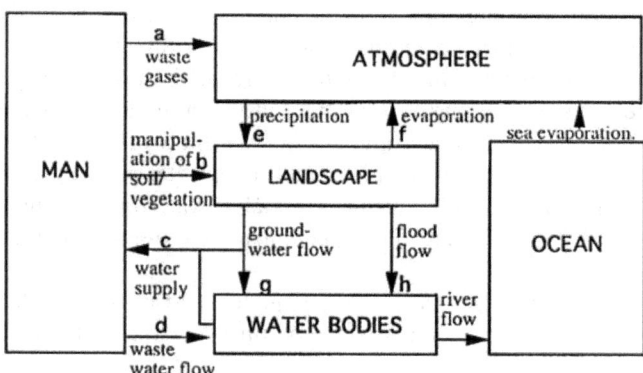

Figure 1. Introduction of water pollutants into the cycling water system: conceptual framework.

the atmosphere transports the evaporated seawater in a pattern of aerial water vapour fluxes. Over the continents the landscapes are being wettened by precipitation from the atmosphere. Water is being returned to the atmosphere both as part of the plant-production process (transpiration) and as direct evaporation from moist surfaces on foliage, on the ground and from water surfaces in the landscape. The remainder goes to recharge the groundwater aquifers or as rapid floods in rivers. Both modes of water flows arrive in the water bodies, defining their habitats. The water bodies are finally drained through larger rivers back to the sea.

Figure 1 also indicates the four principal ways in which man chemically intervenes in the system:

- through waste gas output to the atmosphere;
- through manipulation of the soil and vegetation in the landscape both physically and chemically: to stimulate biomass production by drainage, clearing and agricultural chemicals; and by dry waste deposits and spreading in the landscape of consumer products that have gone out of use, exposed to precipitation;
- by withdrawal of water from aquifers and water bodies for the purpose of water supply of households, municipal activities, industry and irrigation; and
- by returning the used water back to the water bodies as wastewater.

Two key processes are involved in which pollutants reach the water in the landscape: water as a leaching agent and water as a transport agent. The passing water may meet pollutants almost everywhere along its pathways: in the groundwater recharge area, during the transport phase and in the discharge area. Pollutants can be introduced into the mobile water in two main ways: directly through the introduction of waste products or other anthropogenic polluting substances, and indirectly from leaching or by altering the dissolution capacity, for example through acidification and changes in the water table. Pollutants can accumulate in soils and sediments for later release when geochemical conditions change for some reason. Chemical bombs constitute a real risk for unexpected water pollution and are understood only in their generalities.

Multiple Sources Operating in Parallel

In the real world, many causes will be operating in parallel. Measures to mitigate the ecological effects of stream water pollution may therefore have to address a whole set of parallel causes:

- The first type of cause is due to air exhausts of different origin caught by the water cycle and producing a wide cascade of ecological effects: initially producing effects on terrestrial ecosystems as water composition in the root zone is being influenced; the next step is the altered composition of groundwater in the recharge areas and of the flood water running towards the water bodies. When the composition of the water bodies changes, there will be disturbances in the aquatic ecosystems. In the last step changes will follow in the coastal waters and in the marine ecosystems.
- The second type of cause is due to land-based chemicals of different types: surplus fertilizers and pesticides, on the one hand, and dry waste deposits and landfills on the other. The pollutants are caught and carried by the water cycle producing pollution of soil moisture and therefore damage on terrestrial ecosystems; pollution of

groundwater with consequences for water supply and for wetlands fed by that water; pollution of water bodies and therefore damage to aquatic ecosystems; and finally pollution of coastal waters and therefore damage to marine ecosystems.
- The third type of water pollution cause is accumulations of human waste in the landscape in areas where the sanitation system has not yet been developed. The pollutants get caught by the water cycle and reach human populations which obtain their household water from local wells or water bodies. The consequences produced are water-related diseases.
- The fourth type of pollution source is the direct outlets of wastewater from households, villages and cities, on the one hand, and from industries, on the other hand. The effects are damage to aquatic ecosystems, but also to human health to the degree that the water supply depends on polluted water bodies.

Pollution Originating from Many Societal Sectors

Disturbances that impact on water quality have their origin in many sectors of society. Water flowing through a landscape's systems is also influenced by polluting activities upwind or upstream, e.g. urbanization, industry, agricultural intensity or tourism. The main relations between components that determine flow and human inputs are shown in Table 1 and compared with fields of human activities.

Due to water's many parallel functions, health risks to humans originate from multiple ways of human contact with polluted water: drinking water, polluted food and bathing. This is a reflection of water's many parallel functions:

- The carrier function by which water carries pollutants along, transmitting them to humans when humans come in contact with polluted water bodies.
- The health function, which makes water essential for human survival and an ingredient of everyday household life, and transmitting pollutants to humans through drinking water.
- The socio-economic production function which involves huge amounts of water being used in industry, transmitting pollutants to humans when using industrial products.
- The habitat function which makes biota vulnerable to water pollution, later transmitting pollutants to humans when eating seafood and fish.
- The biomass production function in agriculture which may transmit pollutants to irrigated crops and on to humans eating those crops.

Table 1. Result of human activity on the physical and chemical determinants of water flow.

Human activity	Altered physical sector flow determinants			Input of chemicals		
	Relief	Vegetation soil cover	Drainage density	Air	Land	Water
Urbanization	●	●	●	●	●	●
Industry	●			●	●	●
Agriculture	●	●	●	●	●	●
Forest management			●		●	
Tourism		●	●		●	●

Multiple Sources Complicate Water Pollution Mitigation

The conclusion from this principal discussion of pollution sources is that the same water body may suffer from effects from a whole set of pollution causes. Stream restoration is in other words an issue of addressing these and of possible sources of change in the chemical composition of the water. A systematic approach to stream restoration would therefore be needed in order to decide on precise objectives and priorities in relation to the desired state of the water body in question.

The natural *mitigation measures* to these different sources are to cut down altogether or at least to minimize the outputs from the first and fourth types of sources, benefiting from technologies with the effect of hindering the output of waste, and turning to new products or new production methods that leave less waste. The third cause may be mitigated by organized *sanitation* so that human waste is not left in the landscape. The most difficult type to find a solution to is the second type because of the importance of chemicals to stimulate biomass production to supply a growing world population with food, fodder, fibre, fuel-wood and timber. The goal must, however, be to minimize the output by adopting more environmentally sound agricultural practices, degradable pesticides and closing the cycles of consumer products so that dry waste deposits and landfills can be avoided as far as possible.

Key obstacles to water pollution mitigation are, however, linked to great complexity and numerous water quality determinants: river habitat degradation (damming, canalization, navigation locks, etc.); inputs of nutrients, toxic pollutants and biodegradable organic matter; and seasonal or complete streamflow depletion, i.e. loss of aquatic and riparian biota. Besides issues linked to land use, water use and urbanization within the river basin, climate change (e.g. rainfall reduction) and global warming (e.g. temperature increase in rivers and the melting of glaciers in high altitude headwaters) will add another set of constraints on the aquatic biota.

The timing of river impacts is specific to each river basin: some rivers have been impacted for hundreds years, others have just started. It is generally not possible to return to a pristine state and original biodiversity, since endemic species are lost and alien species introduced. The *time delays* to pollution mitigation efforts (Figure 2) may, however, be quite long; the period needed to get rid of a pollution problem is composed of two major components: the social response time, from the first suspicion to taking the necessary technical measures; and the hydrophysical response time, first the time it takes for pollution to build up by accumulation in a water body, and second the time it takes to flush away the accumulated pollutants once that their input has stopped—time delays which are both related to the renewal time of the water mass.

Water on the Move—An Aggressive Liquid Interacting with Surrounding Media

Main Ways of Interaction with the Physical Environment

Water interacts with the surrounding environment during its transport through the landscape above and below the ground surface. A set of dominating processes and reaction systems will determine the concentrations of major inorganic species in natural freshwater (Falkenmark & Allard, 1989):

- The pH of the water is determined through protolysis of the carbonic acid and the exchange with carbon dioxide from the atmosphere and biosphere.

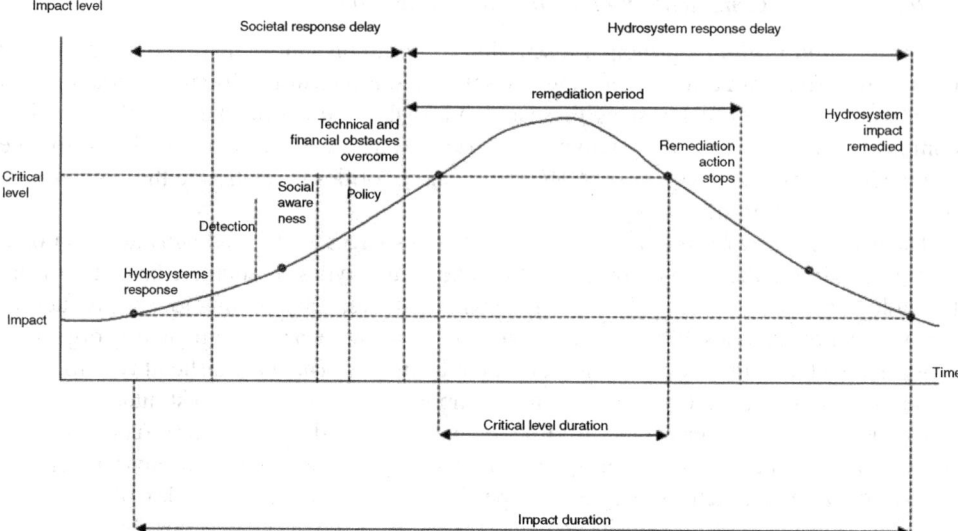

Figure 2. Response-time delays involved in water pollution abatement: societal response delay and hydrosystem response delay. *Source*: Adapted from Meybeck (2002).

- The redox processes are largely determined by the Fe(II)/Fe(III)-O_2-H_2 = system or possibly also the sulphur and nitrogen-related systems as well as by the degradation of organics.
- The production of cations (Ca, Mg, Na) is linked to the slow weathering of silicate minerals and the fast ion-exchange reactions in clay minerals with high cation-exchange capacities.
- The degradation of biogenic materials yields humates and eventually simple organics.

The water/ground system is in other words a *multi-component multiphase system* with a fairly limited number of major components. Reactions in the rootzone are generally fast and have a large impact on water quality. The natural system generally never reaches equilibrium. The interdependence of the various processes involved produces a buffering effect on the pH of the water, counteracting changes in redox potential and cation balance due to the presence of solid phases. The complex interrelationships between the multitude of processes in water/soil systems can explain why the variations in concentrations of the main components are quite small in most freshwater systems.

The major factor governing the transport of solutes in an aqueous system is the physical movement of the water itself. Since water follows different pathways representing various residence times and exchanges with different geological environments, the *mixing of water fractions* is an important factor influencing the resulting water quality at a given point. Solutes in the mobile system undergo various chemical and biological reactions along their pathways, or transformations related to chemical as well as biological and physical conditions in the system. The mere presence of solid surfaces tends to have a large and direct influence on the transport properties of the solute.

At the same time several processes are active in removing dissolved matter from the water: physical absorption, electrostatic absorption, chemisorption, chemical substitution and precipitation/co-precipitation.

The Stream Tube Perception

Water moves through a landscape in a manner defined by gravity, soil permeability and topography. There are basically *two directions of water throughflow*. On the hilltops and upper slopes, water moves away from the surface: it infiltrates the ground and feeds the roots while the surplus continues downwards to form groundwater, recharging the aquifers. On the bottom slopes, in local hollows and valley bottoms, water moves back to the land surface. This flow pattern divides the landscape into *recharge* as opposed to *discharge areas*. These areas vary in size as the groundwater table moves up and down between wet and dry seasons.

Three landscape elements have therefore to be distinguished: those constantly in recharge or discharge areas, respectively, and those where the upper profile shifts between these two conditions. At low runoff rates, the discharge area is only a small fraction of the total basin area, while at higher discharges it increases to a much larger fraction.

The subsurface flow can be thought of as taking place in stream tubes with their entrances in the recharge area and their mouths at the ground surface in the discharge area (Eriksson, 1985). The stream tubes of a basin can be seen as a rather well-ordered bundle of tubes, describing a flow pattern that is fixed for a given topography of the groundwater table. The time taken for a water quantity to flow through such a tube is its transit time.

There may be different categories of stream tubes (Figure 3):

- shallow pathways with short transit times of a few weeks, infiltrated on the slopes;
- intermediate pathways with transit times of months to a year, passing through the upper levels of the ground and infiltrating higher up on the slopes; and
- deeper pathways with long transit times of up to several decades, passing through less conductive layers, and infiltrating on the hilltops.

The fraction of the total discharge belonging to these respective categories depends on factors determining the flow capacity of the ground, and those determining the rate by which water has to leave the basin. When glacial till is located on crystalline rocks, part of

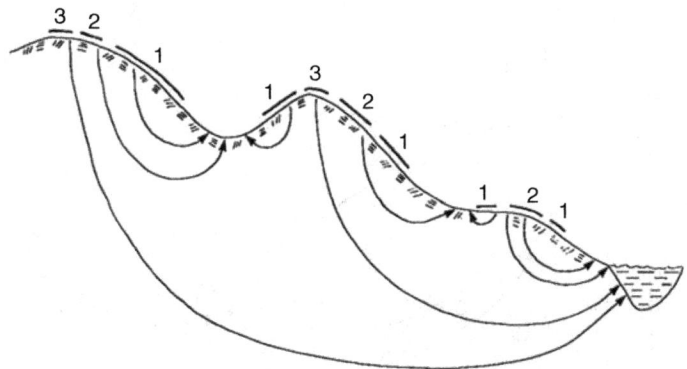

Figure 3. Main types of groundwater with different transit times. *Source*: Falkenmark & Allard (1989), figure 2, with kind permission of Springer Science & Business Media.

the groundwater flow may take place in the upper part of the rock, where conductivity may be considerable due to vertical tension fractures and horizontal fractures in the upper rock.

Water Quality-related Processes in a Stream Tube

Aspects of importance for spatial and temporal quality changes include the following:

- pathways of the water in the ground;
- mixing of water fractions arriving along different pathways; and
- transit times of the various contributions prior to mixing.

While a certain amount of water moves along a stream tube, a gradual natural change takes place of some water quality parameters (Figure 4). (For more details, see Falkenmark & Allard, 1989.)

Processes in the recharge zone. As indicated above, water acts as an aggressive solvent that interacts with geological media on its way through the underground landscape. Figure 5 simplifies the key processes during infiltration and percolation to groundwater. The water reacts with two main systems: the organic components adding CO_2 and humates; and the mineral system adding weathering and ion-exchange products. When reaching the water table, the entering rainwater has achieved a certain pH/alkalinity and redox potential, which is decisive for the processes during the continued transport phase.

Key parameters with large impacts on the mobility and distribution of dissolved components are the composition of the water itself (pH, redox potential, dissolved salts, etc.); the nature of the geological environment; and biological activity going on in the immediate surroundings.

Acidity (pH) is very important for the mobility of heavy metals due to a pH-dependent formation of a particulate phase that can serve as a metal carrier. This was illustrated in a mine tailing area in Sweden where increasingly more of the metals appear in the suspended material as pH rises. Reversely, as the water gets more acid, increasingly more of the metals are dissolved in the water (Karlsson *et al.*, 1987).

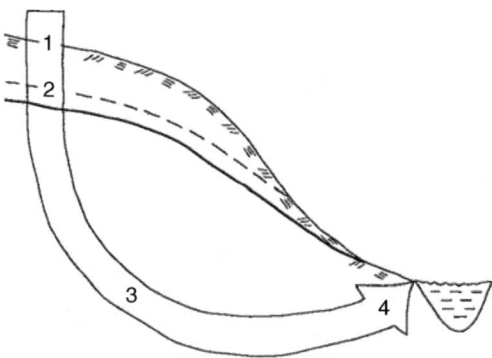

Figure 4. Water quality changes in a stream tube. *Source*: Falkenmark & Allard (1989), figure 14. With kind permission of Springer Science & Business Media.

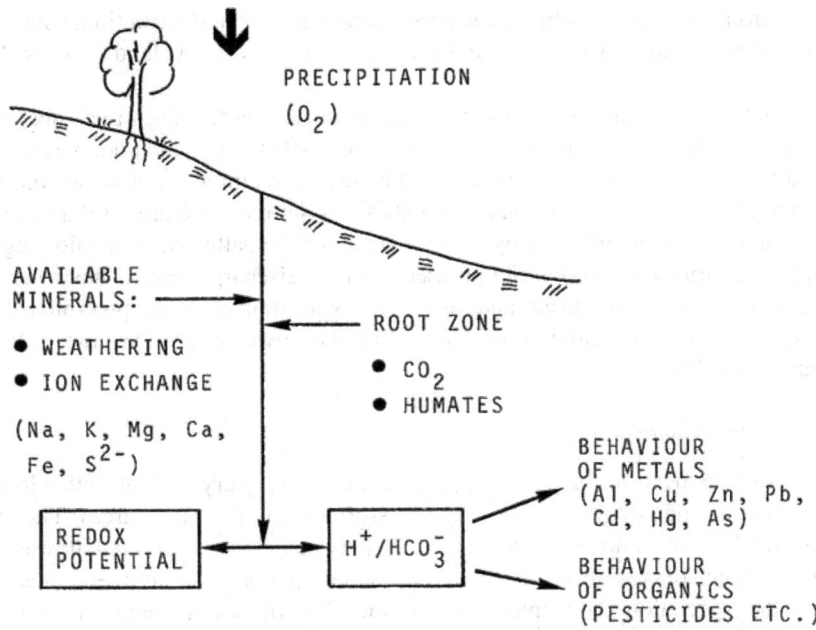

Figure 5. Processes in the recharge zone. *Source*: Falkenmark & Allard (1989), figure 6. With kind permission of Springer Science & Business Media.

Also, the *redox potential* has an impact on the spreading of pollutants, since the formation and existence of the carrier component ferric hydroxide is highly dependent on this potential. There is also an influence on the mobility of redox-sensitive elements such as Cr, Hg, As and U. Also, the degradation of certain organics as well as the chemical state of nitrogen are influenced by this potential.

Processes during the transport phase. Solutes in the mobile aqueous system undergo various chemical or biological reactions along their pathways, or transformations which are related to chemical as well as biological and physical conditions in the system. Moreover, the mere presence of solid surfaces in the system as particulate phases would have a large and direct influence on the transport processes of the solute. The more water that flushes the underground landscape per unit of catchment area, the higher is the amount of solutes carried away with the water draining that basin (Pillsbury, 1981; Walling & Webb, 1983). When the release of contaminants is the rate-limiting step, however, the load can be independent of the runoff.

At the same time, several processes are active in removing dissolved matter from the solute in a freshwater system. Processes involved in these functions include physical adsorption, electrostatic adsorption, chemisorption, and chemical substitution and precipitation/co-precipitation. Changes in pH and concentrations of humic substances may have drastic effects on the adsorption of heavy metals on solid geological matter.

Polluting agents with low mobility such as heavy metals at high pH can under certain conditions interact with soluble organic compounds, drastically enhancing the spreading. There are several cases where the combined deposition of toxic elements and complex organic agents have created mobile toxic species and chemical redistribution. One example is the spreading of radionuclides at the Maxey Flats deposit (Cleveland & Rees, 1981).

Processes in the discharge zone. When groundwater reaches the discharge zone, it enters into contact with the air in the soil (Eriksson, 1984). By evapotranspiration the concentration of dissolved salts is increased, favouring precipitation of some components like calcium. It brings essential plant nutrients from the recharge area, such as calcium, potassium and phosphate, released by weathering along the pathways of the flowing water, encouraging luxurious vegetation rich in species in the discharge area. If the water is poor in oxygen it is easily oxidized and any ferrous iron may form precipitated ferric hydroxides. The carbon dioxide pressure may be lowered so that the calcium is precipitated as calcite.

Processes in the "switching zone"

Certain landscape locations may act as recharge areas during dry periods with a low water table, whereas during wet periods they are transformed into discharge areas. The effect is that the direction of water movements alters between downwards and upwards. The combined effects of variations in redox potential, pH and sometimes organic load influence the speciation, concentration and mobility of heavy metals (Herrmann & Neumann-Mahlkau, 1985).

Water Pollution from Land Use

The links to land use are fairly evident:

- Chemical waste sites: buried chemical waste on waste dumps, factory sites and remote hollows have been a hot issue since the 1980s. Some 27,000 sites have been reported for Denmark alone (Korkman, 1996). The tentative list of contaminant sites in the Dobris Assessment (1997) in Western Europe included more than 55,000 sites, out of which 22,000 were in "critical conditions". These figures did not include Central and Eastern Europe where conditions were probably at least as bad. The combined result of underground water pathways and chemical activity is manifested in groundwater contamination under dry waste disposals (Figures 6 and 7).
- Overexploitation of groundwater: this is well known in many parts of Europe where around 60% of industrial and urban centres are overexploiting their aquifers: this causes water table changes which may alter the hydrochemical situation from an oxygen-free to an oxygen-rich status. The oxidation generated of, for example, pyrites around city areas in Denmark caused a rise in the sulphate content of the groundwater (Korkman, 1996). Nickel and probably also arsenic had been mobilized by the water, and the former had become a real threat to some of the Danish waterworks.
- Agriculture: a significant increase has been observed in stream nitrate load in many water systems in Europe over the past 30–40 years in response to the strong increases in fertilizer application to improve crop yields. This increase is manifested as widespread eutrophication of lakes and coastal waters and high nitrate content in groundwater. Agricultural activities covering 226 million ha in Europe generate a

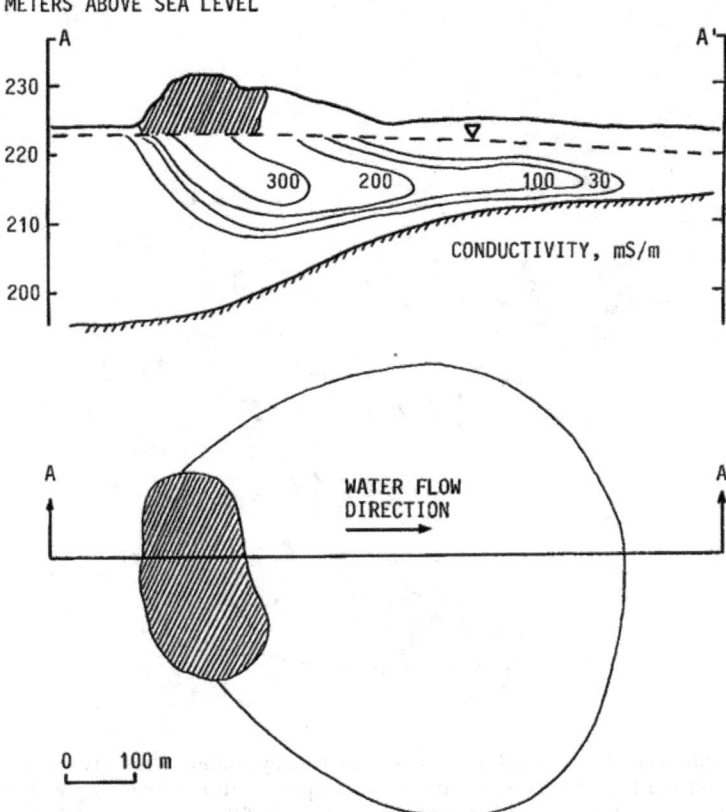

Figure 6. Spreading of pollutants from a land fill. *Source*: Falkenmark & Allard (1989), figure 15. With kind permission of Springer Science & Business Media.

real threat in terms of pollution with nitrate and pesticides (Korkman, 1996). Large parts of Europe are already affected by nitrate levels beyond health limits. Also pesticides are appearing in groundwater. In Denmark pesticides have been found all over the country down to a depth of 80 m, mostly in groundwater recharged 10–40 years ago. Even degradation products have been identified. This is seen as an even greater problem in view of their persistence and mobility.

- Land development: huge hydrological consequences on the Great Hungarian Plain of the continuous deforestation that had already started in the 1000s AD have been reported by Szesztay (1994). He suggests that the vegetation clearing—due to decreased evapotranspiration—has caused an increase in the runoff by some 150–200%. A resulting groundwater rise in discharge areas on the Plain had caused water logging, which had grown into an important challenge of integrated land and water management.
- Mining often results in the dewatering of aquifers and dramatic changes in groundwater quality, as well as surface water being contaminated from the discharge of drainage water into streams and lakes (Korkman, 1996).

Figure 7. Location of sources and stream tubes threatened by pollution from those sources. Dots = point sources in the terrain (dry waste deposits); stripes = urban polluting activities. *Source*: Falkenmark & Allard (1989), figure 3. With kind permission of Springer Science & Business Media.

Long Transit Times May Produce Chemical Time Bombs

The fact that water pollution occurs does not mean that it is possible to detect. It may be hidden due to long transit times. But pollutants may also be accumulated in soils and sediments through chemical adsorption processes going on in the ground that may be building up a chemical time-bomb. The possibility of detecting water pollution is in other words closely linked to delaying processes both in the subsurface transport phase through the drainage basin and in the surface transport phase due to flow retardation in intervening lakes.

Groundwater pollution may be detected when a well penetrates the polluted strata. In Sweden such a case—where pollutants from a waste deposit in a sand pit higher up in the drainage basin had polluted a well in a downstream village—was taken to court, attracting much media attention. The invisible and slow character of the phenomena involved made it extremely difficult to produce strong enough proof for the court, however. The result in the Swedish case was a negotiated solution where the water supply system of a neighbouring city was extended to the village.

As already indicated, pollution of groundwater and surface water may be delayed also where pollutants accumulate in soils and sediments (chemical time-bombs) for later release when geochemical conditions change for some reason (Hekstra, 1995). This means that regardless of the speed of reducing emissions at the source, there will remain a huge legacy of old pollutants stored in soils and sediments. They are, however, not put away for eternity but can be remobilized. The release of the hidden form may be triggered by land-use changes and climate changes that influence the capacity controlling properties, i.e. alter the geochemical conditions in such a way that the contaminants can be leached away by moving water.

Chemicals of concern are heavy metals, halogenated organics, aliphatic and aromatic hydrocarbons, and synthetic fertilizers (N and P). Lead is an example of a contaminant that has been accumulating since the Bronze Age. Persistent chemicals of course cause the greatest concern. The retention time in soils may vary from one to 1 million years.

River Quality Outcome

River water that is a mix of water fractions with different hydrochemical water signatures. Our understanding of the importance of pathway history and chemical signature for stream water quality has been facilitated by the problems around surface water acidification (Bishop, 1991). Acidity was found to be linked to the flow fraction that dominates during flow periods, i.e. to rapidly passing spate flow rather than older "pre-event" water in the catchment soils with longer transit times. The proportions between water fractions with different hydrochemical signatures vary with time. This is reflected in a hysteresis effect in the concentration/flow diagram (Figure 8). The hysteresis effect is also well documented for storm events mirroring different compositions during the rising and falling limbs of a storm hydrograph (Edwards *et al.*, 1984).

Basically, the variation of concentrations depends on many parameters and factors such as land uses, hydrometeorological conditions, waste applications, soil types, etc. Unfortunately, measurement data are frequently erroneous in that they do not include

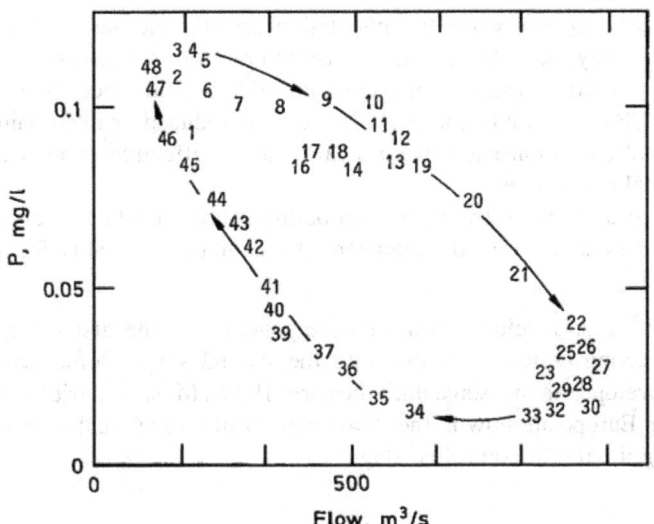

Figure 8. Concentration of phosphate in weekly composite samples as a function of water flow: Rekingen, Rhine. *Source*: Falkenmark & Allard (1989), figure 16. With kind permission of Springer Science & Business Media.

extreme runoff events that could carry the larger part of the total load (Jolankai, 1992). Stable isotopes, especially O_{18}, have been used as natural tracers to clarify both groundwater recharge pathways (Saxena, 1984) and runoff generation (Bishop, 1991).

Inherited water pollution. The major pressures behind water quality degradation have developed step by step in different eras: land use in the first millennium AD, river engineering after 1000 AD, followed by urban population and mining impacts in the 1800s, atmospheric pollution in the first half of the 1900s, and agrochemicals after the Second World War (Meybeck, 2003).

But in addition each region of the world has had a different Earth system history such as glaciation–deglaciation, land cover change, etc. and a different human development history. Also in the future both human drivers and Earth system controls, in terms of global warming and climate variability, will be different from region to region. Also natural sensitivity differs: carbonated rocks, for instance, limit the risk for acidification, semi-arid regions are sensitive to salinization and river depletion, mountain areas and loess-covered regions are more sensitive to sediment imbalance, etc.

In the history of human development, river quality has gradually changed from very limited impacts for hunter–gatherer societies to maximum degradation found in impounded rivers, combining irrigation, mega-cities, industrial and/or mining activities. Limited wastewater collection and limited or unenforced environmental regulations and limited citizen awareness are contributing factors. Meybeck (personal communication October 2008) distinguishes four stages:

- The *population driven situation*, where water use and pollution loads are proportional to population.
- *Accelerated degradation*, where the pollution pressure increases faster than the population increase.
- *Environmental recovery* when the population growth rate has decreased or is stabilized; society has developed infrastructures for wastewater treatment; industrial and mining wastes are recycled; use of toxic substances (e.g. mercury, cadmium, PCBs, DDT, atrazine pesticide, etc.) is reduced or even banned; and environmental education and citizen awareness are effective enough to accept environmental regulations.
- A *fully managed river quality* when water pollution abatement has been achieved; and the water system is directly engineered to maintain the quality for users and aquatic biota.

The slow and least developed countries are generally in the first stage, while the fast developing countries tend to move into the second stage. Some Eastern European countries developed in this stage until the early 1990s. Most countries in North America and Western Europe are now in the third stage, while some local river basins are even experiencing the fourth controlled stage.

River quality changes in a multi-millennia perspective. Table 2 shows the dominant origin in terms of human-caused alterations and the effects in terms of pollutants (Meybeck, 2003). The principal impacts of the different river symptoms tend to be negative when concerning water quality (microbial contamination and development of

WATER QUALITY MANAGEMENT

Table 2. River symptoms and their main causes.

Human pressure/ alteration	Sedimentation	Depletion	Salinization	Chemical contamination	Historic pollution	Acidification	Eutrophication
Land use	●				●	●	●
Mining	●			●	●	●	●
Irrigation		●	●				
Industry				●	●	●	
Urban				●	●		

Source: Meybeck (2003).

pathogens, primarily in the developing world; chemical contamination and nutrients causing problems also in the developed world).

The multiple forms of riverine quality changes caused by different forms of human pressures can be clustered and organized in a set of *river syndromes* in terms of typical worldwide patterns of problematic interaction of people and environment. Based on efforts of reconstruction of past riverine evolution, Meybeck (2003) has shown regional differences in the evolution of some symptoms of river system changes between Western Europe and South America (Figure 9). In Europe organic and faecal and metal contamination grew more or less in parallel up until about the 1970s when they rapidly

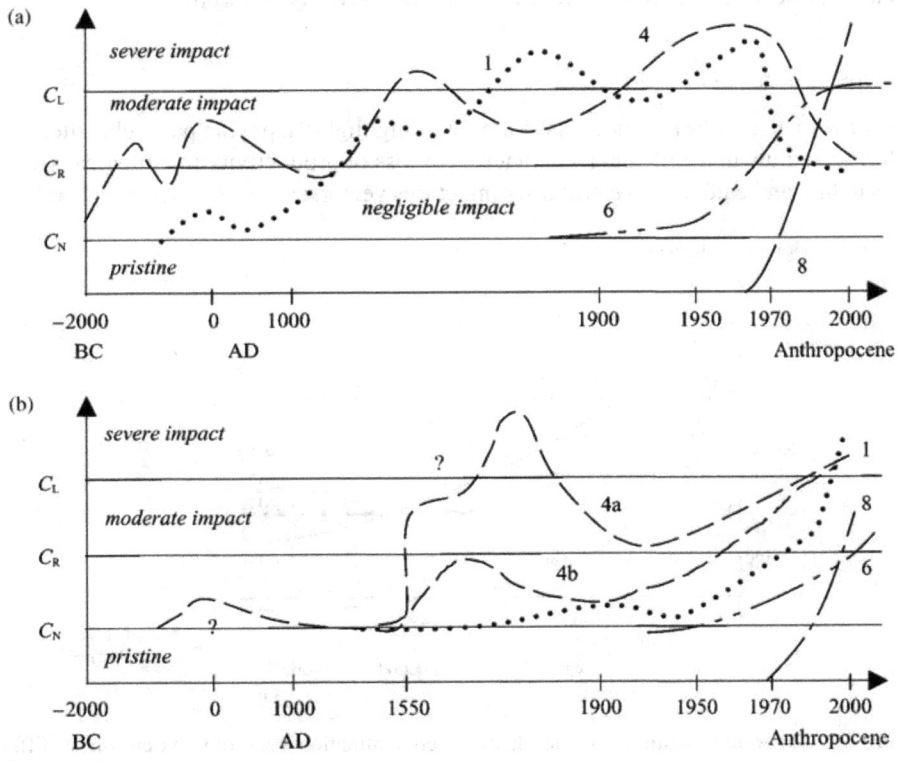

Figure 9. Suggested evolution of some symptoms of river system changes. *Source*: Meybeck (2003), figure 2. With kind permission of The Royal Society, London.

decreased, but nitrate and pesticide pollution more or less exploded. In South America, however, metal contamination grew around 1500 AD with the arrival of European settlers, the gold and silver mining and mercury amalgamation; decreased in the 1800s but increased again after the Second World War in parallel with nitrate and pesticides.

Response to Future Water Pollution Risks

Projections

This paper has clarified that water pollution is extremely widespread and of a highly complex character. Severe impacts have been reported. Pollution is foreseen to expand further. Meybeck (2003) has indicated a schematic evolution of the chemical contamination in terms of space scales of syndrome development from Roman times, when effects may have been severe only very locally; till the year 1800 when effects had expanded to regional scale but still remained negligible. In the 1950s effects had intensified in severity and expanded to continental scale. In the year 2000 the scale was global, the continental syndrome now moderate and the local syndromes severe. He also showed plausible scenarios for 2050 (Figure 10) distinguishing between three cases: business as usual, when the regional scale syndrome had become severe; priority reduction in the most polluted sites, which would have reduced local severity but not much influenced the global scale syndrome; and general application of precaution principle, when both local, regional and global scale severity would be reduced to moderate severity syndromes.

Key Risks To Be Addressed

To start with, particularly serious risks should be highlighted, paying particular attention to both risks to human health and risks in terms of loss of productivity in biological systems. Risks to human health involve primarily infectious vectors; toxic chemicals such as Hg, Pb

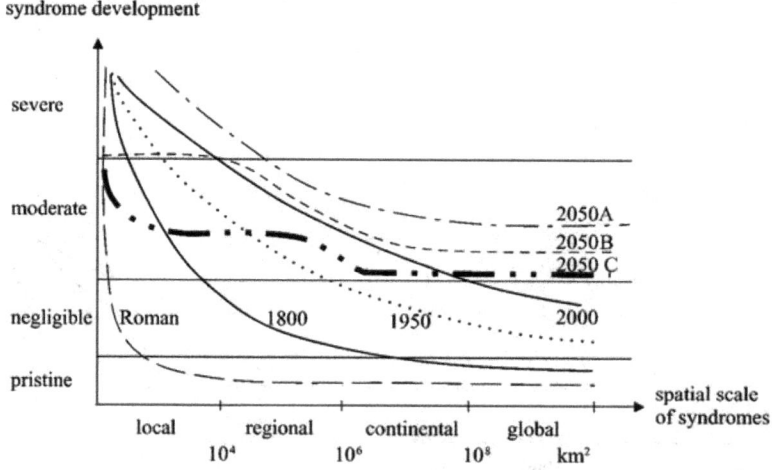

Figure 10. Suggested evolution of the chemical contamination syndrome extension at different space scales from Roman times to present. Three scenarios for the year 2050: A, business as usual; B, priority reduction of the most polluted sites; and C, general application of the precaution principle.
Source: Meybeck (2003), figure 4. With kind permission of The Royal Society, London.

and Cd; pesticides and hormone-mimicking compounds; and nutrients. These substances involve *risks in terms of human health* and water-related diseases, skin diseases, accumulation in human tissues, and accumulation in fish eaten by humans. *Risks in terms of biological systems* may involve the collapse of ecological services of various kinds and include, for instance, soil and water salinization impacting agricultural production, or toxic substances in lake and coastal waters, impacting on fish and seafood, etc.

A key question is of course what can realistically be done in terms of response to these risks and what are the critical issues? Basically highest priority should be given to the most threatening pollutant, the least allowable pollutant, or the one with the most far-reaching effects. For modes of control there are different options: to focus on concentration, by adding dilution; to address the pollutant entry into the water system by wastewater treatment making, for instance, nutrient reuse possible; isolation of the source as is practised with radioactive pollutants; blocking the escape of the pollutant by limiting the maximum allowable amount; or prohibiting a contaminant altogether, as in the case of DDT and other toxic chemical bans.

Pollution-abatement Principles

Principally, pollution-abatement principles may be *reactive* to a pollutant already out, or *proactive* in hindering a potential pollution by banning particular pollutants from use altogether. In terms of the former, there are different options: to address the concentration by securing enough dilution; by addressing entry into the hydrosystem and by wastewater treatment, by isolating the source, by blocking contaminant escape, or by altogether prohibiting its use. The ultimate possibility is to focus on the point of use by applying some user-level treatment technique to protect the user from being damaged during use.

Abatement possibilities differ, in terms of the possibility to identify the particular contaminant source and the responsible polluter, between so-called point and diffuse pollution sources. Diffuse sources are therefore particularly complex to address, as evident from several decades of efforts in industrialized countries to come to grips with an increasingly serious eutrophication of lakes and coastal waters.

The success of a pollution-abatement efforts depends on the ability to manage a long list of sequential activities:

- Awareness and understanding of risks and vulnerabilities involved.
- Identification of the particular source and polluters.
- Policy development in terms of legislation, enforcement, incentives and sanctions.
- Availability of a feasible technical solution.
- Financing mode identified, of which there are many options: polluter-pays principle, user fees, taxation, government funding, etc.
- Political will to act, supported by societal awareness and leading to decision to act, involving financing mode.
- Selection of responsible actor and institution in charge.
- Securing implementation through awareness campaigns to generate societal acceptance.

Although few, there are fortunately a number of success stories to encourage action. In terms of *top-down* approaches, the most basic is of course the cholera explosion in

European cities in the mid-1800s, which launched the transition into organized sanitation to reduce the spreading of bacteriological pollution from human waste. Also the ban of DDT use is an example which at the same time demonstrates the implications of a long response time in hydrosystems with slow water exchange, like the slow recovery of the biota in the Baltic Sea area. There are also examples of *bottom-up* driven approaches like the abolishment of Cl_2 in the paper industry for paper bleaching, where a very quick industrial response was achieved in response to a general public outrage.

Conclusions

What has been shown in this paper is that soluble chemical components are being caught and carried by water, an aggressive liquid on continuous move through the landscape. Its movement through the land can be thought of as taking place in a large set of stream tubes following shallow, intermediate or deep pathways. Dominant processes during this flow are pH, redox processes, and those related to weathering and biodegradation. River water can be seen as a mix of water fractions with a different pathway history and different chemical signatures in response to those pathways. Pollutants have been added to the water from multiple sources: air exhausts in the atmospheric phase, land-based chemicals, human waste and wastewater during the continental phase. Difference in hydroclimate and human history are reflected in large regional differences in terms of current water quality profiles. It can thus be concluded that water pollution is not only extremely widespread, but also at the same time highly complex. It originates from the interaction between natural processes and human activities, altering physical water flow determinants and adding chemical components to the flowing water. Some constituents are more threatening than others: infectious vectors, As, Hg, hazardous chemicals, hormone-mimicking substances, etc. They may be impacting directly through intake or by contact with polluted water, or indirectly through agricultural products and fish. They may also involve loss of productivity in biological systems, sometimes even causing a flipping of the ecological systems or collapse. A particularly threatening phenomenon is salinization of agricultural soils. A considerable part of reported biodiversity decline is linked to aquatic ecosystems. Their habitats tend to reflect the integrated river water response to all human activities in the upstream catchment.

Water quality will continue to change in response to human activity. Water pollution abatement has been a seriously neglected sector—most interest in the past has been on water quantity-related phenomena and issues. In cases where certain action has been decided upon, it may have met problems of implementation (Narain, 1999). While waiting for decisive action, pollution has continued to increase, expand and intensify. The concept of *hydrocide* (Lundqvist, 1998) has been used to clarify the degree of seriousness, and scenarios have been presented to show the potential implications for human fertility of, for instance, persistent hormone-mimicking pollutants (Simonovic, 2002). Severe impacts have been reported for humans in terms of not only water-borne diseases, but also increasingly cancer, skin diseases, hormone disturbances and fertility decline; for ecosystems the effects are impacts such as biota hormone disturbances, Baltic ecosystem flip or Mexican Gulf hypoxy.

Water pollution-abatement efforts are evidently now called for, primarily addressing the most threatening substances, the most polluted sites, the water systems with the largest biodiversity degradation, and the most far-reaching threats in terms of both human health and ecosystem productivity. In the selection of issues the long response-time delays have to be taken into account: both societal response before action starts and hydrosystem response time in both lakes, semi-enclosed coastal seas and groundwater aquifers. Even if the issues are complex, many pollution sources tend to be local, involving more limited challenges. Diffuse sources are more difficult to address, particularly if economic interests are involved, such as the farmer economy in the case of nutrients, which calls for ways of trade off striking.

More than three decades of biological community argumentation and problem formulations in terms of environmental effects and problem formulations have had a poor impact, in terms of convincing power, on economic government advisors. This suggests that it is now time for the water community to take a much stronger part in future water pollution efforts (Falkenmark, 2005):

- explaining the creeping water-related processes behind water pollution;
- contributing with scenarios to make time scales and the outcome of current policy paths visible and understandable;
- contributing with governance models, allowing integrated approaches to water quantity and quality, to land and water, and to humans and ecosystems;
- proposing technically sound, socially acceptable, economically viable, and legally and institutionally feasible solutions; and
- securing water resources' literacy among the general public and its representatives among decision-makers and politicians.

References

Bishop, K. V. (1991) Episodic increases in stream acidity, catchment flow pathways and hydrograph separation, Dissertation, University of Cambridge, Cambridge.

Cleveland, J. & Rees, T. F. (1981) Characterization of plutonium in Maxey Flats radioactive trench leachates, *Science*, 212 (4502), pp. 1506–1509.

Dobris Assessment (1997) *Europe's Environment* (Copenhagen: European Environment Agency).

Edwards, A. C., Creasey, J. & Cresser, M. S. (1984) *The Conditions and Frequency of Sampling for Elucidation of Transport Mechanisms and Element Budgets in Upland Drainage Basins.* No. 150 (Wallingford: IAHS Press).

Eriksson, E. (1984) *Hydrochemical Processes in Groundwater Discharge Areas.* No. 150 (Wallingford: IAHS Press).

Eriksson, E. (1985) *Principles and Applications of Hydrochemistry* (London: Chapman & Hall).

Falkenmark, M. (2000) A European perspective on sustainable consumption of freshwater. Is getting rid of pollutants realistic and affordable?, in B. Heap & J. Kent (Eds) *Towards Sustainable Consumption. A European Perspective*, pp. 49–58 (London: Royal Society).

Falkenmark, M. (2005) Water useability degradation. Economist wisdom or societal madness?, *Water International*, 30, pp. 136–146.

Falkenmark, M. & Allard, B. (1989) Water quality genesis and disturbances of natural freshwaters, in: O. Hutzinger (Ed.) *The Handbook of Environmental Chemistry*, Vol. 5, pp. 45–78 (Part A: Water Pollution (Berlin: Springer).

Falkenmark, M. & Mikulski (1994) The key role of water in the landscape system. Conceptualization to address growing human landscape pressures, *GeoJournal*, 33(4), pp. 355–363.

Falkenmark, M., Andersson, L., Castensson, R. & Sundblad, K. (1999) *Water—A Reflection of Land Use. Options for Counteracting Land and Water Mismanagement* (Stockholm: Swedish Natural Science Research Council).

Hekstra, G. P. (1995) *Delayed Effects of Pollutants in Soils and Sediments: Understanding and Handling of Chemical Time Bombs in Europe.* Ecoscript No. 56 (Amsterdam: Stichting Mondiaal Alternatief).

Herrmann, R. & Neumann-Mahlkau, P. (1985) The mobility of zinc, cadmium, copper, lead, iron and arsenic in ground water as a function of redox potential and pH, *Science of the Total Environment*, 43(1–2), pp. 1–12.

Jolankai, G. (1992) *Hydrological, Chemical and Biological Processes of Contaminant Transformation and Transport in River and Lake Systems. A State-of-the-Art Report*. IHP-IV Projects No. H-3.3. Technical.

Karlsson, S., Sandén, P. & Allard, B. (1987) Environmental impacts of an old mine tailings deposit: metal adsorption by particulate matter, *Nordic Hydrology*, 18(4–5), pp. 313–324.

Korkman, T.-E. (1996) Groundwater for the next generation, in: *Proceedings of the Stockholm Water Symposium 1996* (Stockholm: Stockholm Water Company).

Lundqvist, J. (1998) Avert looming hydrocide, *Ambio*, 27, pp. 428–433.

Meybeck, M. (2002) Riverine quality of the Anthropocene: propositions for global space and time analysis, illustrated by the Seine river, *Aquatic Sciences*, 64, pp. 376–393.

Meybeck, M. (2003) Global analysis of river systems: from Earth system controls to Anthropocene syndromes. *Philosophical Transactions B, Royal Society London*, 358, pp. 1935–1955.

Narain, S. (1999) We all live downstream: urban industrial growth and its impact on water systems, in: *Proceedings of the Stockholm Water Symposium 1999* (Stockholm SIWI).

Pillsbury, A. F. (1981) *Scientific American*, 7, p. 22.

Saxena, R. K. (1984) *Surface and Groundwater Mixing and Identification of Local Recharge–Discharge Zones from Seasonal Fluctuations of Oxygen 18 in Groundwater in Fissured Rock*. No. 150 (Wallingford: IAHS Press).

Simonovic, S. P. (2002) Global water dynamics: issues for the 21st century, in: *Proceedings of the Stockholm Water Symposium 2001. Water Science and Technology*, 45(8), pp. 53–64.

Szesztay, K. (1994) The role of water in landscape ecology and in crop production, *Periodica Polytchnica, Series Civil Engineering*, 38(3), pp. 315–331.

Walling, D. E. & Webb, B. W. (1983) in: B. W. Webb (Ed.) *Dissolved Loads of Rivers and Surface Water Quantity/Quality Relationships*. No. 141 (Wallingford: IAHS Press).

Impact of Agriculture on Water Pollution in OECD Countries: Recent Trends and Future Prospects

KEVIN PARRIS
Organisation for Economic Co-operation and Development (OECD), Paris, France

ABSTRACT *Agricultural pollution of surface water, groundwater and marine waters relates to the contamination of drinking water, and harmful effects on ecosystems and costs for recreational activities, cultural values and commercial fisheries. After the introductory section, this paper examines the recent trends and economic costs of agricultural water pollution. Subsequent sections discuss recent Organisation for Economic Co-operation and Development (OECD) policy experiences in addressing water pollution in agriculture, and the medium outlook for pollution across OECD countries. The final section explores ways forward toward sustainable management of water quality in agriculture.*

Introduction

Agricultural water quality has been identified as a major environmental issue in Organisation for Economic Co-operation and Development (OECD) countries, and as a topic for policy analysis is an issue of relevance across all OECD countries. The primary agricultural sector is mainly responsible for nitrate, phosphorus, pesticide, soil sediment, salt, and pathogen pollution of water from crop and livestock activities, but it can also play a role under certain farm practices in terms of improving water quality through a water purification function. Water pollution from agriculture has associated costs in terms of removing pollutants from drinking water supplies, as well as damage to ecosystems and commercial fishing, recreational, and cultural values associated with rivers, lakes, groundwater and marine waters.

Agricultural water pollution is a focus of attention for policy-makers in most OECD countries due to the following (the importance of these issues varies within and across countries):

- Reduction in pollution by non-agricultural polluters, which has been more rapid than for agriculture, especially nitrate, phosphorus and pesticide pollution.
- Increase in point pollution from agriculture linked to the intensification of livestock farming, especially in the pig, poultry and dairy sectors.

- Greater public awareness of the damage to aquatic ecosystems from certain agricultural practices.
- Growing concerns related to groundwater and coastal pollution, especially from the leaching of phosphorus and pesticides.
- Uncertainty over the extent and severity of those water pollutants derived from farming that are in general poorly monitored (e.g. pathogens, salts, heavy metals).

Trends in OECD Agricultural Water Pollution Since 1990

Overview

Most OECD countries have monitoring networks to measure the actual state of water pollution of water bodies, while some countries use risk indicators that provide estimates, usually based on models of contamination levels. However, monitoring of agricultural pollution of water bodies is more limited with just over one-third of OECD member countries monitoring nutrient pollution and even fewer countries tracking pesticide pollution. Certain farm pollutants are recorded in more detail and with greater frequency (e.g. nutrients, pesticides), whereas an indication of the overall OECD situation for water pollution from pathogens, salts and other agricultural pollutants is unclear. Moreover, pollution levels can vary greatly between countries and regions depending mainly on soil and crop types, agro-ecological conditions, climate, farm management practices, and policy.

The limitations to identifying trends in water pollution originating from agriculture are in attributing the share of agriculture in total contamination and identifying areas vulnerable to agricultural water pollution. In addition, differences in methods of data collection and national drinking and environmental water standards hinder comparative assessments, while monitoring agricultural water pollution is poorly developed, especially for pesticides, in a number of countries, such as Australia, Italy, Japan and New Zealand. The extent of agricultural groundwater pollution is generally less well documented than is the case for surface water, largely due to the costs involved in sampling groundwater, and because most pollutants take a longer time to leach through soils into aquifers.

The remainder of this section examines agricultural pollution in terms of the main agricultural driving forces impacting on water pollution, especially the use of nutrient and pesticides inputs (Figure 1). In turn depending on farming practices and systems, the use of farm inputs will affect the state of the environment with regard to rates of soil erosion (which affects the leaching of pollutants), water quality and impacts on aquatic ecosystems (either in fresh or marine waters). The fourth section examines the policy responses across OECD countries to the state of water pollution, which in turn act as a driving force on the farm systems, practices and inputs used by agriculture.

Recent Trends in Water Pollution from Agricultural Nutrients and Pesticides

The overall pressure of agriculture on water quality in rivers, lakes, groundwater and coastal waters has eased since the early 1990s due to the decline in nutrient surpluses and pesticide use for most OECD countries (Figures 2 and 3). Despite this improvement, absolute levels of agricultural nutrient pollution remain significant in many cases. With point sources of water pollution (i.e. industrial and urban sources) falling more rapidly

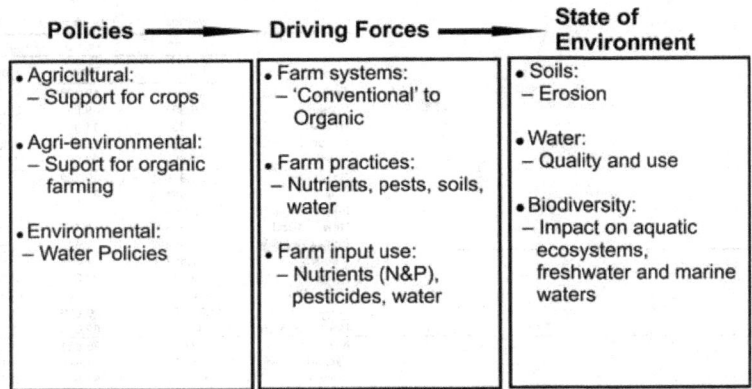

Figure 1. Linkages between policies, driving forces and the state of the environment relevant to water. *Source:* OECD Secretariat, 2010.

		Balance expressed as tonnes of nitrogen (N)				Balance expressed as kg nitrogen per hectare of total agricultural land		
		Average	Average	Change		Average	Average	Change
Change in the nitrogen balance (tonnes N)		1990-92 (000') tonnes N	2002-04 (000') tonnes N	1990-92 to 2002-04 (000') tonnes N	%	1990-92 kgN/ha	2002-04 kgN/ha	1990-92 to 2002-04 %
	Canada[2]	1168	2101	934	80	19	35	85
	Hungary	136	217	82	60	21	37	74
	New Zealand	407	576	169	41	31	46	46
	Ireland	337	360	24	7	76	83	9
	Portugal	168	180	12	7	42	47	13
	United States	14621	15024	402	3	34	37	7
	Spain	966	977	12	1	32	33	5
	Australia[3]	7574	7636	62	1	16	17	5
	Italy	588	582	-5	-1	33	39	16
	Korea	465	456	-9	-2	213	240	13
	OECD	41238	39681	-1557	-4	88	74	-17
	Switzerland	121	116	-6	-5	77	76	-1
	Iceland	17	16	-1	-6	7	7	-5
	Czech Republic	332	300	-31	-9	77	70	-9
	Norway	92	81	-11	-12	92	77	-16
	Japan	935	813	-121	-13	180	171	-5
	Poland	922	797	-125	-14	49	48	-2
	Mexico	2768	2354	-414	-15	27	22	-18
	France	1932	1589	-343	-18	63	54	-16
	EU15	9989	7935	-2054	-21	113	83	-26
	Sweden	193	152	-41	-21	57	48	-16
	Turkey	1493	1148	-346	-23	37	28	-24
	Germany	2515	1926	-589	-23	145	113	-22
	Belgium	344	256	-88	-26	255	184	-28
	Austria	226	161	-65	-29	66	48	-27
	United Kingdom	1022	702	-320	-31	56	43	-23
	Denmark	493	338	-156	-32	178	127	-29
	Netherlands	688	443	-245	-36	345	229	-34
	Finland	211	123	-88	-42	83	55	-34
	Luxembourg	29	16	-12	-43	229	129	-44
	Slovak Republic	197	111	-85	-43	80	46	-43
	Greece	278	130	-149	-53	32	15	-52

Figure 2. Gross nitrogen[1] balance estimates for 1990–1992 and 2002–2004. *Notes*: 1. The gross nitrogen balance calculates the difference between the nitrogen inputs entering a farming system (i.e. mainly livestock manure and fertilizers) and the nitrogen outputs leaving the system (i.e. the uptake of nutrients for crop and pasture production); 2. For Canada, the change in the nitrogen balance is 80%; for Hungary, the change in the nitrogen balance is 60%; and for Greece, the change in the nitrogen balance is −61%; and 3. The average for the period 2002–2004 is an OECD estimate. *Source*: OECD (2008).

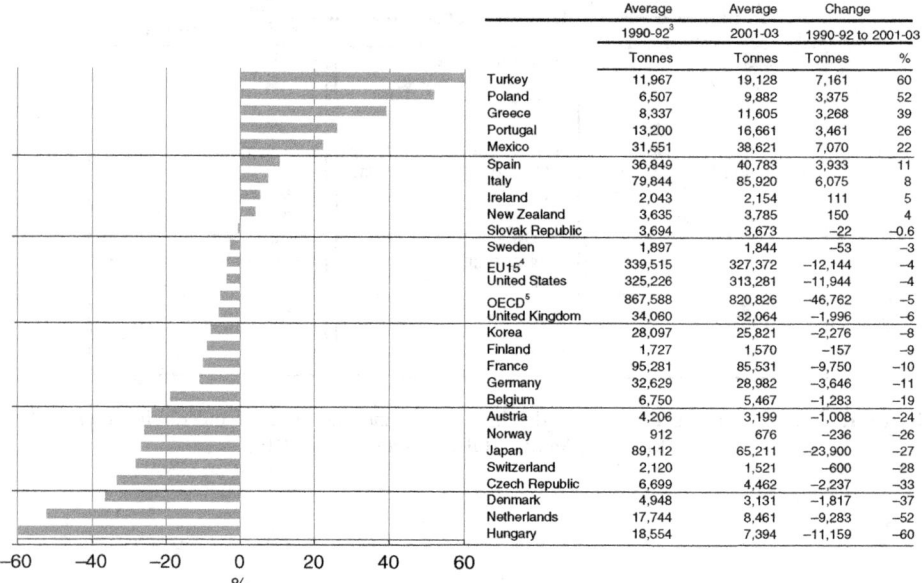

Figure 3. Pesticide use[1] in agriculture[2] for 1990–1992[3] to 2001–2003[4]. Change is in tonnes of active ingredients (%). Caution is required when comparing trends across countries because of differences in data definitions, coverage and time periods. The following countries are not included: Australia, Canada, Iceland (time-series are incomplete), and Luxembourg is included in Belgium. *Notes:* 1. For all countries the data represent pesticide sales except for the following countries: Korea and Mexico (national production data); 2. Pesticide use covers agriculture and non-agricultural uses (e.g. forestry, gardens), except for the following countries which only include agriculture: Belgium, Denmark and Sweden; 3. Data for the 1990–1992 average equal the 1991–1993 average for Greece and Slovak Republic; the 1993–1995 average for Mexico, New Zealand and Turkey; the 1995–1997 average for Italy; and the 1996–1998 average for Portugal; 4. The EU-15 includes the 1996–1998 average for Portugal and OECD Secretariat-estimated values for the following countries and years: Ireland: 2002 and 2003; Greece: 1990; and Italy, Germany and Spain: 2003; 5. The OECD total includes OECD Secretariat-estimated values for the following years and countries: 1990 for Greece and Slovak Republic; 1990–1992 for Mexico, New Zealand and Turkey; 2002 and 2003 for Ireland, Turkey and the United States; and 2003 for Germany, Mexico, Poland and Spain. *Source*: OECD (2008).

than for agriculture over the 1990s and effectively controlled in most situations, the share of agriculture (i.e. nonpoint source of pollution) in nutrient pollution of water has been rising, both in surface (Figure 4) and coastal waters (Figure 7).

Nearly one-half of OECD countries record that nutrient and pesticide concentrations in surface water and groundwater monitoring sites in agricultural areas exceed national drinking water limits for nutrients and pesticides (Figures 5 and 6). But the share of monitoring sites of rivers, lakes and marine waters that exceed recommended national limits or guidelines for environment and recreational uses is much higher, with agriculture a major cause of this pollution in many cases.

With respect to groundwater, however, agriculture is now the major and growing source of pollution across many OECD countries, especially from nutrients and pesticides (Figures 5 and 6). This is largely because other sources of pollution have been reduced

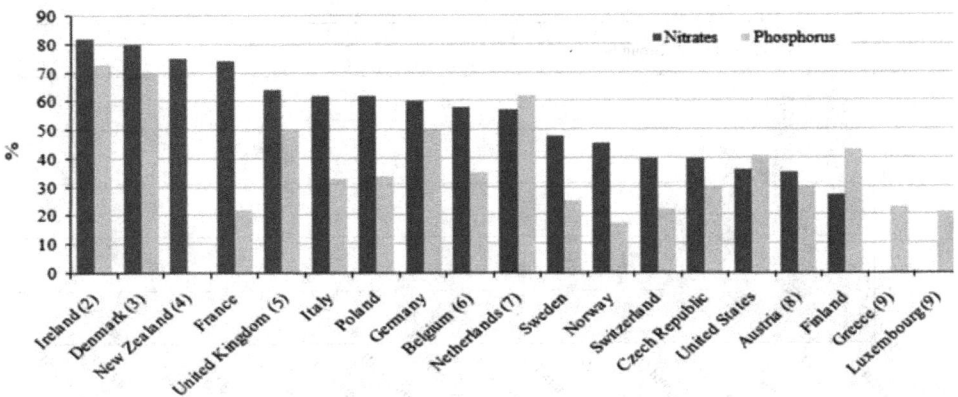

Figure 4. Agriculture's contribution of nitrates and phosphorus in surface water for the mid-1990s to the mid-2000s. *Notes:* 1. Data for the mid-1990s were for Finland, France, Germany, Greece, Italy, Luxembourg, Norway, Poland, Sweden and Switzerland (OECD, 2008, ch. 3); 2. 2004, see Ireland (OECD, 2008, ch. 3); 3. Phosphorus (2002), the percentage refers to Danish lakes only; 4. Data for nitrate contamination of rivers and streams, total input to surface waters from agriculture nonpoint source pollution. Data for phosphorus are not available. *Source:* New Zealand Government (2004, p. 98); 5. United Kingdom, OECD (2008, ch. 3); 6. Flanders only, 2001; 7. Netherlands, 2002 (OECD, 2008, ch. 3); 8. The value for 2000; 9. Data for nitrate emissions are not available. *Sources*: OECD (2008).

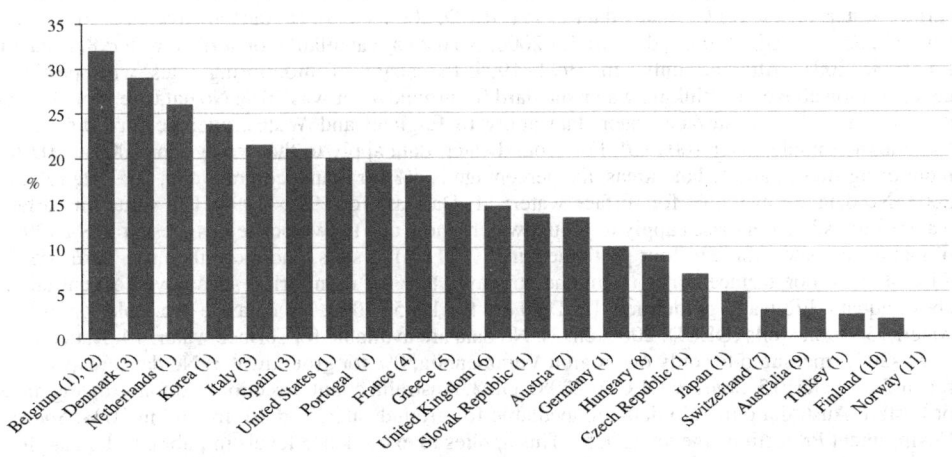

Figure 5. Share of monitoring sites in agricultural areas that exceed recommended drinking water threshold limits for nitrates in groundwater for 2000–2004. *Notes:* 1. Data refer to the average for 1995–2005; 2. Belgium (Flanders only); 3. Data refer to the average for 2002 and 2003; 4. Data refer to 2001; 5. Data refer to the average for 2001–2002, with a range of 10–20% (OECD, 2008 ch. 3); 6. Data refer to 2004; 7. Data refer to 2002; 8. Data refer to the average for 2000–2002 (OECD, 2008, ch. 3) and apply to all surface water-monitoring points; 9. For Australia, see OECD (2008), ch. 3. Groundwater in intensively farmed areas of north-eastern Australia; 10. Data refer to 2002, estimated for shallow wells at 2% and for aquifers at 1.5%; and 11. Norway (National Environmental Monitoring Programme) reported 0% for 1985–2002. *Sources:* OECD (2008).

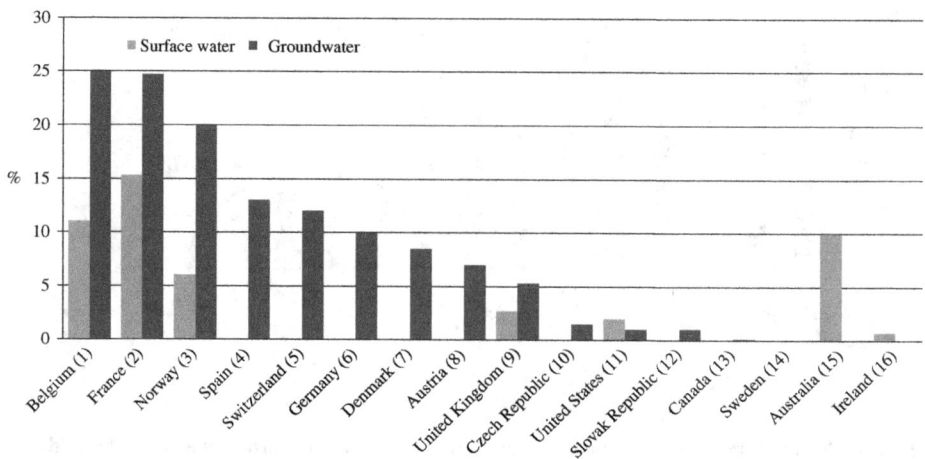

Figure 6. Share of monitoring sites in agricultural areas where pesticide concentrations in surface water and groundwater exceed recommended national drinking water threshold limits for 2000–2002. *Notes:* 1. Data are for 2000–2002. Flanders region only. Atrazine only for surface water. Regional variation shows concentrations ranged between 13% and 32%, with 10% of monitoring sites in excess of 0.5 µg/l compared with drinking water standard of 0.1 µg/l; 2. National data. Average poor and very poor status (OECD, 2008, ch. 3); 3. Data apply only to monitoring locations in high-risk pollution sites. Data are for 1995–2002, with concentration levels for surface water declining in most locations. For groundwater, the percentage share for pesticide presence applies to farmer's drinking water wells, while the pesticide concentration in groundwater is 2% for those aquifers supplying more than 100 people; 4. *Source:* OECD (2004). No data are available for surface water; 5. Data for 2002 apply to water catchments under arable farming. No data are available for surface water; 6. *Source:* Germany data 1995 (OECD, 2008, ch. 3). No data are available for surface water; 7. *Source:* EEA (2005), data are for 2000. No data are available for surface water; 8. Data are for 1990–2001. Atrazine only. In 1992–1994 the share of monitoring sites with pesticide concentration above the drinking water standard for groundwater was 20%. No data are available for surface water; 9. For surface water data apply to England and Wales, average for 2000–2002 for atrazine samples over 100 mg/l. For groundwater, data apply to the average for 2000–2002 for monitoring sites in arable land areas, the percentage is 4% for managed grassland; 10. Data refer to 2003. No data are available for surface water; 11. Data are for 1992–1998. The value for surface water (figures in parentheses apply to groundwater) show one to two pesticides present in 8% (29%) of monitoring sites; three to four pesticides in 18% (11%) of sites; and more than five pesticides in 74% of sites. For surface water (farmland streams), 80% of monitoring sites have concentrations above aquatic life water guidelines; 12. Data are for 1985–2002. No data are available for surface water; 13. Rural wells (OECD, 2008, ch. 3). No data are available for surface water; 14. Data are for 1998–2002, measured for only one region Vemmenhög, 0% for groundwater. No data are available for surface water; 15. *Source:* OECD (2008), ch. 3, Australia country section. Cotton-growing areas of Eastern Australia only. No data are available for groundwater; and 16. Ireland in 2004. *Source*: Environment Protection Agency (2005). This applies to exceedence levels in public water supplies.
Sources: OECD (2008).

more rapidly than for agriculture, although evidence of groundwater pollution is limited. This is a particular concern for countries where groundwater provides a major share of drinking water supplies for both human and livestock populations, and also as natural recovery rates from pollution can take many decades, in particular, for deep aquifers. There is also some evidence of increasing pollution of groundwater from pesticides

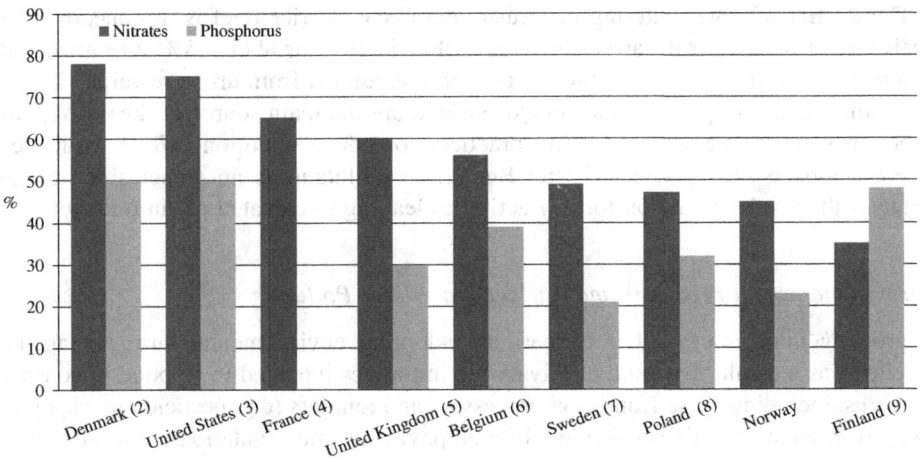

Figure 7. Agriculture's percentage share in total emissions of nitrates and phosphorus in coastal water[1] for 2000–2002. *Notes:* 1. Data on nitrates and phosphorus are from the OECD's 2nd Agri-environmental Indicators Questionnaire 2004 for Belgium, Finland, Norway and the UK; 2. Data refer to 2002; 3. Data refer to 2000; 4. *Source:* OECD (2008, ch. 3), represents nitrates discharged from the River Seine into "la Manche" (the Channel). Data refer to 2000; 5. Nitrate was estimated at between 50% and 70% and between 30% and 40% for phosphorus. *Source*: OECD 2nd Agri-environmental Indicators Questionnaire 2004); 6. Flanders only for the year 2000; 7. Data refer to the year 2000 and to anthropogenic load; agriculture contributed to 21% of the total phosphorus load; 8. Includes a range of 45–50% for nitrates and 30–35% for phosphorus; and 9. Data refer to 1997–2001. *Source:* OECD (2008).

despite lower use in many cases, largely explained by the long delays pesticides can take to leach through soils into aquifers.

Estuarine and coastal agricultural nutrient pollution is also an issue in some regions causing algal blooms (i.e. "red tides" or "dead zones"), damaging marine life, including commercial fisheries in coastal waters adjacent to Australia, Japan, Korea, the United States and Europe, mainly the Baltic, North Sea and Mediterranean. This is evident in the widespread problem of eutrophication reported in surface water across OECD countries, and the damage to aquatic organisms from pesticides.

Two examples of coastal pollution from agricultural activities are found in the Great Lakes of North America and the Australian Great Barrier Reef. The Great Lakes ecosystem is stressed by farm nutrients, pathogens, pesticides and soil sediments from both Canadian and United States sources, which threatens recreational opportunities and raises the costs of treating drinking water and dredging harbours. There has been some improvement in certain areas of the Great Lakes such as the attainment of guideline levels of phosphorus for all lakes (except Lake Erie), due to a reduction of phosphorus inputs from agricultural, municipal and industrial sources. There is evidence that Canadian agricultural nutrient inputs (especially phosphate) to the Great Lakes could be declining as a result of improved farm management practices. Nutrient surpluses are an issue in some key watersheds such as Lake Winnipeg which is showing signs of eutrophication, although farming is not the only source of nutrient pollution.

WATER QUALITY MANAGEMENT

The quality of water entering the Australian Great Barrier Reef is of concern. Water quality entering the Great Barrier Reef has declined, affecting about 25% of its area, partly as a result of farm pollutants, although phosphorus run-off from urban sewerage is also a problem. The dry tropical regions in Queensland are the main source of these pollutants, although some farmers are adopting practices to reduce pollution. While evidence of adverse impacts on the Great Barrier Reef from pollutants is not conclusive, research suggests the need for caution for any activities leading to elevated pollution levels.

Farm Management Practices and Agricultural Water Pollution

A growing number of OECD farmers are adopting environmental farm management practices, as a result of voluntary private-led initiatives intended to respond to consumer concerns, including those from food processors and retailers (e.g. pesticide management) and government incentives provided through payments and regulations. But only around one-third to one-half of OECD member countries are regularly monitoring changes in environmental farm management practices, with the notable exception of organic management where all countries are tracking trends in organic farming.

The adoption of nutrient management practices is widespread across OECD countries, with an increase in their uptake over the period 1990–2004, for around half of the OECD countries monitoring nutrient management practice (Figure 8). For countries with a high

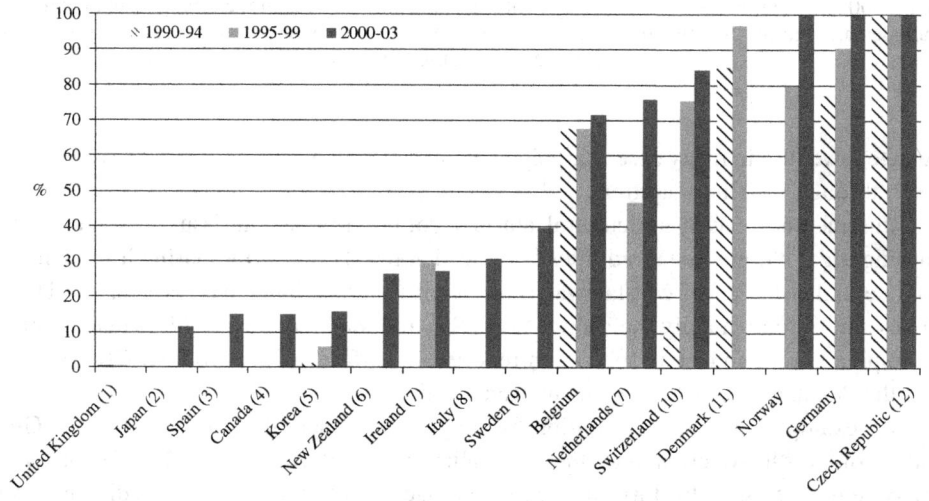

Figure 8. Share of the total number of farms under nutrient management plans for 1990–2003. Nutrient management plans cover nitrogen, phosphorus and potassium, unless stated otherwise. *Notes:* 1. Data for the UK are 0.2% in 1990–1994 and 0.2% in 1995–1999; 2. Chemical fertilizers only; 3. Nutrients are not specified; 4. *Source:* Statistics Canada, 2001 Farm Environmental Management Survey; 5. Data include only arable crops; 6. Nutrients covered by the management plan: nitrogen, phosphorus, potassium sulphur, magnesium and other nutrients; 7. The nutrient management plan covers nitrogen and phosphorus; 8. For 2000 only and the plan applies only to inorganic fertilizer; 9. The number of farms that get support for a crop management plan including a nutrient management plan; 10. Data for 1990–1994 and 1995–1999 refer to the years 1993 and 1999; 11. Data are for 1994 and 1995–1998; 12. The nutrient management plan covers nitrogen, phosphorus, potassium, calcium and magnesium. *Source:* OECD (2008).

and increasing uptake of nutrient management practices they have usually experienced a reduction in nutrient surpluses (Belgium, Czech Republic, Denmark, Finland, Germany, the Netherlands, Norway, Sweden, Switzerland), but for countries where nutrient surpluses have risen or are well above the OECD average (in terms of kilograms of nutrients/hectare of farmland) (Canada, Ireland, Japan, Korea and New Zealand), nutrient management practice adoption rates are generally lower, although increasing in Canada and Korea (Figure 2).

Despite the increase in adoption of environmental integrated pest management practices, the level of uptake across OECD countries is modest, although only about one-third of OECD countries track integrated pest management. But for countries with a high integrated pest management uptake or growth in organic farming (Figure 9) they have also experienced a decrease in pesticide use (Austria, Czech Republic, Denmark, Finland, Germany, Norway, Sweden, Switzerland, United Kingdom, United States) (Figure 3).

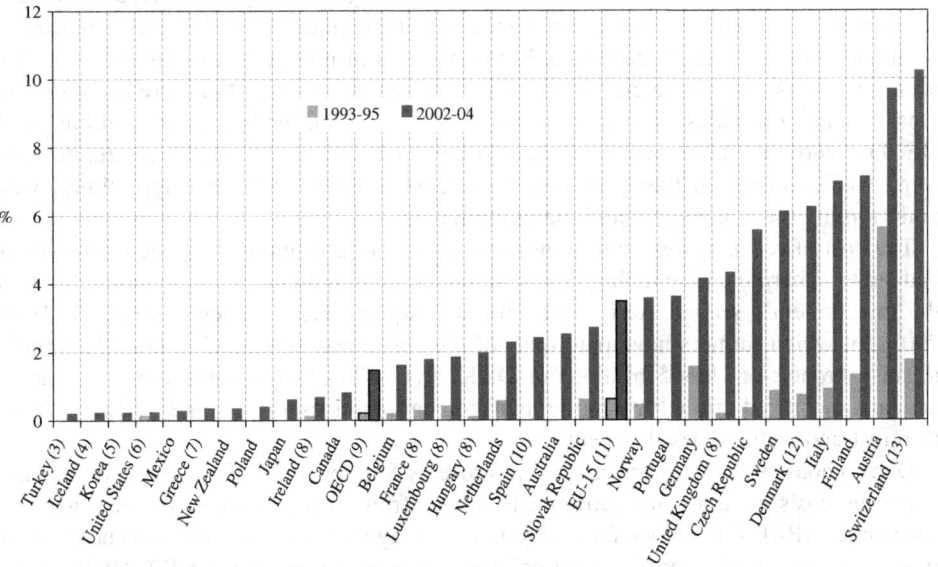

Figure 9. Share of the area under organic farming from the total agricultural land area in 1993–1995[1] and 2002–2004[2]. Notes: 1. Data for 1993–1995 are unavailable for Australia, Canada, Japan, Mexico, New Zealand, Poland and Portugal; 2. Data for 2002–2004 are taken IFOAM; 3. The value for 1993–1995 is 0.01. Data for 1994 are taken from *Environmental Indicators for Agriculture* (OECD, 2001); 4. The value for 1993–1995 is 0.004. Data for 1993–1995 are taken from Environmental Indicators for Agriculture (2001); 5. The value for 1993–1995 is 0.001. Data for 1995 are taken from Environmental Indicators for Agriculture (2001); 6. The value for 1993–1995 equals that for data for 1997 (0.14%) and the value for 2003 equals that for data 2001 (0.25%); 7. The value for 1993–1995 is 0.02; data are taken from Environmental Indicators for Agriculture (2001); 8. Data for 1993–1995 are taken from IFOAM, other data are taken from Environmental Indicators for Agriculture (2001); 9. Australia, Canada, Japan, Mexico, New Zealand, Poland and Portugal are not included in the OECD for 1993–1995; 10. Data for the period 1993–1995 refer to the year 1993, the value is 0%; 11. Portugal is not included in the EU-15 for 1993–1995; 12. Data for 1994 equals the 1993–1995 average; 13. Data for the period 1993–1995 refer to the year 1993. *Source:* OECD (2008).

The area of farmland under soil management practices has remained stable over the past decade, but only one-third of OECD countries monitor changes in soil management practice. Where the rate of soil management practice adoption has risen (Canada, United States), this has led to reduced soil erosion risks and greater provision of habitat for wild species, although where soil management practice uptake rates are low, soil degradation problems remain (Hungary, Italy, Korea, Slovak Republic, Turkey).

The OECD area under certified organic farming has increased substantially between the early 1990s and 2004, even so it accounted for less than 2% of total farmland by 2002–2004 (Figure 9). However, the share is higher in most European countries (around 6% or higher in Austria, Denmark, Finland, Italy, Sweden and Switzerland) but much lower in mainly non-European OECD countries (under 1% in Canada, Japan, Korea, Mexico, New Zealand and the United States).

Economic Costs of Agricultural Water Pollution

The economic cost of agricultural water pollution is high in many cases. Treating water to remove nutrients and pesticides to ensure water supplies meet drinking standards is significant in some OECD countries. Eutrophication of marine waters also imposes high economic costs on commercial fisheries in some cases. To date there are no systematic cross-country estimates of the economic costs of agricultural pollution, although the OECD is currently undertaking a review in this area (OECD, 2011). There are, however, some studies available that give partial insights into the costs that agricultural water pollution imposes on society and the environment.

The cost of reducing agricultural nutrient pollution in Denmark has been considerable and led to a sharp increase in the price of water for household users. The overall cost under the government's second Action Plan for the Aquatic Environment (APAE II, 1998–2003), of which farmers have paid 60% of the costs, was estimated at DKK525 million (Danish kroner; or US$65 million) or DKK15 (US$2)/kg of avoided nitrogen leaching annually, achieved through changing management practices and changes in land use such as forest and wetland development.

Danish household water prices (pre-tax) rose by 58% between 1988 and 1999, in part to cover the costs of removing nutrient discharges from water. A range of measures were used under APAE II to reduce farm nitrogen leaching with their cost-effectiveness varying from an average per kilogram reduction in nitrogen (N) leaching of DKK7 (US$0.9)/kg N for the creation of wetlands to over DKK75 (US$9)/kg N for limits on livestock density. But the economic benefits from lowering farm nutrient loads are currently unknown, although the physical loads of nutrients have been lowered significantly (Figure 2).

In the Netherlands water pollution originating from agriculture is an important environmental concern. While recent trends indicate that the pressure from farming on water quality is diminishing, absolute levels of pollution remain amongst the highest across the OECD (Figures 2–5). Agriculture is the major source of nutrients, pesticides and the only known source of heavy metals in water. Pollution from endocrine disrupters and veterinary medicines in terms of potential impacts on human and wildlife reproductive systems is also a concern. The total external costs of agricultural water pollution are unknown, but in the late 1990s the annual external costs of eutrophication associated with nitrate emissions was estimated at €600 million (US$540 million), and for treating drinking water polluted with nitrates at an annual cost of €23 million (US$21 million).

The UK has undertake some extensive research on estimating the economic costs and benefits of agriculture on the economy and environment. UK agriculture is a major source of water pollution entailing high costs. As urban and industrial water pollution is largely controlled, diffuse pollution is becoming comparatively more important especially farm run-off of nitrates, phosphorus, pesticides and pathogens, mainly of agricultural origin and concentrated in England. The overall cost of water pollution from agriculture was estimated in 2003/04 at around £500 million (€725 million) annually, contributing over 40% of total water pollution costs. Nearly half of the prosecutions for pollution by the agricultural sector in 2002–2003 were related to water pollution incidents, mainly from the dairy sector. Almost 5% of Sites of Special Scientific Interest (e.g. bogs, upland heath) in England in 2005 were in an unfavourable condition because of agricultural water pollution.

UK pesticide use declined by 6% during the period from 1990–1992 to 2001–2003 (Figure 3), but the trend has been variable, linked to changes in cropping patterns and weather conditions. Farming uses almost 90% of pesticides and accounts for most pesticide water pollution incidents. Removing pesticides from drinking water supplies is estimated to cost around £110 million (€160 million) annually. Pesticide incidents involving terrestrial wildlife remain a concern, although the area of cereal field margins, which can help reduce these incidents increased from under 5,000 to over 40,000 hectares between 1997 and 2004, while the area under crop protection management plans is also expanding.

OECD Policies to Address Agricultural Water Pollution

Overview

Water reform programmes are being implemented across the range of national to watershed scales in many OECD countries, while these programmes usually involve, but are not specific to, the agricultural sector. There is a growing recognition that water policies should be coherent across different scales of decision-making, including from the farm to water catchment, national and international levels, and also between the different users (e.g. urban, industry) and uses of water (e.g. aquatic ecosystems, recreational uses). The need for policy coherence is also important across agricultural, environmental and water policies, especially to avoid conflicting signals and incentives to farmers in achieving sustainable water management.

Policy responses to address water quality issues in agriculture need to be part of a policy package that encompasses a range of policy instruments, institutional reforms and broader community engagement. Water policies and institutions need to focus on the public good (e.g. maintaining aquatic ecosystems) and market failure aspects of water resources (e.g. resource depletion and pollution), by facilitating stakeholder involvement, developing information (data) and knowledge (science), and enabling public access to this information. Moreover, given the high level of vulnerability of agricultural systems and water resources to climate change and climate variability, policies will need to be increasingly responsive and flexible in adapting to these changes.

There is a diversity of policy approaches to control agricultural water pollution across OECD countries (as discussed in further detail below) with different emphasis on taxing water pollutants, and payments and regulatory policy approaches to achieve water policy goals. There is also increasing emphasis being placed in many countries in establishing decision support tools and risk-management strategies to improve water management by farmers. Policy focus, however, tends to be on surface water (visible) so attention to the overuse and pollution of groundwater (invisible) also needs to be strengthened.

Understanding the links between agriculture, water use and water quality can help target the appropriate policy responses (Figure 10). Pressure on water quality from agricultural activities can be caused by poor land-management practices (e.g. poorly timed manure spreading, dryland salinity through tree felling, tillage practices exacerbating soil sedimentation run-off). While pressure on water resources (quantity) is largely the result of excessive extractions, the modification of flow regimes through storage, the poor management of irrigation infrastructure and inadequate uptake of efficient water application technologies by irrigators leading to water wastage and inefficiencies.

For countries where water pollution linked to agriculture has been acute, this has prompted them to take action earlier than other countries. Some countries are building on and adapting existing institutional structures to implement water reform programmes and others, at an earlier stage with their reform programmes, are in the process of creating the required institutions.

Some countries are refining, developing and introducing market-based approaches for water pollution, but little evaluation of their economic efficiency and environmental and social effectiveness has been undertaken. Moreover, clearer identification and enforcement of property rights is required if water market approaches are to be developed.

Well-defined and enforceable property rights are the cornerstone of democratic and economic systems in all OECD countries, with most water rights relating to a right to use water or allow discharges into water, both of which provide the foundations of a water trading system. But limits are usually imposed on this right (e.g. drawing water or

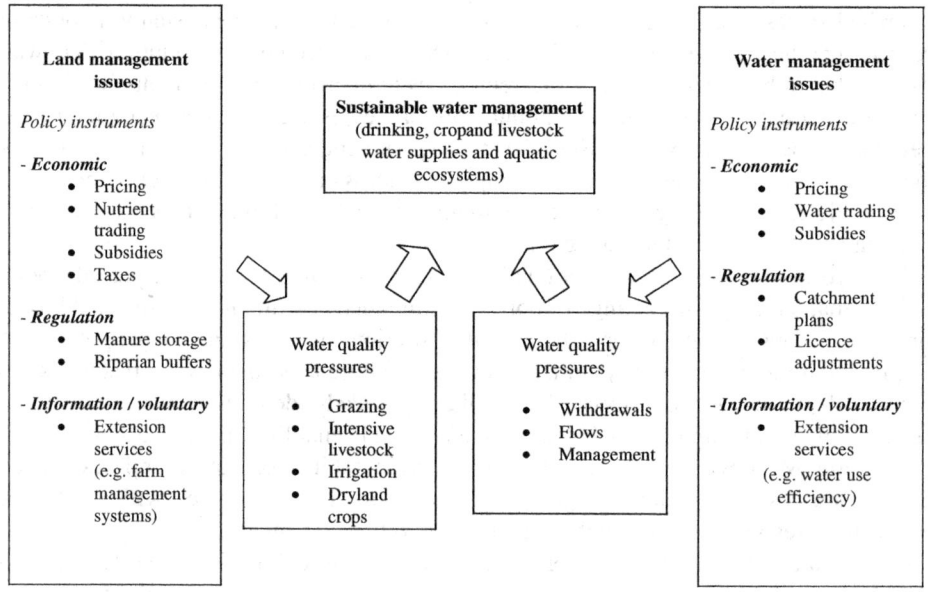

Figure 10. Sustainable water management. *Source:* OECD (2006).

discharging waste into water bodies), and some countries are now engaged in the process of separating water entitlements from land title rights.

The main focus and developments in OECD policies addressing agricultural water pollution can be summarized as follows:

- Policy focus has largely addressed nutrient (nitrogen and phosphorus) and pesticide pollution, with less emphasis on other pollutants.
- A mix of policy instruments are usually used to address water pollution, but the use of payments and regulatory instruments are prominent supported by farmer education and advice.
- Most programmes providing payments to help toward reducing pollution are on a voluntary basis, but in more highly polluted zones programmes tend to be mandatory.
- There are often infringements and poor enforcement of regulatory measures to control water pollution related to agriculture.
- Application of the polluter-pays principle to water pollution in agriculture has proved difficult to apply, although taxes on pollution have been used in some countries
- Use of market based instruments, for example nutrient trading, to address pollution has been very limited but interest is growing across OECD countries.

An OECD (2007) survey of policy objectives and policy instruments addressing agricultural nutrient and pesticide run-off singled out 93 relevant national policy objectives across OECD countries. Of these objectives nearly 50% concerned nutrient run-off, almost 40% pesticide run-off and the remaining 10% addressed both issues. The survey identified 346 policy instruments, of which almost 60% addressed nutrient run-off, nearly 35% addressed pesticide run-off and about 5% addressed both issues. Some 40% of these policy instruments were regulatory instruments, with economic instruments the most common policy instrument used to address both issues, dominated by subsidies. The use of farmer advisory services and other information instruments were also commonly used policy instruments. Overall taxes and charges play a minor role and thus there has been little emphasis on enforcing the polluter-pays principle when governments have addressed agricultural nonpoint source pollution.

Use of Regulatory Policy Instruments to Control Agricultural Water Pollution

Regulatory policy instruments are commonly used in OECD countries to address water pollution from agriculture. These regulations concern those on the use of potentially polluting inputs such as pesticides, industrial fertilizers and manure (storage, management and field application) and land-management measures to prevent the polluting agents from reaching surface waters and/or groundwater. Since the 1980s there has been a growing use in regulatory measures to protect surface water and groundwater, in particular the following (OECD, 2010a):

- *Overall input regulations.* An important aim in all OECD countries is to reduce pollution generated by the use of agricultural inputs, in part by using regulations concerning the marketing and sale of chemical inputs, especially pesticides. Regulations have typically been amended over time such that many countries now approve new pesticides for a limited period only (commonly 5-10 years). Some requirements relating to inputs have been implemented in response to international pressures, e.g. the reduction of nutrients into the Baltic Sea under the Convention to Protect the Baltic Sea (HELCOM Convention).

- *Pesticide regulations.* All OECD countries set regulations concerning the storage, and application of chemical fertilizers and pesticides. The aerial spraying of pesticides is now prohibited in some parts of the European Union and Australia. It is heavily controlled in many other regions and countries, with licences or permits commonly required. In many OECD countries, the use of pesticides is now restricted within a certain distance of watercourses. In the European Union a process is underway to enhance the adoption of integrated pest management.
- *Nutrient regulations.* While regulations prohibiting the direct discharge of livestock waste to surface waters have existed in most OECD countries from the early 1970s, a large number of regulations have since been applied in relation to general farming practices associated with pollution from nutrients. In particular, OECD countries have introduced a variety of requirements relating to manure management in order to limit pollution from livestock farming, including restrictions on the quantity of manure that can be spread; seasonal bans on manure application; manure storage requirements; and limitations on livestock densities and on the expansion of livestock units. Many OECD countries have also tightened regulatory requirements relating to the application of nutrients, either at the national or state/regional level.
- *Regulations concerning the scale of production.* In some OECD countries large-scale livestock production units are controlled through a system of permits, either at the national or regional level. For example, the European Union Integrated Pollution Prevention and Control Directive, which has been applied since 1999 to new facilities (and is to be applied to existing facilities from 2007), requires member states to impose emission limits in environmental permits which are mandatory for potentially polluting plants of a given scale, especially large scale pig and poultry facilities. In Japan, under the Water Pollution Control Law and other associated legislation, upper limits are set for discharges of pollution for specified agricultural facilities, including large-scale pig and cattle facilities.
- *Regulations with regard to buffer strips and catch crops.* Buffer strips around water courses and groundwater sources have become a common requirement to limit nutrient leaching in many OECD countries, including Australia, Canada, France and New Zealand. Some governments have also established regulations requiring farmers to maintain a minimum level of green cover during certain times of the year (catch crops). Requirements for catch crops are notably stringent in Denmark and some parts of Sweden.

Use of Payments to Control Agricultural Water Pollution

Payments to farmers are widely used across OECD countries to control water pollution and encourage the reduction in other polluting emissions (e.g. ammonia), together with payments to promote environmental benefits. There has been a substantial increase across most OECD countries in the use of agri-environmental payments to achieve environmental objectives in agriculture (Table 1). The most common form of payments to control water

Table 1. Total agri-environmental payments[a] in selected OECD countries, 1996–2008.

		1996	1997	1998	1999	2000	2001	2002	2003	2004	2005	2006	2007	2008
EU[b]	EUR million	3,004	3,817	3,931	4,390	5,623	5,828	5,250	5,133	5,527	6,118	6,525	5,620	6,809
	1996 = 100	100	127	131	146	187	194	175	171	184	204	217	187	227
Norway	NOK million	923	922	994	1,043	1,071	1,001	1,198	683	695	712	874	966	998
	1996 = 100	100	100	108	113	116	108	130	74	75	77	95	105	108
Switzerland[c]	CHF million	605	721	689	177	184	193	203	213	224	231	233	239	245
	1996 = 100	100	119	114	29	30	32	34	35	37	38	39	40	40
United States	USD million	2,690	2,731	3,030	2,676	2,751	2,964	3,501	4,093	4,550	4,911	4,946	4,524	4,876
	1996 = 100	100	102	113	99	102	110	130	152	169	183	184	168	181

Notes: [a] Agri-environmental payments used provide support to farmers for undertaking farming practices designed to achieve specific environmental objectives that go beyond what environmental regulation requires. Farm support related to respecting regulations (environmental cross-compliance) and payments to less favoured areas are not included here as agri-environmental payments. Discussion on which payments to less favoured areas can be considered as agri-environmental payments is ongoing in the OECD in the context of the Inventory project.
[b] EU-15 in 1996–2003; EU-25 in 2004–2006; and EU-27 from 2007.
[c] In Switzerland up to 1998 the most important part of agri-environmental payments was those for integrated production. Since 1999, these payments have been abolished and the regulatory requirements for integrated production are compulsory for all direct payments (environmental cross-compliance). However, these payments are not included as part of "agri-environmental payments". This change in policy is reflected by the sharp drop in agri-environmental payments in 1999.
Source: OECD (2009).

pollution are for agricultural production conditional upon reduced use (or no use) of pesticides and fertilizers (such as extensive production, integrated production, organic farming), green cover and buffer strips.

The European Union Nitrate Directive, for example, defines areas vulnerable to nitrates in its member states, and sets guidelines to establish the maximum permitted level of nitrates in water. Moreover, the action programmes developed to implement the Directive, establish the necessary measures to ensure that nitrogen of animal origin spread on the land (manure fertilization) does not exceed 170 kg/hectare. It also makes it mandatory for farmers to ensure that fertilizer use is well balanced to supply the needs of crops. European Union member states have designed and implemented some agri-environmental measures to further reduce nitrogen losses in water that go beyond the statutory obligations. Reduced use of fertilizers, converting arable land to extensive grassland (pasture), green cover and crop rotation are the main instruments implemented by member states to reduce nitrates in water. In addition, the European Union Water Framework Directive imposes the objective of achieving good water status by 2015.

In the European Commission's (2010) third report on the implementation of the Nitrates Directive across the 27 European Union member states for the period 2004–2007, it noted some improvement in nitrate pollution of water bodies. Some 66% of groundwater monitoring stations and 70% of surface water stations showed stable or decreasing nitrate concentrations over this period. However, for about one-third of groundwater monitoring stations an increase in nitrate pollution was observed, while one-third of freshwater stations were defined as eutrophic.

The European Commission noted in its report that action programmes to address nitrate pollution of water were improving across member states. Even so, this appears to be mainly driven by response to infringement procedures. Moreover, some national programmes need further improvements to comply fully with the Nitrates Directive.

Using Taxes to Control Agricultural Water Pollution

Policy measures imposing a tax or charge relating to pollution or environmental degradation include taxes and charges on farm inputs or outputs that are a potential source of environmental damage. The implementation of taxes and charges appears to be rare in agriculture, compared with other sectors. This may at least partly reflect practical problems of measurement, unlike a factory where pollution can normally be monitored at "point", the pollution from agriculture is much more dispersed, as it tends to originate from many different independent farms and in varying intensities.

Nonetheless, some examples of these policy measures do exist, including the Netherlands which has tackled the measurement problem by introducing a range of levies on off-farm nutrient emissions above a set limit. Since 2006, the system directly regulates the maximum amount of fertilizers (animal manure plus maximum amounts of nitrate and phosphate) that may be used on the farm. Similar taxes on the estimated on-farm generation of nutrients over set levels are also in place in Belgium.

In agriculture, environmental taxes are more often applied on the sale of inputs identified as having a potentially adverse impact on the environment. For example, various taxes and charges are currently levied on pesticides in Canada (British Columbia), Denmark, Finland, Italy, Norway and Sweden, while fertilizer levies are applied in Italy, Sweden and some states of the United States. Input-based taxes are generally inexpensive to administer, but may be less effective than a tax on pollution itself, as they do not discriminate on the basis of actual loading on the environment.

Tradable Permits to Control Agricultural Water Pollution

Tradable permits for regulating environmental externalities can often achieve environmental targets at lower social cost than traditional design and performance standards and environmental taxes. Trading offers a mechanism for achieving a cost-effective allocation of environmental effort across alternative sources without environmental regulators knowing the abatement costs of individual agents. To date, however, few countries have utilized this policy instrument to address water pollution in agriculture, although the Netherlands implemented a system of tradable permits in relation to the volume of manure produced by farms.

Water quality trading markets are complex because there is plenty of uncertainty about sources and levels of emissions, about effectiveness of different abatement measures and water quality impacts of effluents originating from different sources. When developing a water quality trading market, the policy-maker has to first define the tradable commodity for nonpoint polluters (e.g. fertilizer use reduction or establishment of buffer strips and green set-asides). A trading ratio has to be determined that takes into account delivery of pollutants and imperfect substitution between point and nonpoint emissions (on the basis of relative uncertainty related to reduction of emissions from these two sources). Finally the aggregate supply of permits has to be limited (capped) so that water quality targets are met.

Other Policy Instruments Used to Control Agricultural Water Pollution

Other policy instruments are also used to control water pollution from agriculture and these mainly concern research and development programmes, technical assistance and farm advisory services, and in some countries community base measures. Agricultural research and development programmes that seek to establish best management practices and develop improvements in technologies, usually provide support to farm advisory services for farmers. These measures provide farmers with on-farm information and technical assistance to plan and implement environmentally friendly farming practices. Most OECD countries have long-established programmes for assisting farmers to adopt technology and improve agricultural practices. Such programmes have traditionally focussed on improving on-farm productivity, but in the past two decades much greater emphasis has been placed on increasing farmers' understanding of resource and environmental issues, in order to induce voluntary changes in farming practices to improve environmental outcomes.

In some countries, notably Australia, Canada and New Zealand, government led information policies are supplemented by the growing use of community-based approaches promoting the exchange and transfer of information, variously known as landcare groups or conservation clubs. These approaches make use of local expertise in solving environmental problems that thereby enhance environmental conservation, and rely upon the self interest of farmers. Such groups seem especially well suited to address issues that are local in nature, but which extend beyond the borders of a single farm. Some of these groups receive administrative or financial support from central or regional authorities, while others are entirely self-financed and independent.

WATER QUALITY MANAGEMENT

Future Prospects for Agricultural Water Pollution

Despite the significant impact of the global financial crisis and economic downturn on all sectors of the economy, agriculture is expected to be relatively better-off compared with the 1997–2006 period, as a result of the recent period of relatively high incomes and a relatively income inelastic demand for food. The OECD-Food and Agriculture Organization (FAO) Agricultural Outlook projections over the next 10 years to 2019 paint a picture of sustained crop prices in nominal and even in real terms (allowing for inflation) that remain well above the levels observed prior to the 2007–2008 price peaks, i.e. during the 1997–2006 period. Most livestock prices, in contrast, are expected to remain close to the average levels for the next 10 years in real terms (Figure 11).

According to recent FAO work using longer term population and income projections, global food production needs to increase more than 40% by 2030 and 70% by 2050, compared with average 2005–2007 levels. There is substantial additional land available for use in agriculture. Some 1.6 billion hectares could be added to the current 1.4 billion hectares of cropland. Over half of the additionally available land is found in Africa and in Latin America. These regions account for most of the available land that has the highest suitability class for rain-fed crop production. But historical expansion of arable land has been slow, and bringing more marginal land into production can involve considerable investment and lower average yields, while possibly incurring social and environmental costs.

Overall these medium- and long-term projections for agricultural production would suggest the following impacts on water pollution across mainly OECD countries:

- Pressures that could lead toward greater water pollution originating from agriculture include:
    - increasing global demand for food leading to growing production levels and intensities in OECD agricultural exporting countries;
    - higher international agricultural commodity prices, including growth in demand for agricultural feedstocks to produce bioenergy, also leading to

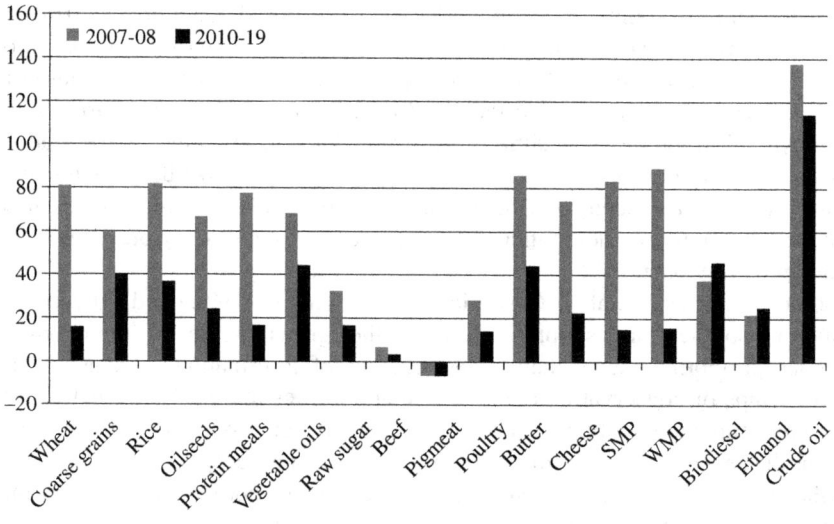

Figure 11. OECD projections for commodity prices in real terms to 2019. *Source*: OECD (2010b).

further expansion in farm production; and
- impact of climate change and climate variability on agriculture.
- Changes that might reduce water pollution originating from OECD agriculture, include:
  - improvements in farm management and related technologies, especially biotechnologies and use of geo-positional systems (GPS);
  - increase in public pressure to reduce the health and environmental costs of water pollution from agriculture; and
  - decline in overall OECD agricultural support and increasing trend toward decoupled support.

Toward Sustainable Management of Water Quality in Agriculture

Policy responses to address water quality issues in agriculture need to be part of a policy package that also encompasses water resource (quantity) issues and a range of policy instruments, institutional reforms and broader community engagement (OECD, 2010c). Water policies and institutions need to focus on the public good (e.g. maintaining aquatic ecosystems) and market failure aspects of water resources (e.g. resource depletion and pollution), by facilitating stakeholder involvement, developing information (data) and knowledge (science), and enabling public access to this information. Moreover, given the high level of vulnerability of agricultural systems and water resources to climate change and climate variability, policies will need to be increasingly responsive and flexible in adapting to these changes.

Water management in agriculture is evolving toward a more holistic and integrated approach by considering water quality and water quantity issues. Consequently, there is increasing integration of land use with water-use management decisions to both help conserve water and enhance water quality, but also to promote the potential of agriculture to provide multiple environmental benefits and services. But further effort is required if policy-makers, ranging from decision-makers at the watershed through to national levels, are to move toward the sustainable management of water quality in agriculture. Among the challenges include:

- seeking opportunities to apply the polluter-pays principle to agriculture, so as to internalize the externality costs from agricultural water pollution;
- using an appropriate mix of instruments and tools aimed at addressing agriculture water pollution issues to ensure the achievement of coherent agricultural, environmental and water policy goals as well as cost-effective implementation, including coordinated policy responsibilities and structures at different levels from the watershed to national level;
- identifying property rights attached to water discharges and ecosystem provision;
- establishing clear lines of responsibility in the institutional framework to manage water—who does what, who pays for what, who monitors and evaluates—underpinned by a long-term commitment from governments to resource the necessary actions, especially with the growing concerns related to climate change and climate variability;
- strengthening water policy reforms to provide a robust regulatory framework to allow, for example, nutrient trading for pollution abatement;
- raising the capacity for farmers, industry, and community groups to participate in the

design and delivery of policy responses for integrated water management, including encouraging and developing greater farmer uptake of practices and technologies to reduce water pollution;
- assessing the economic and environmental trade-offs (manure spreading) and co-benefits (riparian buffers) between water pollution and other environmental policies; and
- integrating and expanding current scientific research and data-collection capacity to underpin improved policy-making.

Acknowledgements

This paper is published under the authorship of Kevin Parris and does not necessarily reflect the views of the OECD or its member countries. The paper largely draws from the OECD publications shown in the bibliography, especially OECD (2011).

References

Environment Protection Agency (EPA) (2005) *The Quality of Drinking Water in Ireland: A Report for the Year 2004* (Wexford, Ireland: Environmental Protection Agency).

European Commission (2010) *Report from the Commission to the Council and the European Parliament on Implementation of Council Directive 91/676/EEC Concerning the Protection of Waters against Pollution Caused by Nitrates from Agricultural Sources Based on Member State Reports for the Period 2004–2007.* SEC(2010)118, COM(2010)47 final (Brussels: European Commission).

European Environment Agency (EEA) (2005) Integration of environment into EU agriculture policy, the IRENA indicator-based assessment report, Copenhagen, Denmark. Available from: http://webpubs.eea.eu.int/content/irena/Latestproducts.htm

Organisation for Economic Co-operation and Development (OECD) (2001) *Environmental Indicators for Agriculture*, vol. 3 (Paris: OECD).

Organisation for Economic Co-operation and Development (OECD) (2004) *Environmental Performance Review of Spain* (Paris: OECD).

Organisation for Economic Co-operation and Development (OECD) (2006) *Water and Agriculture: Sustainability, Markets and Policies* (Paris: OECD). Available from: http://www.oecd.org/water/.

Organisation for Economic Co-operation and Development (OECD) (2007) *Instrument Mixes Addressing Non-point Sources of Water Pollution* (Paris: OECD). Available from: http://www.oecd.org/env/.

Organisation for Economic Co-operation and Development (OECD) (2008) *Environmental Performance of Agriculture in OECD Countries since 1990* (Paris: OECD). Available from: http://www.oecd.org/tad/env/indicators/.

Organisation for Economic Co-operation and Development (OECD) (2009) *PSE/CSE Database* (Paris: OECD). Available from: http://www.oecd.org/tad.

Organisation for Economic Co-operation and Development (OECD) (2010a) *Guidelines for the Design and Implementation of Cost-effective Agri-environmental Policy Measures* (Paris: OECD). Available from: http://www.oecd.org/tad/.

Organisation for Economic Co-operation and Development (OECD) (2010b) *OECD-FAO Agricultural Outlook 2010–2019* (Paris: OECD). Available from: http://www.oecd.org/tad/.

Organisation for Economic Co-operation and Development (OECD) (2010c) *Sustainable Management of Water Resources in Agriculture* (Paris: OECD). Available from: http://www.oecd.org/agr/env/water/.

Organisation for Economic Co-operation and Development (OECD) (Forthcoming, 2011) *Sustainable Management of Water Quality in Agriculture* (Paris: OECD). Available from: http://www.oecd.org/agr/env/water/.

New Zealand Government (2004) *Growing for Good, Intensive Farming, Sustainability and New Zealand's Environment*, October. Available from: http://www.pce.govt.nz/.

Regulating Nonpoint Source Water Pollution in a Federal Government: Four Case Studies

SUSAN GRAHAM, ADAM SCHEMPP & JESSICA TROELL
Environmental Law Institute, Washington, DC 20036, USA

ABSTRACT *Without effective regulation, nonpoint source water pollution is likely to increase as growth continues across the globe. This paper explores the history of and policy, legal and regulatory options for addressing nonpoint source water pollution in countries with a federal government. The legal mechanisms for controlling nonpoint sources of water pollution at the national and state level in four countries are identified and analysed. While the forms of control and level of governance at which these pollutants are regulated vary among the countries explored, the effectiveness of their control will depend largely on how strategic rather than patchwork the structure is.*

Introduction

Across the globe, nonpoint sources of pollution continue to pose a significant threat to water quality. "Nonpoint source" water pollution is pollution from one or more diffuse sources, as opposed to a discharge from a discrete "point source" such as an outfall or pipe. Such pollution is often, though not exclusively, the result of agricultural, urban or industrial runoff from land. Because of their diffuse nature, nonpoint sources have proven extremely difficult to regulate.

In most countries, regulatory emphasis is placed on the control of point source discharges, usually through a system that grants permits to dischargers specifying what level of pollution is allowable and what specific technological or other controls must be undertaken to prevent water quality degradation. Point sources are generally easier to regulate because the party responsible is more readily identifiable and it is possible to determine the exact level of pollution being contributed from that source. In many instances, these regulations have been successful, demonstrating improvements in the water quality of numerous countries. As efforts to reduce water pollution from point sources continue, the impact of nonpoint sources on water quality is becoming more evident. In some instances, meeting even minimum water quality standards can only be accomplished by addressing nonpoint sources.

Opportunities for addressing nonpoint sources depend partly on the system of governance in a country. Those countries with a federal form of government have the added complexity of finding the proper balance of national and state or provincial authority to control nonpoint source pollution most effectively, particularly when watersheds transcend state or provincial boundaries. This paper identifies the variation in law concerning nonpoint sources of water pollution across four countries with a federal style of government. It identifies the relevant statutory and policy mechanisms at the national level of each country for regulating nonpoint sources of water pollution and provides examples of how those sources are being controlled at the state or provincial levels through federal legislative frameworks or independent regulatory regimes.

The four case-study countries: the United States, Australia, the Republic of South Africa[1] and India, were chosen based on several criteria, including general similarities in the structure of government and variation in the levels of development, structure of laws and experiences addressing water quality problems. Each country has had limited success in reducing water pollution from nonpoint sources, and this paper attempts to identify the potential role of the existing legal and regulatory measures to control these sources more effectively. Where possible, the paper highlights examples of implementation of these provisions. It concludes by highlighting key considerations that appear to be consistent across all four case study countries and might therefore guide attempts to address nonpoint source water pollution in a federal context.

Nonpoint Source Regulation in the United States

The National Law

The Clean Water Act (CWA) serves as the foundation for water quality regulation in the United States. Originally the Water Pollution Control Act of 1958, significant amendments were passed in 1972 that instituted many of the most notable provisions of the law as it is now constructed. The CWA has the express purpose of restoring and maintaining the chemical, physical and biological integrity of United States waters (33 U.S.C. § 1251). While the Act is federal, much of the authority for implementing it is delegated to the states, with oversight and assistance from the national Environmental Protection Agency (USEPA).

The states must conduct water quality monitoring, assess the health of waters in the state, make a list of "water quality-limited" bodies of water, and develop total maximum daily loads (TMDLs) that identify pollutant-specific loadings consistent with meeting applicable water quality standards (33 U.S.C. § 1313). The USEPA has published recommended water quality criteria, but each state is responsible for establishing its own required water quality criteria and indicating the categories of, or specific waters to which, each criterion applies (33 U.S.C. § 1313(a)(3)). The USEPA is tasked with reviewing the activities of the states as they fulfil these requirements, and in some cases must perform those tasks for the state if it fails to do so.

> A point source is defined in the CWA as any discernible, confined and discrete conveyance, including but not limited to any pipe, ditch, channel, tunnel, conduit, well, discrete fissure, container, rolling stock, concentrated animal feeding operation, or vessel or other floating craft, from which pollutants are or may be discharged. (33 U.S.C. § 1362(14))

A nonpoint source is any source not falling within this definition. Additionally, the CWA specifically states that return flows from irrigated agriculture and agricultural stormwater discharges are not point sources (33 U.S.C. § 1362(14)).

Point sources are regulated by permits under the National Pollution Discharge Elimination System (NPDES). In most cases, the discharge of a pollutant or combination of pollutants from a point source requires an NPDES permit (33 U.S.C. §§ 1311(a), 1342(a)). The CWA vests this permitting authority in the USEPA (33 U.S.C. § 1342(a)(5)), but this authority has, in most cases, been delegated to the states. There is no complementary federal permitting programme for nonpoint sources.

In 1987, the CWA was amended to address nonpoint sources of pollution. Section 319 of the amended Act requires each state to develop and implement a management programme for nonpoint sources with the objective of improving the quality of state waters (33 U.S.C. § 1329(b)(1)). Without a permitting authority or some other federally granted power to force the reduction of pollution from nonpoint sources, these state nonpoint programmes must primarily rely on promoting best-management practices through technical assistance, financial support and other non-regulatory tools (33 U.S.C. § 1329(b)(2)). A significant portion of the financial support for these efforts comes from the federal government. Federal funding for these programmes is currently approximately US$200 million per year (USEPA, n.d.).

Effect on the States

The structure of the CWA has created some notable complications for the states as they try to implement the law. Most states have the authority to mandate and enforce pollution reductions by point sources, but most states lack an enforceable mechanism for reducing contributions by nonpoint sources. They must instead rely on incentive-based initiatives. The lack of requirements in the CWA to link nonpoint source programmes funded under Section 319 to other aspects of state water quality control has resulted in a disconnect between other water quality protection measures, such as TMDLs, and those activities funded under the 319 programme.

Pursuant to the CWA, associated regulations and USEPA guidelines, state water quality programmes must undertake an ongoing process of water quality monitoring and assessment that leads to continuous updating of the state's list of "water quality-limited" waters. Water quality-limited waters are those bodies of water that fail to meet at least one of the state's water quality standards. A list of these waters must be reported biennially to the USEPA. For each listed water, the state must subsequently develop a TMDL. The TMDL designates how much of this pollutant load may come from specific point sources covered by individual NPDES permits, and how much may emanate from various categories of nonpoint sources.

To reduce point source contributions to a level that meets the amount identified in the TMDL, states may modify NPDES permits so that the sum of all the permits does not exceed the TMDL-mandated number. Achieving reductions in contributions by nonpoint sources is often more difficult. Without the ability to simply modify permits, the state must rely on non-regulatory tools. The incentives that states can offer with funding from the USEPA (pursuant to Section 319 of the CWA) can help in this process, but 319 programmes are not required to meet TMDLs. In many states, the 319 and TMDL programmes are not closely coordinated, with 319 programmes focused on a variety of other priorities. In recent years, the USEPA has issued guidance that attempts to foster linkages between state 319 programmes and other elements of the CWA, such as water quality standards and TMDLs. But the problem, and the effort toward a solution, is ongoing.

States have been able to achieve notable reductions from point sources, which have significantly improved water quality in the United States. However, there still is much to be accomplished. In order to achieve further significant improvements in water quality, nonpoint source controls must play a significant role. With existing point source controls, a larger percentage of the water quality problems originate from nonpoint sources (Copeland, 2001a). According to the USEPA, agriculture alone is responsible for as much as 70% of water quality-limited rivers and streams and 49% of water quality-limited lakes (Copeland, 2001b). Some of the environmental problems caused largely by agricultural runoff, such as the dead zone in the Gulf of Mexico, have even made national headlines.

State Solutions

Under these circumstances, states are employing a variety of methods to address nonpoint sources of pollution in a more coordinated way. Some states have even managed to pass legislation establishing permitting authorities over nonpoint sources. Others are actively directing funding from the 319 Program toward TMDL priorities or physically and/or procedurally integrating their state 319 and TMDL programmes. The following examples are a few of the more notable strategies devised by the states.

In 1969, the California Legislature passed the Porter–Cologne Water Quality Control Act, which provides for the regulation of nonpoint sources of water pollution in the State of California. The Act covers all waste discharges to land, surface water and groundwater regardless of source. It allows for waivers from the Waste Discharge Requirements (essentially a permit), but the waivers are conditional, may be terminated and may not exceed five years without renewal (Cal. Water Code § 13269). The temporary nature of the waivers pushed agricultural producers to choose simply to comply with Waste Discharge Requirements, and thus be voluntarily regulated.

In the State of Washington, the Water Pollution Control Act prohibits any discharges into waters of the state that cause or tend to cause pollution (Wash. Rev. Code § 90.48.080). The Washington Department of Ecology (DOE) has interpreted this statute to apply to nonpoint as well as point sources. In addition, the state's water quality standards require those activities that cause pollution from nonpoint sources to use best-management practices to abate such pollution (Wash. Admin. Code § 173-201A-510(3)(a)). Based on this language, the DOE views the failure to use a best-management practice to be a legal violation and authorizes the DOE to use orders, directives, permits, or civil or criminal sanctions to achieve compliance. While these provisions have not elevated nonpoint source permitting to the level of control possible with an NPDES point source permit, they have given the state significant authority to manage nonpoint sources for the purpose of water quality improvement.

With or without direct authorities over nonpoint sources of pollution, several states, including Washington, Florida and Indiana, have sought to address water quality limitations better by integrating their 319 and TMDL programmes. The Indiana Department of Environmental Management, for example, restructured its Watershed Planning Branch in 2005 in a manner that placed its 319 and TMDL programmes under the authority of a single manager. It also moved the location of the two programmes so that they are now housed in a single building. In addition to this integration, the department aligned the priorities and funding of the two programmes. The 319 Program now focuses its funding on watersheds with approved TMDLs and water segments listed as water quality limited. The TMDL Program now focuses its efforts in watersheds with active citizen groups, tries to anticipate the needs of those groups and supply them with the information they need and provides grantees of federal 319 funding with technical support for their programmes addressing nonpoint source pollution. While Indiana's water quality programme is still not among the better in the country, these changes have improved its productivity and effectiveness and helped it to overcome some of the obstacles stemming from the structure of the CWA.

Even some states with separated TMDL and 319 programmes have managed to match funding to water quality priorities. For example, the Massachusetts Department of Environmental Protection has a system for ranking applications for 319 federal grant funding to support nonpoint source reductions. This system was revised so as to give higher priority for funding to those projects proposing to implement best-management practices in response to a TMDL or watershed action plan or on lands affecting a water segment listed as water quality-limited.

Other state efforts for addressing nonpoint sources of pollution have included social incentives and direct coordination with farmers. The Virginia Department of Conservation and Recreation (VDCR) has made a specific effort to work with influential individuals in farming communities to adopt best-management practices on their lands and serve as an example of success for others in the community. Their experience shows that as more people in the community adopt the practices, the streams improve, the community receives accolades and non-adopters are pressured into cooperating. The VDCR also works with farmers and other stakeholders on regional water quality planning, finding them to be a valuable source of feedback regarding implementation.

Summary

The United States faces significant challenges with regard to addressing nonpoint sources of pollution. The country's primary water quality law, the Clean Water Act, addresses the issue with a funding programme rather than providing enforcement authority to the states tasked with controlling nonpoint sources of water pollution. To complicate matters, the funding programme is not inherently linked with enforceable means of identifying and improving water quality limitations. In varying ways, many states have attempted to mitigate these shortcomings of the national law. And while there have been some successes, the problem remains one of the most pressing water quality issues facing the country.

Nonpoint Source Regulation in Australia

National Laws and Policies

Under the Australian Constitution, the federal government (the Commonwealth) has responsibility for a limited set of matters, including taxation, defence and foreign affairs (Australian Constitution §§ 51, 52). The states retain control over those matters not specifically mentioned as being the responsibility of the Commonwealth or otherwise regulated by the Commonwealth in effectuating international treaties (Legal Services Commission of South Australia, n.d.). As there is no mention of the environment in the list of Commonwealth responsibilities, matters such as water quality have been left for the individual states to regulate.

Despite the lack of a direct constitutional authority, some environmental issues have been addressed through national policy initiatives, by agreement between the Commonwealth and the states. It was through one such national policy initiative that the National Environment Protection Council (NEPC) was established. The NEPC is comprised of representatives from the Australian Government and each state and territory and is now incorporated within the Environment Protection and Heritage Council (EPHC). While the NEPC has the power to make "National Environment Protection Measures" (standards, goals, guidelines or protocols) on issues including water quality (NEPCA § 14), it has not implemented measures relating directly to the control of nonpoint sources of water pollution (Department of Sustainability, Environment, Water, Population & Communities (SEWPAC), Government of Australia, n.d. a). The EPHC and the Natural Resource Management Ministerial Council, however, have developed a National Water Quality Management Strategy (NWQMS) (SEWPAC, n.d. b) that sets out uniform policies and guidelines for water quality management. Guidelines have been issued on topics relevant to nonpoint sources such as "Rural Land Uses" and "Water Pollution and Urban Stormwater Management" (SEWPAC, n.d. a). The NWQMS also establishes a framework for the development, in collaboration with the states and territories, of Water Quality Improvement Plans (WQIPs) (SEWPAC, n.d. c), which seek to improve water quality in identified "hotspots" (SEWPAC, n.d. d). A WQIP may include water quality monitoring, agricultural best-management practices, and water sensitive urban design, each with the potential to address nonpoint sources of pollution.

In a more recent development, innovative cooperation between the states and the federal government is leading to the development of a Basin Plan for the integrated management of water resources in the Murray–Darling Basin, which covers more than 1 million km^2 and includes parts of Queensland, New South Wales, Victoria and South Australia and the whole of the Australian Capital Territory (Murray–Darling Basin Authority, Government of Australia, n.d.). The Murray–Darling Basin Authority was established by the Commonwealth under the Water Act of 2007, which requires the development by the Authority of a Basin Plan for adoption on behalf of the Australian Government. Schedule 1 to the Act contains an agreement between the Commonwealth and the various Basin states, the purpose of which is

> to promote and co-ordinate effective planning and management for the equitable, efficient and sustainable use of the water and other natural resources of the Murray–Darling Basin, including by implementing arrangements agreed between the Contracting Governments to give effect to the Basin Plan, the Water Act and State water entitlements. (Water Act—Schedule 1 § 1)

The Basin plan is currently in the process of development and it remains to be seen how, if at all, it will seek to address nonpoint source water pollution. It is relevant to note, however, that the Water Act of 2007 provides that the Basin plan may not include a provision which directly regulates the control of pollution, land use or planning in relation to land use (Water Act § 22(10)).

In the agricultural sector in Australia, particular emphasis is placed on the education of farmers and the adoption of best-management practices. For example, the dairy industry initiative "Dairying for Tomorrow" (DfT) has developed region-specific action plans and "encourages collaborative partnerships between the dairy industry and catchment managers to set on-farm targets for change that will contribute to healthy catchments and communities" (DfT, n.d. a). It also promotes the use of an environmental self-assessment tool for dairy farmers known as DairySAT (DfT, n.d. b). Under the DfT programme, there has been a 25% increase in the number of farmers adopting best-management practices for nonpoint source control (DfT, n.d. c).

State Regulation: The New South Wales Approach

In the absence of overarching federal controls relating to water quality, each state has been left to develop its own regulatory framework. A good example of state action to address nonpoint source pollution comes from New South Wales (NSW). The key NSW statute is the Protection of the Environment Operations Act of 1997 (POEO Act). Section 120 of the POEO Act provides that it is an offense to pollute any waters or cause or permit any water to be polluted. This is a strict liability offense, and the NSW Environment Protection Authority, which is now incorporated within the NSW Department of Environment, Climate Change, and Water (DECCW), has successfully used it to prosecute emitters of nonpoint source water pollution (Environment Protection Authority, 2009). However, the requisite criminal standard of proof has been difficult to meet in some cases because of the potential for more than one contributing source.

The POEO Act also contains provisions that enable action to be taken to prevent or forestall pollution, which will likely prove to be a more effective mechanism for nonpoint source control than legal prosecution after the pollution has already been emitted. For example, it is an offense to pollute any land or cause or permit any land to be polluted (POEO Act § 142D). Also, the wilful or negligent disposal of waste or causing any substance to leak, spill or otherwise escape in a manner that harms (or is likely to harm) the environment are offenses (POEO Act §§ 115, 116). In addition, the POEO Act includes comprehensive licensing provisions with respect to certain prescribed "activities", including large-scale agricultural processing and intensive livestock production; specific provision is made to enable a licence to be issued to control the carrying out of any activity for the purpose of controlling water pollution (POEO Act §§ 42–48, Schedule 1).

In addition to the POEO Act, there are a number of other statutes in NSW that provide mechanisms for the control of nonpoint source water pollution. For example, the Native Vegetation Act of 2003 (NVA) and the Soil Conservation Act of 1938 (SCA) both provide indirect mechanisms for the control of water pollution caused by surface runoff. The NVA specifically recognizes the contribution made by native vegetation to water quality (NVA § 3(c)). Subject to certain exemptions, prior approval is required for the clearing of native vegetation (NVA § 12(1)). A "stop work" order may be made for any actual or anticipated contravention of the NVA (NVA § 37) and remediation of any unauthorized clearing may be ordered, including "work to ensure that specified land, or any specified river or lake, will not be damaged or detrimentally affected by the clearing" (NVA § 38). If such an order is not complied with, the authorities may act and recover the costs of doing so (NVA § 38).

The SCA provides for the conservation of soil and farm water resources and for the mitigation of erosion. A notice may be issued prohibiting or requiring acts in order to avoid or mitigate soil erosion and land degradation (SCA § 15A(1)). If the person upon whom the notice is served does not act, the authorities may do so and recover from that person any costs incurred (SCA §§ 15E–F).

Historic contamination of land is also a potential source of nonpoint source water pollution in NSW, and the system established in the Contaminated Land Management Act of 1997 (CLMA) for the investigation and remediation of contaminated land may prove to be a relevant tool for its control. Although at first glance this Act would seem likely to be of more relevance to industrial rather than agricultural nonpoint pollution, there are examples of land in Australia having been contaminated by historic agricultural activities. For example, the land on which many cattle dips were established in north-eastern NSW in the first half of the 20th century became contaminated by the use of chemicals such as arsenic and DDT in the dipping process (Kempark Pty Limited 1998).

The DECCW may declare land to be "significantly contaminated" and issue a management order with respect to it on the owner or person responsible for the contamination (CLMA §§ 11–16). Failure, without reasonable excuse, to comply with a management order is a criminal offense (CLMA § 14(6)). If an order is not complied with, the DECCW may step in and carry it out or order another public authority to do so (CLMA § 30). The costs incurred in connection with the issue or carrying out of a management order may be recovered (CLMA §§ 33A–42).

In June 2009, what was then the NSW Department of Environment and Climate Change published the NSW Diffuse Source Water Pollution Strategy (DECCW, 2009). It stresses the economic as well as the environmental and social consequences of nonpoint sources of water pollution. It also highlights the potential impact of climate change through more frequent and intense storm events and through the degradation of vegetation.

The Strategy focuses on three priority problems: sediment, nutrients and pathogens (DECCW, Government of New South, 2009). The Priority Action Plan (2009–2013) lists a number of activities that are designed to address these three problems. Among the actions are to develop an approach to identify hotspots for sediment, nutrients and pathogens; to create best-management practices guidelines and provide incentives to landowners in hotspots to follow them; to develop new land management plans and amend old ones to reduce sediment, nutrient and pathogen loads; to scope the application of a legal framework for implementing market-based instruments or incentives to address nonpoint sources; to gauge community awareness and deliver targeted programmes to motivate stakeholders to take action; and to promote demonstration sites (DECCW, 2009).

Summary

In Australia, environmental protection is not a responsibility of the national government, leaving states to legislate individually on matters such as water quality. There are a number of collaborative national initiatives and guidelines to assist states in developing policy responses to environmental issues. However, there is little national influence on the regulation of nonpoint sources of water pollution. While some states, most notably NSW, have passed laws that directly and indirectly control aspects of nonpoint source water pollution, significant gaps remain, both in law and implementation, as demonstrated by the 2009 NSW Diffuse Source Water Pollution Strategy.

Nonpoint Source Regulation in the Republic of South Africa

National Laws and Policies

The National Water Act (NWA or the Act) serves as the national basis for water quality regulation in the Republic of South Africa. The purpose of the NWA is to ensure that the nation's water resources are "protected, used, developed, conserved, managed, and controlled" in a manner guided by the principles of sustainability and equity and recognize the basic human needs of present and future generations (NWA ch. 1(2); Department of Water Affairs & Forestry (DWAF), 2007). This is all to be done in an integrated manner, taking into account the need for coordinated management and the sustainable development of South Africa's water resources (DWAF, 2004).

The NWA designates the Department of Water Affairs and Forestry (now the Department of Water Affairs, or DWA) as the agency responsible for water resources management at the national level. However, the Act also stipulates that the country should be divided into 19 water management areas, which have since been established, that will eventually come under the direction of newly established Catchment Management Agencies (CMAs). The CMAs are required to develop a Catchment Management Strategy (CMS) that is consistent with the National Water Resource Strategy (NWRS), a legally binding policy that sets forth the "strategies, objectives, plans, use, development, conservation, management and control of water resources". DWA acts as an interim CMA as the Agencies are established.

The NWA does not differentiate between point and nonpoint sources of pollution. As such, all relevant provisions of the Act presumably apply to nonpoint source regulation. However, there is little guidance as to how to apply the Act's provisions specifically to nonpoint source control and regulation. To address this gap, DWA is currently developing a Strategy for the Management of Pollution from Non-point Sources for inclusion in the NWRS, and is developing licence application guidelines for certain water uses that commonly lead to nonpoint source contamination. Still, as explored below, the NWA as currently written provides significant opportunities for addressing nonpoint sources of pollution.

The NWA sets up a system of both source- and resource-directed measures for controlling discharges into South Africa's waters. Resource-directed measures focus on the condition of the water resource itself, including both quantity and quality. Pursuant to Section 12 of the NWA, DWA is currently developing a comprehensive resource classification system that will define the level of protection required for each water body in the country and guide the allowable uses of and discharges into those resources. The classification system also will establish Resource Quality Objectives (RQOs) for each water resource. RQOs are objectives based on the acceptable level of risk appropriate for the water resource class and include the enforceable requirements for the maintenance of the Reserve.[2] All water management practices must "give effect" to these RQOs, and thus the RQOs act as an overall marker to guide water quality management.

The NWA provides that the classification system may "in respect of each class of water resource—set out water uses for instream or land-based activities which activities must be regulated or prohibited in order to protect the water resource" (NWA § 12(2)(b)(iii)). The classification system is thus a potential mechanism for regulating nonpoint sources by linking various land use activities requiring source-directed management to the level of protection required by the RQOs (Pegram *et al.*, 1999). Under the draft regulations establishing the classification system, this is done explicitly by requiring the creation of water use-related water quality requirements for each user of a classified resource (DWAF, 2008). Additionally, the NWA establishes pollution-prevention measures applicable to nonpoint sources. Under Section 19 of the Act, an "owner of land, a person in control of land, or a person who occupies or uses the land on which" activities or processes occur that are likely to cause pollution to a water resource are required to take "all reasonable measures" to prevent pollution from occurring or recurring. If such measures are not taken, the relevant CMA (or DWA in cases where a CMA has not yet been established) may act and recover its reasonable costs in so doing (NWA § 19).

In addition to the resource-directed measures, the NWA also stipulates a number of source-directed measures that define the limits and constraints that must be imposed on the use of water resources to ensure the desired level of protection for the resource (DWAF, 2008). These requirements are implemented through a system of water-use licences (permits) and general use authorizations (for specific water resources, categories of people, or for a defined geographic area or period of time). The NWRS specifies a hierarchy of source-directed responses. First, wherever possible, source-directed controls should be promoted that prevent the pollution or degradation of water resources (DWAF, 2004). Once discharged, such pollution must be minimized by, for example, institution of the polluter-pays principle, but certainly by adhering to the regulatory requirements for meeting the relevant RQOs. Finally, for degraded water bodies, stricter measures may be implemented.

Under the NWA, water "use" is defined broadly, encompassing all activities that might cause detrimental impacts to a water resource, as well as waste discharges and disposals (NWA § 21). In addition, there is a category of water use defined in Section 38(2) of the Act as "controlled activities", which can be declared by the minister for any activity that is likely to affect a water resource detrimentally. Certain water uses must be authorized only in accordance with other pertinent regulations, and must be consistent with the relevant Catchment Management Strategy. The provisions of each water licence are also subject to public consultation procedures, which may in turn impact the final contents and structure of a particular authorization. As such, nonpoint source management practices are likely to be defined in licence conditions or in regulations for "controlled activities" (Pegram & Görgens, 2001).

In addition to DWA, the Department of Agriculture, Forestry, and Fisheries (DAFF) also has a mandate, pursuant to the Conservation of Agricultural Resources Act of 1983, to prescribe control measures relating to the flow pattern of run-off water and the protection of water resources against pollution on account of farming practices. Currently, DWA is working with DAFF and related government departments (provincial and national) to develop a strategy aimed at managing the impacts on water quality of agricultural activities. This will entail the drafting of guidelines for the use of pesticides and fertilizers, as well as for diffuse pollution caused by certain agricultural practices, such as feedlots (DWA, n.d. a).

Finally, and potentially relevant to nonpoint sources of water pollution, the Department of Environmental Affairs and Tourism (DEAT) has responsibility for disposal of wastes on land. Any permitting for such disposal requires approval of DWA with respect to water quality aspects. However, no specific regulations regarding the identification of nonpoint sources of water pollution from land-based waste disposal have been developed.

Catchment-level Responses

As noted above, the NWA requires that each of the country's 19 water management areas (WMAs) develop a Catchment Management Agency (CMA) that, in turn, creates a Catchment Management Strategy (CMS) for the relevant basin or sub-basin. Ultimately, the CMAs will be the primary institution responsible for water resources regulation. Until such time, the nine regional DWA offices are responsible for implementing water quality management at the operational level. The WMAs do not follow political boundaries, but were established on the basis of four considerations: (1) catchment boundaries; (2) social and economic development patterns; (3) efficiency considerations; and (4) communal interests within the area in question. Some WMAs thus coincide with catchments, or portions of a catchment, and some do not. Similarly, some WMAs coincide with local and provincial political boundaries, and others do not. This clearly complicates the picture in terms of implementation of nonpoint source management, as multiple jurisdictions of varying levels of responsibility will likely need to coordinate efforts to ensure effective regulation. It is also notable that eleven WMAs are located in trans-boundary river basins (Karar, 2008).

So far, nine CMAs have been established, but none has yet completed a CMS. Prior to completing a CMS, a CMA must undertake a catchment assessment study, which entails an assessment of the activities and impacts on the water resources of the catchment, as well as the current status of the physical environment, the hydrology, the land-use activities, the demands and the various impacts of those activities on water resources. In addition, the assessment is meant to project future developments and growth and estimate the impacts on the water resources within the catchment (DWA, n.d. b). The assessment will thus provide the basis for the CMS and for nonpoint source regulation within each catchment. With respect to water quality, the CMS must establish resource water quality objectives (RQOs) for use of the resource; determine the appropriate source management objectives to meet these objectives; formulate a sectoral water quality management plan indicating the management requirements and responsibilities to achieve these objectives; and develop a single source and sector-specific water quality management plan to give effect to this framework, and introduce single source interventions (water use licences) to ensure that resource water quality objectives are met (DWA, n.d. b).

It is expected that, once a CMS has been finalized, authority to govern technical water resource management functions, including water quality management, will be delegated from DWA to the relevant CMA.[3] Under the constitutional principle of cooperative governance, the CMA is responsible for fostering relationships with local government authorities that could impact nonpoint source regulation in their catchment. Municipal governments are tasked with water services delivery and have the constitutional mandate of controlling water pollution and waste management in their jurisdiction, so the institutional mechanisms set up to foster cooperation will be extremely important in ensuring effective nonpoint source control. These mechanisms could include memoranda of understanding, or even delegation of authority by the CMA of certain regulatory functions.

Summary

While the lack of distinction between point and nonpoint sources of water pollution in the South African National Water Act provides a unique opportunity to create comprehensive frameworks at the catchment level for regulating nonpoint sources, there is little guidance as to how this should be implemented. The newly established Catchment Management Agencies and other government agencies with the authority to address nonpoint source pollution are left to identify and develop the mechanisms for "cooperative governance" necessary to coordinate the regulation of these sources. The Department of Water Affairs is attempting to fill this gap by developing a Strategy for the Management of Pollution from Non-point Sources for inclusion in the National Water Resource Strategy, and is developing licence application guidelines for certain water uses leading to nonpoint source contamination. This entire process is early in its development stages, meaning that the true extent of the obstacles and realistic opportunities for managing nonpoint sources of water pollution under the 1998 Water Act in South Africa are still unfolding.

Nonpoint Source Regulation in India

National Laws and Policies

The Water (Prevention and Control of Pollution) Act of 1974 (the Act) is the primary water quality management law in India. It requires the establishment of a Central Pollution Control Board (CPCB) and individual State Pollution Control Boards (SPCBs) (Act §§ 3, 4). The main function of the CPCB is to "promote cleanliness of streams and wells in different areas of the States" (Act § 16(1)). The SPCBs are responsible for routine water quality monitoring and setting effluent standards (Act § 17), but they are bound by the directions of the CPCB (Act § 18). The CPCB has established a national network of monitoring stations, at which samples are taken monthly, quarterly, or (in the case of groundwater) every six months. Quality objectives have been set according to the use to which a water resource is put (Bhardwaj, 2005). However, a number of operational constraints have been identified, including inadequate manpower and resources, quality control, and lack of training (Bhardwaj, 2005).

The Act does not explicitly address nonpoint source pollution, but several of its provisions could be used to control pollution from nonpoint as well as point sources. For example, it is an offense under the Act to knowingly cause or permit any "poisonous, noxious or polluting matter" on land or to enter into any stream, well, or sewer (Act § 24(1)(a)). However, this does not appear to be a strict liability offense and a conviction for "knowingly" causing or permitting nonpoint source water pollution is likely to be particularly difficult to obtain.

Where any poisonous, noxious, or polluting matter is present in any stream or well or on land and the SPCB is of the opinion that it is necessary or expedient to take immediate action, the SPCB is authorized to carry out operations to remove the matter, remedy or mitigate any pollution caused, or issue a restraining order (Act § 32). In addition, the SPCB may apply to the court for a restraining order where it is determined that the water in any stream or well is likely to be polluted by reason of the disposal, or likely disposal, of any matter in such stream or well or in any sewer, or on any land, or otherwise (Act § 33).

Under the Environment (Protection) Act of 1986 (EPA), the central government is empowered to take measures to protect and improve the quality of the environment and prevent, control and abate environmental pollution (EPA § 3(1)). These measures may include a national programme for preventing, controlling and abating environmental pollution; standards for environmental quality; emission or discharge standards; and area-based restrictions on industrial or other processes (EPA § 3(2)). Rules may be made in connection with all or any of these matters (EPA §§ 6, 25). Examples of these include the Environment Protection Rules of 1986, the Municipal Solid Wastes (Management and Handling) Rules of 2000 and the Hazardous Wastes (Management and Handling) Rules of 1989. In the exercise of its powers and the performance of its functions under this Act, the central government may also give directions. Such directions may include the closure, prohibition or regulation of any industry, operation or process and the stoppage or regulation of the supply of electricity or water or any other service (EPA § 5). All of these authorities have the potential to be used to address nonpoint sources of pollution.

It was recognized by the CPCB in its 2000/01 annual report that nonpoint water pollution sources were becoming increasingly prominent and were likely to include farming, mining, rural hamlets, leaks and spillovers from point pollution sources, leachates and deposition of air-pollutants (CPCB, 2001). More recently, in its 2007 report on the Status of Groundwater Quality in India—Part I, the CPCB talked of the "alarming picture" of declining groundwater quality. It was observed that "all activities carried out on land have the potential to contaminate groundwater, whether associated with urban, industrial or agricultural activities" (CPCB, 2007). Diffuse sources of groundwater pollution were listed, including the leaching of agrochemicals and animal wastes, subsurface discharges from latrines and septic tanks and infiltration of polluted urban run-off and sewage where sewerage does not exist or is defunct. The Report concluded that "the only solution to diffuse sources of pollution is to integrate land use with water management" (CPCB, 2007).

The 2007 National Policy for Farmers contains reference to water quality problems caused by "over exploitation and the indiscriminate use" of fertilizers, pesticides and toxic chemicals (Department of Agriculture & Cooperation (DAC), Ministry of Agriculture, Government of India, 2007). By comparison, however, the 2008 Draft Indian Standard Requirements for Good Agricultural Practices (IndiaGAP) (Part 1 Crop Base) (Bureau of Indian Standards, 2008), which is primarily concerned with the quality of produce, considers environmental safety a "minor" criterion for consideration in the certification process (Bureau of Indian Standards, 2008). However, in recent years, Conservation Agriculture (CA) has begun to be introduced into the Indo-Gangetic Plain. CA is a system for the cultivation of crops without tilling the soil. Other characteristics include leaving and managing crop residues on the surface of the soil and the adoption of crop rotations that minimize adverse environmental impacts (Abrol & Sangar, 2006). In addition to water conservation and reducing run-off and erosion, environmental benefits may include a reduction in the use of fertilizers and pesticides (Jat *et al.*, 2006).

Promulgated under the auspices of the Ministry of Environment and Forests (MoEF), a National Environment Policy was adopted by the Union Cabinet in 2006. It sets out a number of policy objectives, principles and strategic themes. In particular, it proposes the revisiting of the legislative framework with a view to, among other things, establishing an integrated approach to the management of environmental and natural resources (Jat *et al.*, 2006). With regard to water pollution abatement, it states that action plans for addressing water pollution in major cities should be prepared and implemented, groundwater pollution should be considered in the pricing policies for agricultural inputs and measures should be taken to prevent other sources of water pollution (Jat *et al.*, 2006).

In September 2009, the MoEF issued a discussion paper seeking comments on a proposal for the establishment of a National Environment Protection Authority (NEPA)

(MoEF, 2009). It is recognized in the paper that existing institutional structures are "inadequate for responding to the emerging environmental challenges" (MoEF, 2009). The NEPA would be an independent statutory body responsible for the "national stewardship" of enforcement and compliance (MoEF, 2009). Four options are outlined for the overall structure and responsibilities of the NEPA, as well as its relationships with the MoEF, the CPCB and SPCBs. In addition to enforcement and compliance, other functions that could be discharged by the NEPA include certain of the regulatory, research and development responsibilities currently vested in the MoEF and/or the CPCB. This proposal is still in the early stages of development.

State Regulation

With little emphasis from the CPCB and other national authorities on nonpoint source pollution as well as no clear mandates from the Act to address it, SPCBs have focused their resources on monitoring compliance with point source authorizations issued under the Act. A 2006 study of India's environmental compliance and enforcement programmes found there was an "over emphasis" on permitting, monitoring and inspection of the activities of large industry while the "significant cumulative pollution impacts" from small and medium-sized enterprises, municipal sources, transport and agriculture are "virtually disregarded" (Organisation for Economic Co-operation & Development (OECD), 2006). The same study highlighted various existing enforcement challenges faced by the SPCBs (OECD, 2006) and it seems likely that, in the absence of national initiatives, nonpoint source pollution will remain a low priority at the state level unless and until industrial and municipal point sources of pollution are regulated effectively.

Summary

Water quality regulation is driven nationally, primarily through the Water (Prevention and Control of Pollution) Act of 1974, but responsibility for enforcement is vested in the individual SPCBs. Emphasis is placed on the regulation of point source discharges, and there are no laws that specifically address nonpoint sources of water pollution. Although it would appear that the central government has the power to regulate nonpoint sources under that Act and the Environment (Protection) Act of 1986, no substantive action has been taken. However, a number of policy documents and initiatives have drawn attention to the problem over the course of several years.

Conclusion

As with nearly any issue in a federal government, the level of government at which regulation occurs, as well as how it is enforced and the way in which various authorities are coordinated, are critical to successfully addressing nonpoint sources of pollution. The national government may need to play a greater role in developing authorities to regulate nonpoint sources and creating structure and guidance for doing so when it traditionally has led developments in water pollution control.

In the United States, South Africa and India, the foundation for water pollution control has come from the national government. A few states in the United States have established regulations that seek to fill the gap regarding nonpoint sources of pollution left by the Clean Water Act, but perhaps the best example, the Porter–Cologne Water Quality

Control Act in California, was passed before enactment of the national Clean Water Act, and thus predates the leadership of the national government in water quality issues. Similarly, few developments in nonpoint source control have occurred in the states of India without national influence on the matter.

By contrast, the case study of Australia suggests that when water quality control is not historically driven by the national government, and the states have the capacity to regulate such matters, nonpoint source pollution controls may develop at the state level. While far from perfect, the myriad laws and Priority Action Plan in New South Wales demonstrate the potential for managing nonpoint source pollution at the state level. The Porter–Cologne Water Quality Control Act in California is another example of state activity on nonpoint sources in the absence of national action in water quality management generally.

Regardless of the level of national involvement in water pollution control writ large or on nonpoint source pollution specifically, the national government likely will need to play a role in trans-boundary nonpoint pollution issues because the states or provinces cannot address those problems internally. The national government could be a more efficient level of governance at which to establish such policies. In addition, a national government should be cognizant of the political, financial, structural and scientific capacity of state or provincial governments to control nonpoint sources of pollution adequately. A related issue is that of enforcement capacity since, as the case studies have shown, there may be operational constraints and inherent evidential difficulties in controlling nonpoint source pollution by direct regulation. The level of support needed could range from simple guidance to developed infrastructure and technical support, but the more support provided by the national government, the more aware of implementation obstacles the national government should be. The South African government has created a detailed water quality management structure, but the Catchment Management Agencies that are to be adopted and implemented by local parties face a difficult challenge coordinating with other governance bodies with authority over aspects of water quality management.

Growing populations and expanding interest in biofuels, among other influences, will have a significant impact on how we use the land and what it means for water. Water quality standards, monitoring and national policies and guidelines are an essential first step. However, a coordinated structure of regulatory authorities, local buy-in, proper enforcement and other implementation measures will be necessary to achieve in nonpoint source pollution even just those reductions thus far accomplished from point sources. The level of government at which those authorities and programmes are created and implemented will necessarily be unique to each country, but should be strategic rather than patchwork to be effective.

Notes

1. While South Africa was established as a unitary state, it has a number of federal characteristics that lend themselves towards inclusion in this analysis. First, the regulation of water, while governed at the national level, is meant to be implemented at the catchment level with the establishment of Catchment Management Agencies that have significant regulatory control over water management (National Water Act Schedule 3). Additionally, the South African Constitution established a number of areas of concurrent national and provincial legislative competence, including environment and nature conservation (South African Constitution Schedule 4).

2. The Reserve is the quality and quantity of water required to provide for basic human needs (the Basic Human Needs Reserve) and sustainable ecosystem functioning (the Ecosystem Reserve).
3. It is envisioned that delegation of authority over various management functions will be done on an ongoing basis, as the CMA demonstrates the necessary capacity (financial and technical) to undertake those functions.

References

Abrol, I. P. & Sangar, S. (2006) Sustaining Indian agriculture—conservation agriculture the way forward, *Current Science*, 91(8), pp. 1020–1025. Available from: http://www.ias.ac.in/currsci/oct252006/1020.pdf (accessed 4 October 2010).

Bhardwaj, R. M. (2005) *Water Quality Monitoring in India—Achievements and Constraints*. Available from: http://unstats.un.org/unsd/environment/envpdf/pap_wasess5a2india.pdf (accessed 4 October 2010).

Bureau of Indian Standards (2008) *Draft Indian Standard Requirements for Good Agricultural Practices—IndiaGAP Part 1 Crop Base*. Available from: http://www.bis.org.in/sf/fad/FAD22(1949)c.pdf (accessed 4 October 2010).

Central Pollution Control Board (CPCB) (2001) *Annual Report 2000–2001*. Available from: http://cpcbenvis.nic.in/ar2001/annual_report2000-01-15.htm (accessed 4 October 2010).

Central Pollution Control Board (CPCB) (2007) *Status of Groundwater Quality in India – Part I*. Available from: http://www.cpcb.nic.in/upload/NewItems/NewItem_47_foreword.pdf (accessed 4 October 2010).

Clean Water Act, 33 United States Code Sections 1251–1387 (2009).

Commonwealth of Australia Constitution Act 1900.

Conservation of Agriculture Resources Act, Act 43 of 1983 (South Africa).

Contaminated Land Management Act 1997 (New South Wales).

Copeland, C. (2001a) *Clean Water Issues in the 107th Congress: An Overview*. Available from: http://ncseonline.org/nle/crsreports/water/h2o-43.cfm (accessed 4 October 2010).

Copeland, C. (2001b) *Water Quality: Implementing the Clean Water Act*. Available from: http://ncseonline.org/nle/crsreports/water/h2o-15.cfm (accessed 4 October 2010).

Dairying for Tomorrow (DfT) (n.d. a) *Best Management Practices*. Available from: http://land.vic.gov.au/dpi/vro/vrosite.nsf/3d36e5863b7b2b3f4a25692e000848d2/251d33460001ababca257439007b7144?OpenDocument&ExpandSection=6 (accessed 4 October 2010).

Dairying for Tomorrow (DfT) (n.d. b) *DairySAT*. Available from: http://www.dairyingfortomorrow.com/index.php?id=57 (accessed 4 October 2010).

Dairying for Tomorrow (DfT) (n.d. c) *About Dairying for Tomorrow*. Available from: http://www.dairyingfortomorrow.com/index.php?id=9 (accessed 4 October 2010).

Department of Agriculture & Cooperation (DAC), Ministry of Agriculture, Government of India (2007) *National Policy for Farmers*. Available from: http://www.indg.in/agriculture/rural-employment-schemes/npff2007.pdf (accessed 4 October 2010).

Department of Environment & Climate Change and Water (DECCW), Government of New South Wales (2009) *NSW Diffuse Source Water Pollution Strategy*. Available from: http://www.environment.nsw.gov.au/resources/water/09085dswp.pdf (accessed 4 October 2010).

Department of Sustainability, Environment, Water, Population & Communities (SEWPAC), Government of Australia (n.d. a) *National Environmental Protection Council (NEPC)*. Available from: http://www.environment.gov.au/about/councils/nepc/index.html (accessed 4 October 2010).

Department of Sustainability, Environment, Water, Population & Communities (SEWPAC), Government of Australia (n.d. b) *National Water Quality Management Strategy*. Available from: http://www.environment.gov.au/water/policy-programs/nwqms/index.html (accessed 4 October 2010).

Department of Sustainability, Environment, Water, Population & Communities (SEWPAC), Government of Australia (n.d. c) *Water Quality Improvement Projects*. Available from: http://www.environment.gov.au/water/policy-programs/nwqms/wqip/projects.html (accessed 4 October 2010).

Department of Sustainability, Environment, Water, Population & Communities (SEWPAC), Government of Australia (n.d. d) *Water Quality Hotspots*. Available from: http://www.environment.gov.au/water/policy-programs/nwqms/wqip/hotspots.html (accessed 4 October 2010).

Department of Water Affairs & Forestry (DWAF) (2004) *National Water Resource Strategy*. Available from: http://www.dwaf.gov.za/Documents/Policies/NWRS/Default.htm (accessed 4 October 2010).

Department of Water Affairs & Forestry (DWAF) (2007) *Guidelines for the Development of Catchment Management Strategies: Towards Equity, Efficiency and Sustainability.* Available from: http://www.dwaf.gov.za/Documents/Other/CMA/CMSFeb07/CMSFeb07Ed1.pdf (accessed 4 October 2010).

Department of Water Affairs & Forestry (DWAF) (2008) *Draft Regulations for the Establishment of a Water Resource Classification System.* Available from: http://www.pmg.org.za/files/gazettes/080919water-regulations.pdf (accessed 4 October 2010).

Department of Water Affairs (DWA) (n.d. a) *Project Details: 1.2 Agriculture.* Available from: http://www.dwaf.gov.za/Dir_WQM/projectsDetails.asp#proj01b01 (accessed 4 October 2010).

Department of Water Affairs (DWA) (n.d. b) *Water Quality Management in South Africa.* Available from: http://www.dwa.gov.za/Dir_WQM/wqm.asp (accessed 4 October 2010).

Environment (Protection) Act, no. 29 of 1986 (India).

Environment Protection Authority v. Ross [2009] N.S.W. Land & Env't Court 36.

Jat, M. L., Sharma, S. K. & Singh, K.K. (2006) *Conservation Agriculture for Sustainable Farming in India.* Available from: http://www.rwc.cgiar.org/pubs/145/CASustFarmIndia.pdf (accessed 4 October 2010).

Karar, E. (2008) The institutional reforms in South Africa. Paper presented at the International Conference on Water Management in Federal and Federal-type Countries. Available from: http://www.forumfed.org/en/global/thematic/water_papers/Eiman%20Karar.pdf (accessed 4 October 2010).

Kempark Pty Limited v. State of New South Wales [1998] N.S.W. Land & Env't Court 205.

Legal Services Commission of South Australia (n.d.) *The Law Handbook Online: Commonwealth Responsibility.* Available from: http://www.lawhandbook.sa.gov.au/ch18s02s01.php (accessed 4 October 2010).

Ministry of Environment & Forests (MoEF) (2009) *Towards Effective Environmental Governance: Proposal for a National Environmental Protection Authority.* Available from: http://moef.nic.in/downloads/home/NEPA-Discussion-Paper.pdf (accessed 4 October 2010).

Murray–Darling Basin Authority, Government of Australia (n.d.) *FAQs: About the Murray–Darling Basin.* Available from: http://mdba.gov.au/basin_plan/faqs/the-basin (accessed 4 October 2010).

National Environment Protection Council Act 1994 (Australia).

National Water Act, Act 36 of 1998 (South Africa).

Native Vegetation Act 2003 (New South Wales).

Organisation for Economic Co-operation & Development (OECD) (2006) *Environmental Compliance and Enforcement in India: Rapid Assessment.* Available from: http://www.oecd.org/dataoecd/39/27/37838061.pdf (accessed 4 October 2010).

Pegram, G. C., Görgens, A.H.M. & Guibell, G.E. (1999) *A Framework for Implementing Non-point Source Management Under the National Water Act: A Discussion Paper for the Department of Water Affairs and Forestry & Water Research Commission.* DWAF Report No. WQP 0.1 (Pretoria: Department and the Commission).

Pegram, G. C. & Görgens, A. H. M. (2001) *A Guide to Non-point Source Assessment: To Support Water Quality Management of Surface Water Resources in South Africa.* Available from: http://www.wrc.org.za/Knowledge%20Hub%20Documents/Research%20Reports/TT-142-01.pdf (accessed 4 October 2010).

Porter–Cologne Water Quality Control Act, California Water Code Division 7 (2009).

Protection of the Environment Operations Act 1997 (New South Wales).

Soil Conservation Act 1938 (New South Wales).

United States Environmental Protection Agency (USEPA) (n.d.) *Clean Water Act Section 319(h) Grant Funds History.* Available from: http://www.epa.gov/owow/nps/319hhistory.html (accessed 4 October 2010).

Washington Water Pollution Control Act, Revised Code of Washington Chapter 90.48 (2009).

Water (Prevention and Control of Pollution) Act, no. 6 of 1974 (India).

Water Act 2007 (Australia).

Introduction to Environmental and Economic Consequences of Hypoxia

ROBERT J. DÍAZ* & RUTGER ROSENBERG**

*Virginia Institute of Marine Science, College of William and Mary, Gloucester Pt., Virginia, USA;
**Kristineberg Marine Research Station, University of Gothenburg, Gothenburg, Sweden

ABSTRACT *Low dissolved oxygen environments (known as hypoxic or dead zones) occur in a wide range of aquatic systems and vary in frequency, seasonality and persistence. While there have always been naturally occurring hypoxic habitats, anthropogenic activities related primarily to organic and nutrient enrichment related to sewage/industrial discharges and land runoff have led to increases in hypoxia and anoxia in both freshwater and marine systems. As a result, over the last 50 years there has been a rapid rise in the areas affected by hypoxia. The future status of hypoxia and its consequences for the environment, society and economies will depend on a combination of climate change (primarily from warming, and altered patterns for wind, currents and precipitation) and land-use change (primarily from expanded human population, agriculture and nutrient loadings). The overall forecast is for hypoxia to worsen, with increased occurrence, frequency, intensity and duration. The consequences of eutrophication-induced hypoxia can be reversed if long-term, broad-scale and persistent efforts to reduce nutrient loads are developed and implemented.*

Introduction

By the early 1900s dissolved oxygen (DO) was a topic of interest in research and management, and by the 1920s it was recognized that a lack of DO was a major hazard to fishes (Jones, 1952). It was not obvious that DO would become critical in estuarine and shallow coastal systems until the 1970s and 1980s when large areas of low-DO started to appear with associated mass mortalities of invertebrates and fishes. From the mid-20th century to today there have been drastic changes in DO concentrations and dynamics in marine coastal waters (Díaz & Rosenberg, 2008; Rabalais *et al.*, 2010). Díaz & Rosenberg (1995) noted that no other environmental variable of such ecological importance to estuarine and coastal marine ecosystems as DO has changed so drastically in such a short time. Currently there are over 500 hypoxic areas or dead zones around the world related to human activities (Figure 1).

Accounts of environmental problems related to low DO predate our ability to measure oxygen concentration in water. For example, the Drammensfjord in Norway appears to have been persistently hypoxic and anoxic since at least the 1700s based on foraminiferan proxies (Alve, 1995). Even in this small fjord with an extended residence time of deep water, historic anoxia has been made worse over the last two centuries by organic enrichment related to sewage/industrial discharges and land runoff. Improvements were

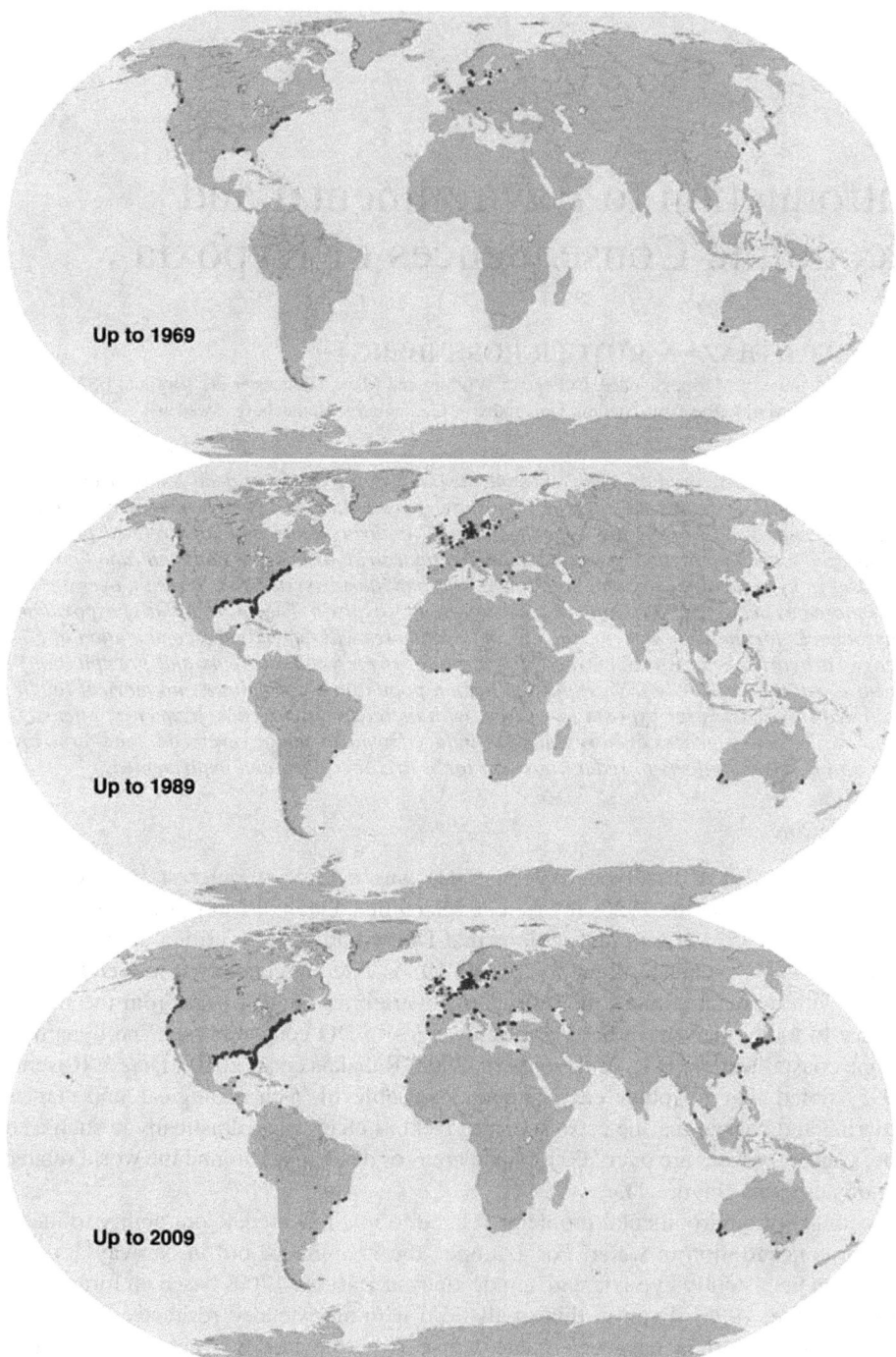

Figure 1. Global distribution of documented case of hypoxia related to human activities (dots). The number of hypoxic areas is cumulative for the successive time periods. *Source:* Rabalais *et al.* (2010).

observed only after reductions in organic loading primarily through improved water treatment. Other examples would be the Mersey Estuary in the UK which has had poor water quality and hypoxia since at least the 1850s (Jones, 2006), and Mobile Bay in Alabama, USA, since at least the 1860s (May, 1973).

Recently there have been a series of summary and review articles that have examined the key factors leading to the development of low DO and the subsequent environmental consequences (Díaz & Rosenberg, 2008; Rabalais & Gilbert, 2008; Vaquer-Sunyer & Duarte, 2008; Keeling *et al.*, 2010; Gilbert *et al.*, 2010; Gooday *et al.*, 2009; Levin *et al.*, 2009; Middelburg & Levin, 2009; Peña *et al.*, 2010; Rabalais *et al.*, 2010). The present paper will summarize the key environmental and economic aspects of hypoxia on coastal ecosystems.

Environmental Consequences of Hypoxia

There is a similarity of faunal response across systems to varying types of hypoxia that range from beneficial to mortality (Díaz & Rosenberg, 1995; Vaquer-Sunyer & Duarte, 2008). The consequences of low DO are often sub-lethal and can affect growth (Stewart *et al.*, 1967), immune responses (Thomas *et al.*, 2007), and reproduction (Wu *et al.*, 2003). In fish, short-term exposure to hypoxia from 0.5 to 96 hours can reduce bactericidal activity, antibody levels and production of disease-fighting reactive oxygen species, although some of these effects may be reversible upon restoration of normoxic conditions (Valenzuela *et al.*, 2005; Welker *et al.*, 2007). Similarly, shrimp and crabs exposed to hypoxia are immunocompromised and may suffer increased susceptibility to disease and mortality from bacterial infections (Le Moullac *et al.*, 1998; Holman *et al.*, 2004).

When a system becomes hypoxic, mobile fauna have to contend with two simultaneous problems: loss of habitat as they are forced to migrate into higher DO waters, and reduced or changed prey resources for demersal feeders. Sessile fauna initiate a graded series of behaviours to survive and will eventually die as DO declines or extends through time (Díaz & Rosenberg, 1995; Rabalais *et al.*, 2001). The resulting habitat compression or squeeze occurs when hypoxia overlaps with essential habitat such as nursery areas or deeper, cooler water during the summer (Coutant, 1990). Hypoxic habitats avoided are functionally lost from the system (Breitburg, 2002; Rabalais & Turner, 2001). As a result, habitat loss due to hypoxia is far greater than would be estimated by calculations based on species recruitment or survival tolerances.

Loss of habitat to hypoxia is not limited to demersal species. The 1976 New York Bight hypoxia blocked the northward migration of bluefish. Fish that encountered the hypoxic zone did not pass through or around it, but stayed to the south waiting for it to dissipate and then continue their migration (Azarovitz *et al.*, 1979). In the northern Gulf of Mexico hypoxia interfered with the migration of brown shrimp from inshore wetland nurseries to offshore feeding and spawning grounds. Juvenile brown shrimp leaving nursery areas migrate farther offshore when hypoxia was not present and were compressed inshore when hypoxia was present (Gazey *et al.*, 1982). In avoiding hypoxia brown shrimp aggregated both inshore and offshore of low DO areas losing about 25% of their shelf habitat (Craig *et al.*, 2005). If it is important for fish and shrimp to reach critical nursery or feeding areas at certain times in their life cycle then hypoxia may have indirect effects on populations dynamics by delaying arrival to spawning or feeding grounds. In such cases the cost of delayed migration in terms of population mortality and production is not known.

The elimination of benthic prey and compression of habitat by hypoxia also have profound effects on ecosystem energetics as organisms die and are decomposed by microbes. Under certain circumstances demersal feeding fishes are able to consume low DO-stressed benthic prey. Over a narrow range of conditions, hypoxia can therefore facilitate the upward trophic transfer as physiologically stressed benthic fauna forced to the sediment surface during hypoxia may be exploited by predators (Nestlerode & Díaz, 1998; Pihl *et al.*, 1992). Aggregation of demersal predators on the edge of dead zones may be a combination of responses that include flight from physiological stress and trophic advantage (Craig & Crowder, 2005). Thus, short-lived and mild hypoxia may not have a net negative effect on trophic transfer as does severe seasonal hypoxia. Increased trophic transfer is most prevalent under diel cycling hypoxic conditions. Diel cycling hypoxia may be a common phenomenon to which fishes respond in an opportunistic manner (Smith & Able, 2003). In Pepper Creek, Delaware, fish emigrated only when DO dropped to very low values and returned quickly with improving DO (Tyler *et al.*, 2009). While it may be physiologically stressful for juvenile fishes to remain in or near low DO water, the gains in trophic resources and added protection from predation may be greater and diel hypoxic areas can still serve as important nursery habitats. In areas where hypoxia is intermittent, and does not cause substantial mortality of the benthos, the behaviour of benthos may facilitate trophic transfer to oxygen-tolerant bottom-feeding fish, which are physiologically capable of withstanding short-term exposure to DO levels to take advantage of weakened benthic prey and receive an energetic gain.

Upward trophic transfer is inhibited in areas where hypoxia is severe as either benthic resources are killed directly and/or predators capable of detecting low DO would avoid the area. Systems reporting mass mortality provide primary examples of degradation in trophic structure. The recently developed hypoxia along the Oregon coast has caused mass mortality over large areas and will likely lead to an altered trophic state if it persists (Grantham *et al.*, 2004; Chan *et al.*, 2008). If it occurs, the increase in the proportion of production transferred to predators is temporary and as mortality of benthos occurs, microbial pathways quickly dominate energy flows (Baird *et al.*, 2004). This energy diversion tends to occur in ecologically important places and at the most inopportune time for predator energy demands, and causes an overall reduction in an ecosystem's functional ability to transfer energy to higher trophic levels and renders the ecosystem potentially less resilient to other stressors (Díaz & Rosenberg, 2008).

Economic Consequences of Hypoxia

Economic effects attributable to hypoxia are subtle and difficult to quantify even when mass mortality events occur. Much of the problem is related to the multiple stressors acting on targeted commercial populations (habitat degradation, over-exploitation, pollution) and also factors that stress fisher's economics (aquaculture, imports, economic costs of fishing, fisheries regulations). From assessing ecological effects of hypoxia it is known that populations can experience a range of problems that at some point will negatively affect economic interests in their populations (Table 1). Direct effects of hypoxia on stocks are related to reduced growth, movement to avoid low DO, aggregation and predation pressure. Direct effects of hypoxia on fishers are related to increased time in fishing grounds, the cost of searching for stocks and market forces that control dock side prices.

Table 1. Generalized response of populations to hypoxia and potential economic effect.

Factor	Result	Economy
Mortality	Loss of stock, may take years to recover	Lower landings Increased time fishing
Reduced recruitment	Smaller populations, effect may be long lasting	Lower landings Increased time fishing
Reduced growth	Smaller individuals	Lower individual value
Poor body condition	Weaker individuals	Lower value
Increased migration	Energy resources diverted to movement	Smaller individuals Increased time fishing
Aggregation	Exposure to increased risks of predation and exploitation	Less time fishing But easier to catch
Altered behaviour	More/less susceptible to fishing gear	Increased or decreased catch-ability

Fish kills are some of the most dramatic manifestation of environmental problems and result from a combination of severe physical conditions (many times including low DO) and the failure of mobile animals to avoid or escape lethal conditions. The frequency or magnitude of fish kills in the United States has increased as nutrient-related hypoxia has worsened (Thronson & Quigg, 2008), but quantifying costs of these events has not been undertaken. By far the majority of fish that die along the Atlantic and Gulf coasts are low-value menhaden. From the 1950s to the 2000s an estimated 383 million fish were killed along the Texas coast, about 70% of which were menhaden (Thronson & Quigg, 2008). For sessile species, estimates of economic loss from hypoxia-related mortality have been estimated. For oysters in Mobile Bay, hypoxia in the early 1970s killed over US$500,000 (1970 prices) in stocks (May, 1973), but a greater economic cost was associated with the declining stocks and poor recruitment of oysters associated with severe hypoxic years.

A lack of identifiable economic effects in fisheries data does not imply that the effects would not occur should conditions worsen. In the northern Gulf of Mexico brown shrimp landing appears to be inversely related to the area of hypoxia (O'Connor & Whitall, 2007). Whether this relationship will remain linear or transform into catastrophe function (Jones & Walters, 1976) at some critical point is unknown. Experience with other hypoxic zones around the globe shows that both ecological and fisheries effects become progressively more severe as hypoxia worsens (Caddy, 1993; Breitburg et al., 2009b). Several large systems around the globe have suffered serious ecological and economic consequences from seasonal hypoxia; most notable are the Kattegat and Black Sea. The consequences range from the localized loss of catch and recruitment failure to complete system-wide loss of fishery species (Karlson et al., 2002; Mee, 1992).

The valuation of recreational fishing relative to DO in the Patuxent River, Maryland, showed that as DO declined losses to recreational fishers increased (Lipton & Hicks, 2003). Total losses for striped bass fishing in the Patuxent River for DO not exceeding 3 mg/l was about US$10,000 with a net-present value of about US$200,000. If the same water quality were allowed to occur in the entire Chesapeake Bay, the net-present value of the losses due to hypoxia would be more than US$145 million (Lipton & Hicks, 2003). A similar analysis of recreational fishing in north-east and middle Atlantic regions, and for other species, found that overall as bottom DO declined, the capture rate of fish declined (Bricker et al., 2006). Finding effects of hypoxia on recreational fishing for one species would led one to believe that similar effects are likely being experienced by commercial fishers.

The direct connection of hypoxia to fisheries' landings at large regional scales is weak because of a number of factors that include confounding effects of overfishing, and compensatory mechanisms that alter or mask effects of hypoxia on landings (Breitburg et al., 2009a). Both mobile species and fishers can distribute themselves to avoid low DO and consume prey-enriched areas (Craig & Crowder, 2005). Also, by the time a system experiences seasonal hypoxia many changes have occurred in populations and energy flows and direct mortality of fisheries' stocks tends to be low, making economic assessment difficult or impossible. Because of their devastating effects on stocks and fishers, it is possible to estimate the effects of mass mortality events. The estimated losses, both actual and to the resources, to marine-related industries from the New York Bight hypoxic event in the summer of 1976 were over US$570 million (Figley et al., 1979). Much of this loss was in surf clams that accounted for more than US$430 million (Table 2). Factored by the area of hypoxia (987 km^2), the 1976 hypoxia event cost about US$580,000/km^2 for resources and fisheries-related activities.

Climate Change and Hypoxia

Climate change, whether from global warming or from microclimate variation, will have consequences for eutrophication-related oxygen depletion (Rabalais et al., 2009, 2010). The influence of multiple climate drivers needs to be considered to understand what future

Table 2. Estimated losses (US$, millions) to New Jersey's commercial and recreational marine fisheries related industries as a result of the 1976 hypoxia event in the New York Bight.

Fishery	Type of loss	1976 Loss	Future loss
Commercial			
Finfish			
Dockside	Actual	1.9	?
Processing/marketing	Actual	4.6	?
Surf clam			
Dockside	Resource	17.8	106.5
Processing/marketing	Resource	44.4	266.4
Ocean quahog			
Dockside	Resource	–	34.4
Processing/marketing	Resource	–	86.0
Sea scallop			
Dockside	Resource	0.1	0.3
Processing/marketing	Resource	0.2	0.7
Lobster			
Dockside	Actual	0.4	1.2
Processing/marketing	Actual	1.0	3.1
Total for commercial			
Dockside		20.1	142.4
Processing/marketing		50.2	356.2
Recreational			
Party boats	Actual	1.0	?
Charter boats	Actual	0.7	?
Dive boats	Actual	0.03	?
Private boats	Actual	2.0	?
Grand totals		74.0	498.6

Source: Modified from Figley et al. (1979).

change to expect (Table 3). Climate change may make systems more susceptible to development of hypoxia through direct effects on water column stratification, the solubility of oxygen, metabolism and mineralization rates. This will likely occur primarily thought warming, which will lead to increased water temperatures and subsequent decrease in oxygen solubility, increase in organism metabolism and remineralization rates. Indirect effects on the quality and quantity of organic matter produced will also be important. All factors related to climate change will progressively lead to an onset of hypoxia earlier in the season and possibly extend it through time (Boesch et al., 2007). Earlier warmer surface waters would also extend and enhance water column stratification intensity.

Along with warming, future climate predictions include large changes in precipitation patterns, but what will happen with precipitation within any given watershed is highly uncertain (Christensen et al., 2007). If changes in precipitation lead to increased runoff to estuarine and coastal ecosystems, stratification and nutrient loads are likely to increase and worsen oxygen depletion (Justić et al., 2007). Conversely, if stratification decreases due to lower runoff or is disrupted by increased storm activity or intensity, the chances for oxygen depletion should decrease (Table 3). For the Mississippi River Basin associated with the northern Gulf of Mexico seasonal dead zone, climate predictions suggest a 20% increase in river discharge (Miller & Russell, 1992) which would lead to elevated nutrient loading, a 50% increase in primary production and expansion of the oxygen-depleted area (Justić et al., 1996).

Much of how climate change will affect hypoxia in estuarine and coastal systems will depend on coupled land–sea interactions with climate drivers. The future pervasiveness of hypoxia will also be linked with land management practices and the expansion of agriculture. Climate change will affect physical and biological processes of water column stratification, organic matter production, nutrient discharges and the rates of oxygen consumption. Land management will affect the nutrient budgets and concentrations of nutrients applied to land through agriculture. If in the next 50 years humans continue to modify and degrade coastal systems as in previous years (Halpern et al., 2008), human population pressure will likely continue to be the main driving factor in the persistence and spreading of coastal dead zones. The expansion of agriculture for the production of crops to be used for food and biofuels will result in increased nutrient loading and expand eutrophication effects (Rabalais et al., 2007), and contribute to greenhouse gases (Searchinger et al., 2008). Overall, climate drivers will tend to magnify the effects of an expanding human population.

Climate-related changes in wind patterns are of great concern for coastal system as they control Eckman transport and up-/downwelling strength, which would affect the stratification strength and delivery of deep-water nutrients into shallow coastal areas. Even relatively small changes in wind and current circulations could lead to large changes in the area of coastal seabed exposed to hypoxia, particularly along the Pacific coast (Helly & Levin, 2004; Keeling et al., 2010). Recent changes in the pattern of upwelling in the Pacific off the Oregon and Washington coasts due to shifts in winds that affected the California current systems appeared to be responsible for the recent development of severe hypoxia over a large area of the inner continental shelf.

Future Directions

Much of the scientific interest and management concern about hypoxia includes a focus on the general problem of eutrophication caused by the over-enrichment of coastal waters

Table 3. Influence of climate drivers on the extent and severity of hypoxia.

Climate driver	Direct effect	Secondary effect	Influence on hypoxia
Increased temperature	More evaporation	Decreased stream flow	+
		Land-use and cover changes	±
	Less snow cover	More nitrogen retention	−
	Warmer water	Stronger stratification	+
		Higher metabolic rates	+
More precipitation	More stream flow	Stronger stratification	+
		More nutrient loading	+
	More extreme rainfall	Greater erosion of soil P	+
Less precipitation	Less stream flow	Weaker stratification	−
		Less nutrient loading	−
Higher sea level	Greater depth	Stronger stratification	+
		Greater bottom water volume	−
		Less hydraulic mixing	+
	Less tidal marsh	Diminished nutrient trapping	+
Summer winds and storms	Weaker, less water column mixing	More persistent stratification	+
	Stronger, more water column mixing	Less persistent stratification	−
	Shifting wind patterns	Weaker/stronger upwelling potential	±

Source: Modified from Boesch *et al.* (2007).

from nutrient runoff (Selman & Greenhalgh, 2009). Two primary examples of attempts to understand the causes and manage a reduction in hypoxia can be found in the Northern Gulf of Mexico (United States Environmental Protection Agency (USEPA), 2008) and the Baltic Sea (Helsinki Commission (HELCOM), 2007). In both systems reduction in both nitrogen and phosphorus are recommend, but primarily reductions in nitrogen. Nitrogen enrichment generally increases biological production, while hypoxia reduces biomass, and habitat quality and quantity. The future effects of hypoxia and nutrient enrichment on food webs and fisheries may be strongly influenced by the extent to which these two factors co-occur. Unless the leakage of nutrients from land to sea can be reduced, the future for estuarine and coastal resources looks bleak. The combination of stressors associated with eutrophication has and continues to degrade coastal systems. Nutrient management has the reverse effects of hypoxia in some systems and expanded effort will be needed to allow more systems to recover. The development of nutrient trading programmes is a promising example of what can be done with market-based approaches (Selman et al., 2009).

There is a growing literature on the ecological consequences of hypoxia, but economic evaluations are few. The lack of economic analysis is related to a lack of data to account for the many sources of variability that influence population dynamics and landings, respectively. To understand how any subset of factors, like nutrient enrichment and hypoxia, affect populations, a complete picture is needed of what role each factor plays in regulating the three basic elements of population dynamics: mortality (all forms of loss), recruitment and growth. Along with the complexity of interactions among physical, chemical and biological processes, two key elements hinder assessment of the effects of hypoxia: (1) a lack of detail on the actual occurrence and distribution of hypoxia; and (2) the accessibility of data on various ecosystem components. To resolve these critical issues, long-term monitoring programmes are needed to measure dissolved oxygen at proper spatial and temporal scales. To manage these data and make them useful for the management of resources, a database system is needed that would facilitate future synthesis of existing data on fishery and non-fishery species.

References

Alve, E. (1995) Benthic foraminiferal distribution and recolonization of formerly anoxic environments in Dramrnensfjord, southern Norway, *Marine Micropaleontology*, 25, pp. 169–186.

Azarovitz, T. R., Byrne, C. J., Silverman, M. J., Freeman, B. L., Smith, W. G., Turner, S. C., Halgren, B. A. & Festa, P. J. (1979) Effects on finfish and lobster, in: R. L. Swanson & C. J. Sindermann (Eds) *Oxygen Depletion and Associated Benthic Moralities in New York Bight, 1976*. NOAA Professional Paper 11, pp. 295–314 (Washington, DC: US Government Printing Office).

Baird, D., Christian, R. R., Peterson, C. H. & Johnson, G. A. (2004) Consequences of hypoxia on estuarine ecosystem function: Energy diversion from consumers to microbes, *Ecological Applications*, 14, pp. 805–822.

Boesch, D. F., Coles, V. J., Kimmel, D. G. & Miller, W. D. (2007) Ramifications of climate change for Chesapeake Bay hypoxia, in: Pew Center (Ed.) *Regional Impacts of Climate Change, Four Case Studies in the United States*, pp. 57–70 (Arlington, VA: Pew Center on Global Climate Change).

Breitburg, D. L. (2002) Effects of hypoxia, and the balance between hypoxia and enrichment, on coastal fishes and fisheriesq, *Estuaries*, 25, pp. 767–781.

Breitburg, D. L., Craig, J. K., Fulford, R. S., Rose, K. A., Boynton, W. R., Brady, D. C., Ciotti, B. J., Díaz, R. J., Friedland, K. D., Hagy, J. D. III, Hart, D. R., Hines, A. H., Houde, E. D., Kolesar, S. E., Nixon, S. W., Rice, J. A., Secor, D. H. & Targett, T. E. (2009b) Nutrient enrichment and fisheries exploitation: interactive effects on estuarine living resources and their management, *Hydrobiologia*, 629, pp. 31–47.

Breitburg, D. L., Hondorp, D. W., Davias, L. W. & Díaz, R. J. (2009a) Hypoxia, nitrogen and fisheries Integrating effects across local and global landscapes, *Annual Review Marine Science*, 1, pp. 329–350.

Bricker, S., Lipton, D., Mason, A., Dionne, M., Keeley, D., Krahforst, C., Latimer, J. & Pennock, J. (2006) *Improving Methods and Indicators for Evaluation Coastal Water Eutrophication: A Pilot Study in the Gulf of Maine*. NOAA Technical Memorandum No. NOS NCCOS 20 (Silver Springs, MD: National Ocean Survey).

Caddy, J. F. (1993) Towards a comparative evaluation of human impacts on fishery ecosystems of enclosed and semi-enclosed seas, *Review of Fisheries Science*, 1, pp. 57–95.

Chan, F., Barth, J. A., Lubchenco, J., Kirincich, A., Weeks, H., Peterson, W. T. & Menge, B. A. (2008) Emergence of anoxia in the California current large marine ecosystem, *Science*, 319, pp. 920–923.

Christensen, J. H., Hewitson, B., Busuioc, A., Chen, A., Gao, X., Held, I., Jones, R., Kolli, R. K., Kwon, W. T., Laprise, R., Magaña Rueda, V., Mearns, L., Menéndez, C. G., Räisänen, J., Rinke, A., Sarr, A. & Whetton, P. (2007) Regional climate projections, in: S. Solomon, D. Qin, M. Manning, Z. Chen, M. Marquis, K. B. Averyt, M. Tignor & H. L. Miller (Eds) *Climate Change 2007: The Physical Science Basis*. Contribution of Working Group I to the Fourth Assessment Report of the Intergovernmental Panel on Climate Change (Cambridge: Cambridge University Press).

Coutant, C. C. (1990) Temperature–oxygen habitat for freshwater and coastal striped bass in a changing climate, *Transactions of the American Fisheries Society*, 119, pp. 240–253.

Craig, J. K. & Crowder, L. B. (2005) Hypoxia-induced habitat shifts and energetic consequences in Atlantic croaker and brown shrimp on the Gulf of Mexico shelf, *Marine Ecology Progress Series*, 294, pp. 79–94.

Craig, J. K., Crowder, L. B. & Henwood, T. A. (2005) Spatial distribution of brown shrimp (*Farfantepanaeus aztecus*) on the northwestern Gulf of Mexico shelf: effects of abundance and hypoxia, *Canadian Journal of Fisheries and Aquatic Science*, 62, pp. 1295–1308.

Díaz, R. J. & Rosenberg, R. (1995) Marine benthic hypoxia: a review of its ecological effects and the behavioural responses of benthic macrofauna, *Oceanography and Marine Biology*, 33, pp. 245–303.

Díaz, R. J. & Rosenberg, R. (2008) Spreading dead zones and consequences for marine ecosystems, *Science*, 321, pp. 926–928.

Figley, W., Pyle, B. & Halgren, B. (1979) Socioeconomic impacts, in: R. L. Swanson & C. J. Sindermann (Eds) *Oxygen Depletion and Associated Benthic Moralities in New York Bight, 1976*. NOAA Professional Paper No. 11, pp. 315–322 (Washington, DC: US Government Printing Office).

Gazey, W. J., Galloway, B. J., Fechhelm, R. C., Martin, L. R. & Reitsema, L. A. (1982) Shrimp mark and release and port interview sampling survey of shrimp catch and effort with recovery of captured tagged shrimp, in: W. B. Jackson (Ed.) *Shrimp Population Studies: West Hackberry and Big Hill Brine Disposal Sites Off Southwest Louisiana and Upper Texas Coasts, 1980–1982*, Vol. 3. Department of Commerce NOAA/NMFS Final Report to the US Department of Energy (Washington DC: US Government Printing Office).

Gilbert, D., Rabalais, N. N., Díaz, R. J. & Zhang, J. (2010) Evidence for greater oxygen decline rates in the coastal ocean than in the open ocean, *Biogeosciences Discussion*, 6, pp. 9127–9160.

Gooday, A. J., Jorissen, F., Levin, L. A., Middelburg, J. J., Naqvi, S. W. A., Rabalais, N. N., Scranton, M. & Zhang, J. (2009) Historical records of coastal eutrophication-induced hypoxia, *Biogeosciences*, 6, pp. 1–39.

Grantham, B. A., Chan, F., Nielsen, K. J., Fox, D. S., Barth, J. A., Huyer, A., Lubchenco, J. & Menge, B. A. (2004) Upwelling-driven nearshore hypoxia signals ecosystem and oceanographic changes in the northeast Pacific, *Nature*, 429, pp. 749–754.

Halpern, B. S., Walbridge, S., Selkoe, K. A., Kappel, C. V., Micheli, F., D'Agrosa, C., Bruno, J. F., Casey, K. S., Ebert, C., Fox, H. E., Fujita, R., Heinemann, D., Lenihan, H. S., Madin, E. M. P., Perry, M. T., Selig, E. R., Spalding, M., Steneck, R. & Watson, R. (2008) A global map of human impact on marine ecosystems, *Science*, 319, pp. 948–952.

Helly, J. J. & Levin, L. A. (2004) Global distribution of naturally occurring marine hypoxia on continental margins, *Deep-Sea Research Part I*, 51, pp. 1159–1168.

Helsinki Commission (HELCOM) (2007) *The Baltic Sea Action Plan* (Helsinki: HELCOM).

Holman, J. D., Burnett, K. G. & Burnett, L. E. (2004) Effects of hypercapnic hypoxia on the clearance of *Vibrio campbellii* in the Atlantic blue crab, *Callinectes sapidus* Rathbun, *Biological Bulletin*, 206, pp. 188–196.

Jones, D. D. & Walters, C. J. (1976) Catastrophe theory and fisheries regulations, *Journal of the Fisheries Research Board of Canada*, 33, pp. 2829–2833.

Jones, J. R. E. (1952) The reactions of fish to water of low oxygen concentration, *Journal of Experimental Biology*, 29, pp. 403–415.

Jones, P. D. (2006) Water quality and fisheries in the Mersey estuary, England: a historical perspective, *Marine Pollution Bulletin*, 53, pp. 144–154.

Justić, D., Bierman, V. J. Jr, Scavia, D. & Hetland, R. D. (2007) Forecasting gulf's hypoxia: the next 50 years? *Estuaries and Coasts*, 30, pp. 791–801.

Justić, D., Rabalais, N. N. & Turner, R. E. (1996) Effects of climate change on hypoxia in coastal waters: a doubled CO_2 scenario for the northern Gulf of Mexico, *Limnology and Oceanography*, 41, pp. 992–1003.

Karlson, K., Rosenberg, R. & Bonsdorff, E. (2002) Temporal and spatial large-scale effects of eutrophication and oxygen deficiency on benthic fauna in Scandinavian and Baltic waters—a review, *Oceanography and Marine Biology*, 40, pp. 427–489.

Keeling, R. F., Körtzinger, A. & Gruber, N. (2010) Ocean deoxygenation in a warming world, *Annual Review of Marine Science*, 2, pp. 199–229.

Le Moullac, G., Soyez, C., Salnier, D., Ansquer, D., Avarre, J. C. & Levy, P. (1998) Effect of hypoxic stress on the immune response and the resistance to vibriosis of the shrimp *Penaeus stylirostris*, *Fish and Shellfish Immunology*, 8, pp. 621–629.

Levin, L. A., Ekau, W., Gooday, A. J., Jorissen, F., Middelburg, J. J., Naqvi, S. W. A., Neira, C., Rabalais, N. N. & Zhang, J. (2009) Effects of natural and human-induced hypoxia on coastal benthos, *Biogeosciences*, 6, pp. 2063–2098.

Lipton, D. & Hicks, R. (2003) The cost of stress: low dissolved oxygen and economic benefits of recreational striped bass (*Morone saxatilis*) fishing in the Patuxent River, *Estuaries*, 26, pp. 310–315.

May, E. (1973) Extensive oxygen depletion in Mobile Bay, Alabama, *Limnology and Oceanography*, 18, pp. 353–366.

Mee, L. D. (1992) The Black Sea in crisis: a need for concerted international action, *Ambio*, 21, pp. 278–286.

Middelburg, J. & Levin, L. A. (2009) Coastal hypoxia and sediment biogeochemistry, *Biogeosciences*, 6, pp. 1273–1293.

Miller, J. R. & Russell, G. L. (1992) The impact of global warming on river runoff, *Journal of Geophysical Research*, 97, pp. 2757–2764.

Nestlerode, J. A. & Díaz, R. J. (1998) Effects of periodic environmental hypoxia on predation of a tethered polychaete, *Glycera americana*: implications for trophic dynamics, *Marine Ecology Progress Series*, 172, pp. 185–195.

O'Connor, T. & Whitall, D. (2007) Linking hypoxia to shrimp catch in the northern Gulf of Mexico, *Marine Pollution Bulletin*, 54, pp. 460–463.

Peña, M. A., Katsev, S., Oguz, T. & Gilbert, D. (2010) Modeling dissolved oxygen dynamics and hypoxia, *Biogeosciences*, 7, pp. 933–957.

Pihl, L., Baden, S. P., Díaz, R. J. & Schaffner, L. C. (1992) Hypoxia induces structural changes in the diet of bottom-feeding fish and crustacea, *Marine Biology*, 112, pp. 349–361.

Rabalais, N. N., Harper, D. E. & Turner, R. E. (2001) Responses of nekton and demersal and benthic fauna to decreasing oxygen concentrations, in: N. N. Rabalais & R. E. Turner (Eds) *Coastal Hypoxia Consequences for Living Resources and Ecosystems*, pp. 115–128 (Washington, DC: American Geophysical Union).

Rabalais, N. N., Díaz, R. J., Levin, L. A., Turner, R. E., Gilbert, D. & Zhang, J. (2010) Dynamics and distribution of natural and human-caused hypoxia, *Biogeosciences*, 7, pp. 585–619.

Rabalais, N. N. & Gilbert, D. (2008) Distribution and consequences of hypoxia, in: E. R. Urban, B. Sundby, P. Malanotte-Rizzoli & J. Milello (Eds) *Watersheds, Bay, and Bounded Seas: The Science and Management of Semi-Enclosed Marine Systems*, pp. 209–225 (Washington, DC: Island).

Rabalais, N. N. & Turner, R. E. (Eds) (2001) *Coastal Hypoxia Consequences for Living Resources and Ecosystems*. Coastal and Estuarine Studies No. 58 (Washington, DC: American Geophysical Union).

Rabalais, N. N., Turner, R. E., Justić, D. & Díaz, R. J. (2009) Global change and eutrophication of coastal waters, *ICES Journal of Marine Science*, 66, pp. 1528–1537.

Rabalais, N. N., Turner, R. E., Sen Gupta, B. K., Boesch, D. F., Chapman, P. & Murrell, M. C. (2007) Hypoxia in the Northern Gulf of Mexico: does the science support the plan to reduce, mitigate, and control hypoxia? *Estuaries and Coasts*, 30, pp. 753–772.

Searchinger, T., Heimlich, R., Houghton, R. A., Dong, F., Elobeid, A., Fabiosa, J., Tokgoz, S., Hayes, D. & Yu, T. -H. (2008) Use of U.S. croplands for biofuels increases greenhouse gases through emissions from land use change, *Science*, 319, pp. 1238–1240.

Selman, M. & Greenhalgh, S. (2009) *Eutrophication: Policies, Actions, and Strategies to Address Nutrient Pollution*. Note No. 3 (Washington, DC: World Resources Institute Policy).

Selman, M., Greenhalgh, S., Branosky, E., Jones, C. Y. & Guiling, J. (2009) *Water Quality Trading Programs: An International Overview*. Issue Brief No. 1 (Washington, DC: World Resources Institute).

Smith, K. J. & Able, K. W. (2003) Dissolved oxygen dynamics in salt marsh pools and its potential impacts on fish assemblages, *Marine Ecology Progress Series*, 258, pp. 223–232.

Stewart, N. E., Shumway, D. L. & Dourdorff, P. (1967) Influence of oxygen concentration on growth of juvenile largemouth bass, *Journal of the Fisheries Research Board of Canada*, 24, pp. 475–494.

Thomas, P., Rahman, M. S., Khan, I. A. & Kummer, J. A. (2007) Widespread endocrine disruption and reproductive impairment in an estuarine fish population exposed to seasonal hypoxia, *Proceedings of the Royal Society, Biological Sciences*, 274, pp. 2693–2701.

Thronson, A. & Quigg, A. (2008) Fifty-five years of fish kills in coastal Texas, *Estuaries and Coasts*, 31, pp. 802–813.

Tyler, R. M., Brady, D. C. & Targett, T. E. (2009) Temporal and spatial dynamics of diel-cycling hypoxia in estuarine tributaries, *Estuaries and Coasts*, 32, pp. 123–145.

United States Environmental Protection Agency (USEPA) (2008) *Hypoxia in the Northern Gulf of Mexico An Update*. Science Advisory Board, Hypoxia Assessment Panel, EPA-SAB-08-004 (Washington, DC: USEPA).

Valenzuela, A., Silva, V., Tarifeño, E. & Klempau, A. (2005) Effect of acute hypoxia in trout (*Oncorhynchus mykiss*) on immature erythrocyte release and production of oxidative radicals, *Fish Physiology and Biochemistry*, 31, pp. 65–72.

Vaquer-Sunyer, R. & Duarte, C. M. (2008) Thresholds of hypoxia for marine biodiversity, *Proceedings of the National Academy of Sciences of the United States of America*, 105, pp. 15452–15457.

Welker, T., McNulty, S. & Klesius, P. (2007) Effect of sublethal hypoxia on immune response and susceptibility of channel catfish, *Ictalurus punctatus*, to enteric speticemia, *Journal of the World Aquaculture Society*, 38, pp. 12–23.

Wu, R. S. S., Zhou, B. S., Randall, D. J., Woo, N. Y. S. & Lam, P. K. S. (2003) Aquatic hypoxia is an endocrine disruptor and impairs fish reproduction, *Environmental Science and Technology*, 37, pp. 1137–1141.

Financing Water Quality Management

CÉLINE KAUFFMANN
Organisation for Economic Co-operation and Development (OECD), Paris, France

ABSTRACT *In a context of greater pressure on water resources, ensuring continued and adequate access to safe water supplies will require the investment of significant funds and expertise in public sewerage and water and wastewater treatment infrastructure. After a short introduction, this paper examines the recent trends in the development of wastewater infrastructure. It then discusses the investment needs and reviews the potential sources of funding, before turning to recent trends in private-sector participation in water and wastewater treatment infrastructure and the framework conditions to make this participation work in the public interest.*

Introduction

Increasing withdrawals and degraded quality are placing greater demands on limited water resources in both Organisation for Economic Co-operation and Development (OECD) and non-OECD countries. While not the greatest contributor to the degradation of water quality, pollution from urban wastewater constitutes a growing threat, notably in emerging economies. In these countries, fast-growing populations and urbanization are outstripping the efforts of governments to develop public sewerage and wastewater treatment infrastructure. Consequently, the volume of sewage produced increases and so does the risk of contamination of fresh water resources. This in turn may jeopardize the efforts of countries to ensure adequate access to a safe water supply for their populations. Bringing these countries to a more acceptable level of wastewater treatment will require significant investment funds and expertise. While OECD countries have made important efforts to treat pollution from wastewater, they also face important financing challenges to renew ageing infrastructure and bring it in conformity with more stringent regulation. To meet these investment needs or bring in the necessary expertise, many countries have sought the involvement of the private sector. Ensuring that such partnerships yield the hoped-for benefits to all constituencies is a necessity for policy-makers.

Water Quality Management: The Importance of Treating Wastewater

The human and economic costs of an inadequate standard of water quality are significant. There is a direct correlation between the quality of water and morbidity of children under five years of age.[1] In a number of countries large portions of the population rely on water from contaminated sources, and diarrhoea from its consumption is widespread. In Indonesia, for instance, 70% of the population suffer from bad-quality water and diarrhoea is the second cause of mortality in children, as is the case in Rwanda.[2] Outbreaks of cholera remain common in a number of countries, including Kenya.[3] According to Marañón-Pimentel (2009), the total monetary cost of diarrhoeas attributable to deficiencies in potable water services amounted to 2.689 billion pesos in Mexico as of December 2005. Working days lost accounted for 5.6% of the total. This can be compared with the annual investment required to provide potable water and sanitation of 776.9 million pesos between 2005 and 2030 (estimated as the amount required to provide total coverage and mitigate the economic costs of diarrhoeas in the 13 states where 75% of the cases occur).

Developing countries have made important efforts to increase population access to drinking water. However, inadequate management of wastewater may jeopardize these efforts. Environmental sustainability is expected to worsen if the water supply target is achieved without accompanying measures for sanitation and effluent management (African Development Bank/Organisation for Economic Co-operation and Development (ADB/OECD), 2007). The volume of sewage produced increases proportionally with the availability of water, exacerbating its threat as a main source of water pollution. In Morocco, for instance, owing to the development of urban water supply systems, the volume of wastewater grew from 48 million to 546 million m^3 between 1960 and 2000 and is expected to reach 900 million m^3 in 2020. Some sanitation solutions adopted by households, such as septic tanks, flush toilets and sewer connections without proper treatment, can also cause water pollution. The excreta disposal situation is also worsening because on-site installations are multiplying but are not properly emptied. Moreover, with rapid urbanization, the number of people per plot is increasing, and the new lands are often poorly drained and shallow. Inadequate management of sewerage risks compounding a situation where already households in many countries need to add treatment to their water for drinking purposes (Biswas & Tortajada, 2009).

Despite the dramatic impacts, the OECD (2008a) estimates that by 2030 an additional 1.1 billion people will not be connected to public sewerage compared with 2000, bringing the total number of unconnected people to 5.4 billion. Globally, the portion of the world population to be connected to advanced (nitrogen-removal) sewerage treatment plants will increase only slightly, from 9% to 16% in 2030. In particular, nitrogen from urban sewerage is expected to increase strongly in fast-developing and emerging economies, notably India, China and the Middle East, as a result of sustained population growth and urbanization outstripping the development of wastewater treatment infrastructure. By 2030 the BRIC countries (Brazil, Russia, India and China) are expected to host half of the those who will not be connected to public sewerage. As a consequence of a strong increase in nitrogen surplus from agriculture and urban sewage, the OECD (2008a) forecasts that nitrogen export to coastal waters in BRIC countries will increase by 11% by 2030.

Efforts to collect and treat wastewater rarely follow the progress of governments in providing safe and sustainable access to drinking water. Information on the proportion of wastewater which is treated and on the level of treatment is relatively sparse for developing countries and emerging economies (for some available information, see Table 1). However, the ADB (2007) notes that the level of wastewater treatment is poor or very poor in a number of Asian countries, including India. By contrast, China has made significant efforts to improve wastewater treatment coverage. As of 2005, half of urban wastewater underwent some sort of treatment. The 2011–2015 plan seeks to increase this

Table 1. Indicators of water supply quality in selected developing and emerging countries

Country	Continuous water supply	Unaccounted for water	Treatment of wastewater
Bangladesh	Mostly up to 4 h/day; 24 h/day in Chittagong and Dhaka (2005)	40–50%	10% in Dhaka
Bolivia	24 h/day in La Paz and El Alto; 15 h/day in Cercado	28% in La Paz and El Alto; 29% in Santa Cruz de la Sierra City; 21% in Sucre (2005)	Complete lack of wastewater treatment in La Paz and minimal amounts in El Alto
Brazil	Regional rationing in dry periods	National average of 39.8%; 32% in Sao Paulo; 44% in Porte Alegre (2006)	17% of the urban population has access to treated wastewater
Cambodia	24 h/day in Phnom Penh and Banteay Meanchey; 25% of Sihanoukville (2003)	6% in Phnom Penh (2003)	Poor
Chile	Yes	National average of 34% (2006); 28% in Santiago (2005)	81.9% in 2006, with a goal of reaching 98.8% of all collected municipal wastewater in 2010
China	24 h/day in many cities	38% in Henan; 18% in Chengdu	In 2004, 90.7% industrial wastewater treatment and 45.2% domestic wastewater treatment. In 2005, 52% of all urban wastewater treated and 46% overall. The 12th Five Year Plan 2011–2015 seeks to increase urban sewage treatment to at least 60%; 70% for urban areas in Beijing
Colombia	Two-thirds of households in large urban areas; no continuity in smaller municipalities; 24 h/day in Cartagena (2005)	National average of 49%; 45% in Los Patios; 38% in Puerto Carreno (2003); 41% in Cartagena (2005)	10% of the wastewater generated in the country undergoes some kind of treatment
India	1/4000 utilities meet 24 h/day; 4–5 h/day in Bangalore, Chennai and Mumbai	From Jamshedpur (12.8%), Mumbai (13.6%), Jabalpur (14.3%) and Chennai (17.0%) to Nashik (59.6%), Amritsar (57.4%), Nagpur (51.9%) and Bangalore (45.1%)	Very poor

Continued

Table 1. Continued

Country	Continuous water supply	Unaccounted for water	Treatment of wastewater
Peru	No. 17 h/day on average in 2005; 37% of operators provide less than 12 h/day; 37% between 12 and 20 h/day and 26% over 20 h/day	National average of 45%; 30% in Tacna; 57% in Loreto; 41% in Lima (2005)	22% in urban zones in 2005, 22% as a national average in 2004
Philippines	No and low water pressure; 88% in Manila (2003)	48% in Manila (2004)	Poor
Singapore	Yes	4.5%	100%
South Africa	Yes for 98% of the population	Average of 31%; 19.2% in Cape Town; 29% in Johannesburg (2006)	84.3% of water supplied
Vietnam	20.2 h/day	37%	Poor

Source: Updated from Organisation for Economic Co-operation and Development (OECD) (2009b).

proportion to 60% (70% in Beijing). Similar patterns can be found in Latin America where Franceys & Gerlach (2008) notes the absence of treatment in the large cities of Bolivia and estimates for other countries such as Brazil, Colombia and Peru range from 10% to 20% (of wastewater treated or of the population with access to wastewater treatment). While Chile stands out as a country that has made significant efforts (with more than 80% of treated wastewater), only 10% of collected sewage in major urban centres of Latin America is treated properly and then disposed of in an environmentally safe way, according to Biswas (2006).

By contrast, point source discharges to surface water from urban wastewater systems have been significantly reduced in most OECD countries (OECD, 2008a). Still, considerable variations continue to exist across OECD countries in terms of coverage and the level of sewerage treatment (OECD, 2009a) (Table 2). Some countries are still completing sewerage systems or implementing the first generation of municipal wastewater-treatment plants, including Belgium, Mexico and Turkey.[4] Other countries, including Japan, Korea, Luxembourg, Spain, and the United Kingdom, exhibit high secondary treatment coverage.[5] The countries that have a particularly high level of tertiary treatment[6] include Austria, Denmark, Finland, Germany, the Netherlands, Sweden and Switzerland. According to Eurostat,[7] 95% of the population connected to urban wastewater had access to tertiary treatment in the Netherlands (in 2006), 93% in Germany (in 2005), and 88% in Austria (2006). By contrast, this proportion reached 10% in Turkey (in 2006). Turkey has nevertheless made important progress over the last few decades to improve the proportion of the population connected to wastewater treatment: connection to any kind of treatment increased from 10% in 1996 to 42% in 2006, while access to secondary treatment improved from 2% to 28% over the same period.

Table 2. Coverage of sanitation services in selected OECD countries

Country (year)	Connected to public sewage		Connected to independent sewerage	Not connected
	Connected to a sewage-treatment plant	Not connected to a sewage-treatment plant		
Canada (1999)	71.7	2.6	25.7	0
Mexico (2005)	35	32.6	15.9	16.5
Japan (2005)	69.3	0	8.6	22.1
Austria (2004)	88.9	0	11.1	0
Belgium (2005)	54.6	31.3	n.a.	n.a.
France (2004)	80.1	2.3	15.7	1.9
Germany (2004)	94.1	1.4	3.4	1.1
Hungary (2004)	59.8	4.1	14	22.1
Netherlands (2005)	99			
Norway (2005)	77.2	4.8	16.3	1.7
Poland (2006)	59.8		25.6	14.6
Spain (2005)	92	8	0	0
Sweden (2005)	86	0	14	0
Turkey (2004)	35.9	32.1	n.a.	n.a.
UK (2005)	97.1	0.7	n.a.	n.a.

Note: n.a., Not available.
Sources: Updated from OECD (2009a).

Investment Needs

Substantial investments are needed if developing countries are to scale up wastewater infrastructure significantly and achieve the Millennium Development Goal (MDG) sanitation target (which "only" aims to reduce by half the number of unserved). For reference, annual investment needs are estimated at US$655 million and US$884 million to reach the sanitation and the water targets, respectively, in the nine African countries considered in OECD (2009b): Burkina, Ghana, Kenya, Mauritania, Mozambique, Senegal, South Africa, Uganda and Zambia. This compares with a level of official development assistance for water and sanitation of US$347 million in 2005. According to the same source, annual investment needs to reach the sanitation MDG target are also significant in Latin America and reach US$141 million in Argentina, US$586 million in Brazil and US$227 million in Mexico.

OECD countries also face significant financial challenges to replace ageing water infrastructure[8] and comply with ever-stringent environmental regulations. Stricter regulations include higher standards of water quality—such as established in the European Union through the Drinking Water Directive,[9] the Urban Wastewater Treatment Directive[10] and the Water Framework Directive,[11] or such as under consideration by the United States Environmental Protection Agency (USEPA),[12] for instance—and growing requirements to separate the sewer system and street drainage. Tightening of regulation in the field of energy efficiency and greenhouse gas emissions is also impacting the water and wastewater industry, traditionally an energy-intensive activity (Palmer, 2010). As an example of costs that non-compliance may induce, the European Commission is imposing on Belgium a fine of €15 million (with a daily penalty of €62,000) for its failure to bring the country's wastewater treatment up to the Urban Wastewater Treatment Directive standards.[13] Consequently, the OECD (2007) estimates that France and the UK would need to increase spending on water by 20% to maintain current services; Japan and Korea by over 40%. In its 2008 Clean Watersheds Needs Survey, the USEPA estimates the investment needs for wastewater-treatment plants, pipe repairs, and buying and installing new pipes at US$192.2 billion, to which US$63.6 billion should be added for combined sewer overflow correction.[14] Overall, the OECD (2007) estimates that annual (current and investment) expenditure on water and wastewater services will reach some US$600 billion in 2025 for OECD countries (half of which is for Mexico and the United States), and US$400 billion for the BRICs (half of which is for China).

Climate change is expected to have a significant impact on water quality and generate substantial additional financing needs to adapt drinking water and sanitation infrastructure. According to OECD (2008b), in regions where precipitations are expected to increase, issues such as flood management and wastewater treatment may become even more problematic and impose substantial additional costs. Indeed, many sewer systems combine sewers and street drainage with the risk that heavy rain may overflow the sewer system and clog wastewater-treatment facilities. The USEPA estimates that many systems have reached their age limit and are plagued by chronic overflows, which are responsible for the discharge of 850 billion gallons (3230 billion litres) of untreated wastewater into surface waters every year. The issue is also becoming prominent in China, where heavy summer rains have led to the congestion of eight wastewater-treatment plants in the Beijing area and to the flushing of the overload of untreated wastewater and storm water directly into rivers (*Global Water Intelligence*, 2010). Regions where water availability will decline on account of glacier retreat and fewer precipitations are not necessarily better off. Investments in enhanced storage, improved efficiency of water allocation and use, and water recycling and reuse are much needed. According to Howard *et al.* (2010), few water and sanitation technologies are resilient to climate change, but those that are need to be prioritized in future investment. Building resilience is also about improving water management (minimizing losses, for instance, and managing demand). Overall, the United Nations Framework Convention on Climate Change (UNFCCC) (2007) estimates the costs of adapting the water sector to climate change at US$225 billion for the period up to 2030, equivalent to some US$11 billion per year.

Where Will the Funds Come From?

Ensuring an adequate supply of finance is essential for meeting policy objectives in both OECD and non-OECD countries. Conventional water and wastewater treatment is very costly. It has led a number of countries to seek sources of finance that could relieve budgets under tight constraint or smooth the necessary expenditure over the long run, including through built–operate–transfer (BOT) arrangements, the typical contract for private sector involvement in this area.

A prerequisite for meeting the financial needs of the sector and attracting more funding is that the sector demonstrates that it operates efficiently. Improving the efficiency of production (e.g. through leakage control and energy-efficiency measures) and consumption (e.g. through demand-side measures) reduces the need for investment and represents a crucial step for establishing realistic and affordable objectives for water and sanitation services. The OECD (2007) highlights the case of New York City, which reorganized the management of the watersheds to protect ecosystems and water quality rather than investing in an expensive new water-treatment infrastructure. This helped the municipality save between US$4 billion and US$6 billion.

In addition, as clarified in OECD (2009a), while commercial funding such as loans, bonds or equity play a critical role in helping the sector to cope with large up-front investment costs and the resulting cash flow needs, they must be repaid through cash flows. Ultimately, ensuring financial sustainability of the sector requires finding the right mix between the ultimate sources of revenue, the so-called "3Ts": tariffs, taxes and transfers.

Tariff, alone, rarely reflects the cost of operating and maintaining the sewerage network. Connection costs also reach prohibitive levels. As highlighted in OECD (2009c), a household's willingness to pay for sanitation services tends to be lower than for water. Cross-subsidization (as in the case of the Tirapur project in India[15]) and spreading the connection costs to all (including to the unserved population as was done in the Philippines) have also shown limitations, in particular in terms of their sustainability over the longer-run. On the other hand, public support is justified because of the staggering investment needs to extend basic sanitation facilities and the public good dimension of wastewater treatment. Typically, the money that businesses or households allocate for improved wastewater treatment will benefit health and economic development for the wider community, e.g. including downstream users who cannot be excluded from the fruition of improved resource quality (OECD, 2009a). The World Health Organization (WHO) estimates that each US dollar invested in water supply and sanitation generates US$4–12 in health benefits alone, depending on the intervention. This raises the need for regulators and policy-makers to look into tariff structures that provide operators with incentives for sewerage connections, and for public authorities to consider targeted subsidies.

Experience in OECD countries shows that tariffs structures and methodologies to compute prices vary widely across and often within countries, although most of the time the basis for charging sewerage and wastewater treatment remains the volume of water consumption. OECD (2009a) highlights the case of Mexico where municipalities have started to charge a percentage of water tariffs to cover sanitation costs. However, practices across municipalities differ widely (between, for instance, Monterrey where 25% of the tariff is for sanitation and Mexico, where sanitation is not considered) and tariffs do not always explicitly indicate the amount that corresponds to sanitation. Denmark imposes a wastewater tax on water consumers—household and industry, but not agriculture—in addition to the tariffs for wastewater treatment services (OECD, 2010b). In the United States, a recent survey of 30 cities showed a wide variety in sewer rates, from US$13.9 for 100 gallons per person per day in Salt Lake City, Utah, to US$208.6 in Atlanta, Georgia. In these cases, the survey authors directly link prices to the investments undertaken by the two cities to upgrade their wastewater-treatment facilities.[16] Large divergences may also arise from the use of different methodologies and basis to compute prices. Wastewater services in the United States are often paid through property taxes, with the result (and this is the case in Salt Lake City) that residential rates are low. Despite the disparities, OECD (2010b) shows a consistent increase in wastewater charges in OECD countries as costs are brought in line with investments needs to comply with more stringent environmental regulations. With regards to the management of wastewater from industrial users, OECD (2010b) shows a trend of increasingly basing wastewater treatment charges on the pollution load of effluents, reflecting a willingness both to align charges with real treatment costs and to apply the polluter-pays principle.

Private Sector Participation

Recent trends show many governments turning towards the private sector to inject much-needed investment capital and improve the performance of public services, with varying degree of success. Investing in wastewater and water-treatment facilities has proven less sensitive than doing so in the provision of water services. In effect, projects that involve water or wastewater-treatment facilities can be more easily ring-fenced. They mainly involve a relationship between a private company and a responsible authority (municipality or other), with limited contact with end-users.

Private sector participation in sewerage is very substantial in some OECD countries, such as France and the UK where 55% and 90% respectively of the population is connected to privately run sewage (Table 3). In other countries it is less so. In particular, private sector participation is absent or quasi-inexistent in Austria, Norway, Sweden and Switzerland. Some non-OECD countries have also substantially involved the private sector as a strategy to extend sewerage networks and treatment facilities. This is for instance the case in Chile, where the involvement of the private sector generated private capital inflows of US$1.5 billion between 1999 and 2009 that enabled wastewater coverage to increase from 12% to 82% in a decade, and allowed improvement in the efficiency of water and sanitation systems. Similarly, in Malaysia sewerage utilities were privatized in 1994. Indah Water Konsortium (IWK), which operates and maintains sewage treatment plants, has gone from serving fewer than 4 million people in 1994 to more than 14 million people in 2001 (OECD, 2009b).

More recently, aware of the undesirable environmental and public health effects that infrastructure development may have if it does not adequately balance water provision and

Table 3. Population served by the private sector

Country	Private sector participation in the provision of drinking water	Private sector participation in the provision of sewerage services
Austria	7%	0%
Belgium	3%	10%
France	74%; Veolia, 39% and Suez, 19%	55%; Veolia, 26% and Suez, 18%
Germany	21% (RWE = 16%)	18%
Hungary	29%	27%
Italy	40% (ACEA = 16%)	29%
Netherlands	0%	10%
Norway	6%	0%
Poland	3%	3%
Sweden	1%	1%
Switzerland	0%	0%
UK	88% (more than 17 private utilities)	90%

Note: RWE, Rhenish-Westphalian Electric power company; ACEA, Azienda Energia e Ambiente
Sources: Pinsent Masons *Water Yearbook 2008–2009*; and Veolia website, Suez Activity Report www.veoliaeau.com/profil/implantations/france.htm and www.suez-environnement.fr/publication/fr/rapport-activite-2009.

wastewater treatment, an increasing number of countries are calling on the private sector to contribute to financing the high upfront costs of wastewater infrastructure development, or to providing the necessary expertise that may be lacking in the short-term. Egypt is one such country where the widening gap between drinking water coverage (almost 100%) and wastewater coverage (55% in Egypt as a whole, 15% in rural areas) is prompting the government to adopt more aggressive strategies to extend the relevant infrastructure. A total of 27% of wastewater collected in the Greater Cairo area is not treated and the renovation need is estimated at 50% of the existing network (MED EUWI, 2009a). The development of a Master Plan for water supply and sanitation infrastructure to 2037 and the successful completion of a tender for the New Cairo wastewater-treatment plant with the consortium Orasqualia have triggered a renewed interest in discussing the potential private involvement in the development and management of wastewater-treatment facilities. The government has five additional projects in the pipeline for the coming 12–24 months and expects more in the future (MED EUWI/OECD, 2010). There are also clear signs that the public–private partnership (PPP) projects are perceived as being attractive to the private sector. This is corroborated by the significant interest raised by both the tender for the "New Cairo" (60 interested parties, seven qualified, five bids) and the pre-qualification stages for the "6th of October" (10 qualified) projects.

More generally, the Middle East is a fast developing market for water and wastewater projects. This is prompted by the relative water scarcity that affects the region and increasing awareness of the need to preserve water quality. As an example of this rising awareness, Abu Dhabi's Regulation and Supervision Bureau has just announced new regulations for the disposal of wastewater and recycling, requiring businesses to obtain consent for the discharge of wastewater in the system or into the environment and, for the first time, setting standards for treated wastewater (*Water21*, 2010). Private companies see the potential of the region for further development of water treatment projects. Veolia Water Solutions & Technologies has for instance opened a new office in Qatar.

China's private participation in the wastewater sector has also grown significantly over the last decade. In 2006 alone, the country claimed two-thirds of the new private participation in water infrastructure contracts reported to The World Bank,[17] mostly in the form of BOT arrangements for water and sewage treatment plants. Transfer–own–transfer (TOT) contracts, a variation of BOTs where the vendor sells an existing facility to an operator for a period of time, are also becoming common. These trends respond to the rapid development of municipal wastewater systems. Just between 2001 and 2004, according to Clark *et al.* (2006), the number of municipal treatment plants increased by half (from 452 to 708), the number of cities served grew from 200 to 364, and the percentage of wastewater receiving treatment went from 35% to 45%. An important driver is the increase in number of cities that charge wastewater tariffs (from 300 to 475).

Increasingly, wastewater is also considered as too precious to waste. Concerns about water resource scarcity and the environmental consequences of untreated discharges in some areas are supporting the development of new technologies such as wastewater reclamation and reuse, desalination plants and advanced filtration membranes. For instance, using a simulation model, Hochstrat *et al.* (2006) find a significant potential for an increased utilization of reclaimed wastewater in European countries, specifically in Spain and the Mediterranean region. The OECD (2008c) notes the new trends in a number of OECD countries, such as Australia, Japan, Spain and the United States, to encourage the use of reclaimed water. Spain, for instance, has set a target of tripling the volume of wastewater reuse by 2015. While these technologies can be adopted by public and private operators alike, they constitute important opportunities for businesses to get involved in the water sector. *Global Water Intelligence*[18] foresees a tripling of water reuse capacity between 2008 and 2016 (from 20 million to 60 million m^3/day).

Hyflux is one of these "new" companies thriving through the development, sale and installation of water-filtration technologies. As highlighted in OECD (2010a), Hyflux launched in 2007 the Hyflux Water Trust responsible for operating and managing the company's BOT contracts. In four years the portfolio of projects grew from 11 to 18 treatment and recycling plants in China, and reached a total design capacity of 580,000 m^3/day.[19]

Windhoek in Namibia was one of the first cities in the world to introduce direct recycling of effluents for drinking purposes. In order to attract technical and operating know-how, the City of Windhoek signed in 2002 a 20-year performance-based operation and maintenance contract with Windhoek Goreangab Operating Company (WINGOC: VeoliaWater, Berlinwasser International and WABAG). Extensive water-quality-monitoring programmes are in place to ensure the required level of water quality after each treatment process, as well as the quality of the water finally supplied to the City of Windhoek.

Probably the best-performing utility in Asia, Singapore's Public Utilities Board (PUB), started engaging private firms in technological advancement programmes, including NEWater,[20] in response to water-scarcity concerns. Four NEWater Factories were developed up to 2007, among which the fourth, the Ulu Pandan NEWater Project, was developed by PUB as a design–build–own–operate and involved the private sector in operating and maintaining the assets.

In addition to conventional arrangements with international players, the treatment of water and wastewater is also attracting numerous smaller-scale private actors responsible for developing on-site facilities and systems (OECD, 2008c). New businesses entering the sector include property developers; big water users such as beverage and mining companies (in the context of increased competition across different uses of water, these companies are concerned with water supply and costs and with acceptance by local communities); and technology providers that provide on-site water and wastewater services for industrial or residential use under design-build-operate agreements. They benefit from new legislation in a number of countries (including India) that allow or may even require on-site treatment to take the weight off overburdened centralized systems. The Chinese government has legislated requirements for major industry to reuse secondary effluents in its 2002 Water Law. The iron and steel industry, for instance, is required to display a water reuse ratio of 90%.

Improving the Framework Conditions for Private Sector Participation

As underlined by the Camdessus Panel and the Gurria task force,[21] shortcomings in the governance of the water and sanitation sector may hamper its ability to mobilize private sector finance and expertise. Past experiences have underlined the importance of better risk management and of strengthening the creditworthiness of the water sector. Even though private sector participation has led to more fruitful collaboration in the development and management of treatment infrastructure than it has in the direct provision of water services, many challenges remain. These relate to the planning of infrastructure development (notably insertion of the treatment facility in the overall chain of provision); the long-term financial sustainability of projects; the quality of the regulatory framework; the incentives put in place to ensure that partners work in the public interest and in a coordinated way; etc.

To respond to the challenges raised by private sector participation in water and sanitation infrastructure, OECD (2009b) proposes a *Checklist for Public Action*, a series of key principles in support of governments' efforts to assess and improve the quality of their investment framework. The OECD emphasizes three crucial issues to make cooperation with the private sector work in the public interest: the financial sustainability of partnerships, the policy and regulatory framework, and accountability mechanisms.

Ensuring the Financial Sustainability of Partnerships

Financial sustainability of projects is a key focus of OECD (2009b) and involves that projects bring value for money, i.e. are assessed through a cost–benefit analysis, are tailored to local needs and characteristics, are affordable for governments, and that their costs can be recovered in the long-term through the ultimate sources of finance.

In particular, adequate insertion of the infrastructure developed and managed by the private sector into the supply chain is a critical element of success. Many wastewater-treatment plants in developing countries are oversized compared with sewerage flows and therefore under-utilized. This often results from inadequate corresponding investments in basic sanitation and wastewater collection, the use of inappropriate technologies, and/or partial assessment of financial sustainability that fails to ensure cost recovery. In Lebanon, for instance, a number of treatment plants are not connected to the wastewater network (MED EUWI, 2009b). Wrong long-term incentives have also undermined the financial sustainability of projects. A case in point was the Tirapur project in India where heavy cross-subsidies combined with new requirements by the Central Pollution Control Board incentivizing wastewater reuse led to a reduction in the demand for the services of the treatment plant and undermined costs recovery.[22]

Tailoring private sector participation involves taking into account local specificities and making the best of private partners' strengths. In particular, the choice of technology is an important step in the determination of affordability. There is a wide range of choices, especially for sanitation: different levels of on-site, conventional and simplified sewerage. Diversifying service provision can help ensure financial sustainability while serving a pro-poor objective. In Latin America, non-conventional systems in the form of simplified sewer designs—or shallow sewerage—have emerged as a technical option that may contribute towards improving sanitation for population groups that do not have access to such services.[23]

It is often the case that the private sector will participate only if national authorities clearly signal that the interests of the private sector will be safeguarded. This often takes the form of sovereign guarantees. As an example, the granting of sovereign guarantees by the Ministry of Finance has been the trigger behind renewed private sector interest in the new stream of PPP projects in Egypt. There is however a trade-off between guarantees to attract the private sector and resulting contingent liabilities bearing on fiscal accounts. There is also a limit to the number of sovereign guarantees that governments can extend without tampering with the credibility of these guarantees. Disclosing future costs of private sector participation and incorporating them in medium-term budgetary projections and debt sustainability analysis has proved important to maintaining fiscal transparency and sustainability.

Developing a Conducive Policy and Regulatory Framework

OECD (2009b) makes clear that private sector participation does not relieve the governments from the essential responsibilities of establishing adequate policies, regulatory frameworks, institutions and contractual arrangements, and overseeing their functioning. In the current context of credit constraint and tighter financial conditions, private developers are likely to be more selective, demanding higher quality, more "bankable" projects, with clearer forms of public support and risk-sharing arrangements. Inadequate framework conditions constitute a risk that the private sector will factor in as a cost.

In particular, strong political commitment remains critical, notably in addressing service affordability. In Morocco, strong political commitment was one of the key factors behind the success of the 30-year concession contract signed in 1997 between the public authority (three main cities, among which was Casablanca, and 14 medium-size cities) and Lydec (owned for 51% by Suez Environment).[24] The contract initially covered drinking water supply, wastewater and electricity distribution, and was further extended in 2007 to include street lighting. While at the beginning the concession agreement did not include extending access to households in informal settlements and slums, the National Initiative for Development launched by the King of Morocco in 2005—with a focus on access to infrastructure and basic social services for the very poor—gave a new impetus to the partnership. Consequently, Lydec and the delegating authority entered an agreement to provide water, sanitation and electricity services to an additional 85,000 households on a social tariff and adapted payment modes. As of 2009, some 30,000 households were (or underway to be) connected to the services.[25] More generally, concession achievements include a quasi-doubling of the population with access to wastewater services (from 440,000 in 1997 to 840,000 in 2007). This is the result of a €6.1 billion investment programme, of which wastewater accounted for 37%.[26]

In addition to political commitment, the establishment of adequate regulation plays a pivotal role in the success of government's cooperation with the private sector. The success of Chile in involving the private sector can largely be attributed to the stability of its water and sanitation policy, and to its high-quality regulatory framework (OECD, 2009b). This involved a clear separation of roles across the different bodies in charge of regulation and supervision, and from the activities of service provision of water operators; a regulation geared towards ensuring efficiency of operation and investment and strong enforcement mechanisms; and a strong focus on sustainable access (notably through a pro-poor subsidy scheme).

Another important challenge is to ensure both policy and administrative coherence. Water and sanitation are segmented sectors, they involve multiple actors with oversight responsibilities for resource management and service provision often split horizontally between different ministries, and vertically across national, regional and local authorities. ADB/OECD (2007) highlights the case of Botswana where, as of 2006, the institutional framework for wastewater management comprised five departments responsible for 11 Acts: Department of Waste Management and Pollution Control, local authorities, Department of Water Affairs, Directorate of Public Service Management, and Department of Local Government. This segmentation may raise important capacity challenges and generate issues of consistency across government levels. Careful allocation of roles and responsibilities is needed across authorities, taking into account existing capacity gaps, and ensuring that resources are allocated in line with duties and distributed in a predictable way. Preserving consistency across government policies also involves strengthening coordination mechanisms across government levels and building common government understanding of the objectives, means and resources for water provision. The establishment of PPP units, as has been done in South Africa and Egypt, can contribute to developing this required capacity. In the case of Egypt, a PPP Unit was established within the Ministry of Finance.[27] Defining itself as a "Centre of expertise", it lists among its lead responsibilities "importing experience from other countries and developing guidance and methodologies that is appropriate to Egypt". Other coordination mechanisms such as contracts, inter-sectoral collaboration, performance indicators, information systems, and financial transfers have proved crucial for efficient and effective joint management of water quality by public and private actors.[28]

The multiplication of smaller-scale providers or non-conventional businesses in the treatment of water and wastewater may compound the administrative capacity and policy coherence challenges. In particular, dispersion and multiplicity of activities and operators may make difficult the mobilization of sufficient financial and human resources to regulate, coordinate, and monitor projects in order to ensure that public health and the environment are protected. Commenting on the recent system of inset appointments trend in England, which allows service providers licensed by Ofwat (customers of more than 50 megalitres of water per year) to replace existing suppliers, Professor Franceys asked in an OECD expert meeting in November 2007: "Who will undertake water quality testing at the customers' taps?" and "What level of services will be provided?".[29] These systems may also challenge the long-term financial sustainability of the sector. They, for instance, may make it difficult to preserve the solidarity mechanisms between different types of users (cross-subsidy). They also potentially expose users to a number of risks, including capture by the provider, if the substitution mechanisms in case of bankruptcy or of a unilateral decision to increase tariffs are not in place.

Rooting the Partnerships in Strong Accountability Mechanisms

OECD (2009b) emphasizes that contractual arrangements with the private sector for the development of wastewater facilities are long-term and unlikely to cover all aspects of the complex relationship between the private and the public sectors. Many past difficulties have also arisen from dispute over the real state of water systems and the quality of baseline data.

Mechanisms exist that may help reduce the uncertainty that comes with long-term incomplete contracts or deal with its consequences. They include adopting performance-based contractual arrangements; providing for clauses and mechanisms to frame discussions on future issues as well as formal dispute-resolution mechanisms; and strengthening competitive pressure and promoting information sharing. Monitoring processes can contribute to reducing uncertainties when they are focused on a small number of key indicators that are clear and easy to measure. In Chile, cooperation with the private sector involves a monitoring process that includes inspection of operators and considerable penalties in case of default. In addition, disputes arising between the regulatory authority and the operators (notably on tariffs adjustment) are dealt with through an innovative mechanism involving expert panels.

Past experiences have shown that partnerships in the water and sanitation sector should not simply be viewed as bilateral relationships between the public and the private sector as they generate strong interest from consumers and communities. Greater involvement of civil society (non-governmental organizations (NGOs) and consumer groups) may grant users and local communities a stronger feeling of ownership, contribute to better protection of consumer rights, and enhance monitoring service provision. One strong point of the concession contract with Lydec in Morocco is that it developed strong partnerships with local communities and NGOs.

Conclusion

Evidence shows that there is large scope for the private sector to contribute to the improvement of water quality. There is a demand and a market for it. Countries need more and better water and wastewater treatment systems and facilities to keep track of the growing population and urbanization or to replace aging infrastructure and comply with ever-stringent regulation. Concerns about water resource scarcity and the environmental consequences of untreated discharges in some areas are providing companies with new business opportunities in technologies such as wastewater reclamation and reuse and advanced filtration membranes. The recent dynamism of built–operate–transfer (BOT) projects in China and the Middle East shows that private sector appetite is there.

However, making sure that the partnerships with the private sector yield the hoped-for benefits to all involves establishing the appropriate mechanisms to ensure the coordination and coherence of efforts and to make the cooperation with the private sector work in the public interest. This has three parts: designing arrangements and establishing framework conditions that support the long-term financial sustainability and affordability of partnerships; ensuring that the policy and regulatory frameworks support the alignment of corporate practices with public goals and reduce risks for investors; and strengthening the accountability mechanisms that can help to overcome the uncertainty that is intrinsic to long-term contracts while preserving the flexibility and capacity of the partnerships to adapt to changing conditions.

Acknowledgements

This paper does not necessarily reflect the views of the OECD or its member countries. Useful comments were provided by Carole Biau and Aziza Akhmouch. The paper largely draws from the OECD publications given in the References.

Notes

1. For the telling slides of Professor Rita Colwell presented at the 2010 World Water Week in Stockholm, see http://www.worldwaterweek.org/documents/WWW_PDF/2010/tuesday/K2/Colwells.pdf/.
2. See http://www.worldwaterweek.org/documents/WWW_PDF/2010/sunday/T6/HWT_Indonesia.pdf and http://www.worldwaterweek.org/documents/WWW_PDF/2010/sunday/T6/HWT_Rwanda.pdf/.
3. See http://www.worldwaterweek.org/documents/WWW_PDF/2010/sunday/T6/HWT_Kenya.pdf/.
4. Primary treatment of (urban) waste water is a treatment by physical and/or chemical processes (such as sedimentation, flotation, etc.), in which the BOD5 and the total suspended solids load of the incoming waste water are reduced by at least 20% and 50%, respectively.
5. Urban waste water is treated by a process generally involving biological treatment resulting in a biochemical oxygen demand (BOD) removal of at least 70% and a chemical oxygen demand (COD) removal of at least 75%.
6. Tertiary treatment comes on top of secondary treatment and is targeted to remove nitrogen and/or phosphorous and/or to tackle any other pollutant affecting the quality or a specific use of water, like microbiological pollution, colour, etc. The following minimum treatment efficiencies define a tertiary treatment: organic pollution removal of at least 95% for BOD and at least 85% for COD, and at least one of the following: nitrogen removal of at least 70%; phosphorus removal of at least 80%; microbiological removal achieving a faecal coliform density less than 1000 in 100 ml.
7. See http://epp.eurostat.ec.europa.eu/portal/page/portal/environment/data/main_tables
8. As an example, the American Society of Civil Engineers ranks the state of wastewater infrastructure D- in a report card developed in 2009, among the least performing infrastructure sectors (http://www.infrastructure reportcard.org).
9. The 1998 Drinking Water Directive sets stringent requirements in relation to a number of chemical substances (http://ec.europa.eu/environment/water/water-drink).
10. Towns and cities across the European Union are required to collect and treat their urban waste water under the Urban Wastewater Treatment Directive adopted in 1991. The main type of waste water treatment envisaged by the directive is biological or 'secondary' treatment. However, for agglomerations over 10,000 inhabitants discharging into water bodies designated as sensitive, more stringent treatment was required by 31 December 1998. For agglomerations of more than 15,000 inhabitants not discharging into sensitive areas, the deadline for secondary treatment infrastructure was 31 December 2000. In smaller agglomerations the deadline for compliance was 31 December 2005 (http://ec.europa.eu/environment/water/water-urbanwaste).
11. The 2000 Drinking Water Directive sets as a requirement to reach "good status" of all surface and groundwater bodies by 2015 (http://ec.europa.eu/environment/water/water-framework).
12. The USEPA is seeking inputs on a new drinking water strategy, which considers extending the list of contaminants requiring treatment (http://water.epa.gov/lawsregs/rulesregs/sdwa/dwstrategy and http://water.epa.gov).
13. See http://europa.eu/rapid/pressReleasesAction.do?reference=IP/10/835&type=HTML/.
14. See http://water.epa.gov/scitech/datait/databases/cwns/.
15. See http://www.oecd.org/dataoecd/36/41/40017113.pdf/.
16. See http://www.circleofblue.org/waternews/2010/world/the-price-of-wastewater-a-comparison-of-sewer-rates-in-30-u-s-cities/.
17. See The World Bank PPI database at http://ppi.worldbank.org/.
18. Reuters, quoting Global Water Market 2008: Opportunities in Scarcity and Environmental Regulation (www.prweb.com/releases/2008/01/prweb660064.htm).
19. See http://www.hyfluxwatertrust.com/.
20. NEWater, branded by Singapore, is a water-treatment process that purifies wastewater using dual-membrane and ultraviolet technologies. The water is mostly utilized for commercial and industrial uses, but is safe to consume. By 2011, NEWater is expected to provide 15% of Singapore's water requirements.

21. See http://www.financingwaterforall.org/.
22. See http://www.oecd.org/dataoecd/36/41/40017113.pdf/.
23. See http://www.irc.nl/page/8193/.
24. See http://www.lydec.ma/.
25. See http://www.unhabitat.org/downloads/docs/7803_90885_ACCESS%20TO%20BASIC%20SERVICES%20IN%20SHANTY%20TOWNS,CASABLANCA,MOROCCO_LYDEC.pdf/.
26. See http://www.riob.fr/IMG/pdf/presentation_lydec_Beyrouth_2010.pdf/.
27. See http://pppcentralunit.mof.gov.eg/.
28. The use and effectiveness of these coordination mechanisms are currently under investigation by the OECD (http://www.oecd.org/gov/water).
29. See http://www.oecd.org/dataoecd/38/37/40015975.pdf/.

References

African Development Bank/Organisation for Economic Co-operation and Development (ADB/OECD) (2007) *African Economic Outlook 2006/2007* (Paris: OECD).

Asian Development Bank (ADB) (2007) *Asian Water Development Outlook* (Manila: ADB). Available from: http://www.adb.org/Water/Knowledge-Center/Books/2007-asian-water-development-outlook.asp/.

Biswas, A. K. (2006) Water management for major urban centres, *International Journal of Water Resources Development*, 22(2), pp. 183–197.

Biswas, A. K. & Tortajada, C. (2009) Changing global water management landscape, in: A. K. Biswas, C. Tortajada & R. Izquierdo (Eds) *Water Management in 2020 and Beyond*. Water Resources Management and Development Series, pp. 1–34 (Berlin: Springer).

Clark, M., Huang, C. R. and Liu, S. B. (2006) *Development of National Wastewater Tariff Guidelines for China* (CDM). Available from: http://www.cdm.com/NR/rdonlyres/586600E8-3539-4B40-9EE4-AB3C282F7230/0/DevelopmentofNationalWastewaterTariffGuidelinesforChina.pdf/.

Franceys, R. & Gerlach, E. (2008) *Regulating Water and Sanitation for the Poor* (London: Earthcan).

Global Water Intelligence (2010) Flooding exposes Chinese infrastructure woes, *Global Water Intelligence*, 11(8). Available from: http://www.globalwaterintel.com/.

Hochstrat, R., Melin, T., Wintgens, T. & Jeffrey, P. (2006) Assessing the European wastewater reclamation and reuse potential—a scenario analysis, *Desalination*, 188, 1–8.

Howard, G., Charles, K., Pond, K., Brookshaw, A., Hossain, R. & Bartram, J. (2010) Securing 2020 Vision for 2030: climate change and ensuring resilience in water and sanitation services, *Journal of Water and Climate Change*, 1(1), pp. 2–16.

Marañón-Pimentel, B. (2009) Economic costs of water-related health problems in Mexico: deficiencies in potable water services and the costs of treatment of diarrhoeas, *International Journal of Water Resources Development*, 25(1), pp. 65–80.

MED EUWI (2009a) *Development of a Financing Strategy for the Water Supply and Sanitation Sector in Egypt* (Athens: Mediterranean Component of the EU Water Initiative).

MED EUWI (2009b) *Organisation of Institutions and Human Resources, Policy Dialogue on Integrated Water Resources Management Planning in the Republic of Lebanon*. Available from: http://www.gwpmed.org/files/090320_MEDEUWI_Rapport_RF-EOA.pdf/.

MED EUWI/Organisation for Economic Co-operation and Development (OECD) (2010) *Framework Conditions for Private Sector Participation in Water Infrastructure in Egypt* (Athens: Mediterranean Component of the EU water Initiative).

Organisation for Economic Co-operation and Development (OECD) (2007) *Infrastructure to 2030. Mapping Policy for Electricity, Water and Transport* (Paris: OECD).

Organisation for Economic Co-operation and Development (OECD) (2008a) *Environmental Outlook to 2030* (Paris: OECD).

Organisation for Economic Co-operation and Development (OECD) (2008b) *Economic Aspects of Adaptation to Climate Change. Costs, Benefits and Policy Instruments* (Paris: OECD).

Organisation for Economic Co-operation and Development (OECD) (2008c) *Alternative Ways of Providing Water Emerging Options and Their Policy Implications* (Paris: OECD).

Organisation for Economic Co-operation and Development (OECD) (2009a) *Managing Water for All. An OECD Perspective on Pricing and Financing* (Paris: OECD).

Organisation for Economic Co-operation and Development (OECD) (2009b) *Private Sector Participation in Water Infrastructure. Checklist for Public Action* (Paris: OECD).

Organisation for Economic Co-operation and Development (OECD) (2009c) *Strategic Financial Planning for Water Supply and Sanitation* (Paris: OECD).

Organisation for Economic Co-operation and Development (OECD) (2010a) *Innovative Financing Mechanisms for the Water Sector* (Paris: OECD).

Organisation for Economic Co-operation and Development (OECD) (2010b) *Pricing Water Resources and Water and Sanitation Services* (Paris: OECD).

Palmer, S. J. (2010) Future challenges to asset investment in the UK water industry: the wastewater asset investment risk mitigation offered by minimising principal operating cost risks, *Journal of Water and Climate Change*, 1(1), pp. 17–35.

Pinsent Masons (2008–2009) *Pinsent Masons Water Yearbook 2008–2009* (London: Pinsent Masons).

United Nations Framework Convention on Climate Change (UNFCCC) (2007) *Investment and Financial Flows to Address Climate Change*. Available from: http://unfccc.int/files/cooperation_and_support/financial_mechanism/application/pdf/background_paper.pdf/.

Water21 (2010) Project progress in the Middle East and Africa, *Water21*, August.

Water Governance in Aragon

RAFAEL IZQUIERDO
The Water Institute of Aragon, Regional Government of Aragon, Aragon, Spain

ABSTRACT *Good management of water resources goes beyond a mere political or socio-economic exercise. It is a responsibility on a global scale, which has to begin with appropriate actions at a local level. The addition of small local projects under unified planning, regulation and management criteria will determine an inflection point in sustainable management of water at a planetary level. It is like a chamber orchestra where every instrument plays its role in a certain location, but always under the unified coordination of a conductor who sets the "tempo", the intensity, the rhythm and the idiosyncrasy of each individual action.*

Water in Aragon

When it comes to breaking down how the responsibility for local governance is exercised in water matters, the key for achieving a stable balance in the ecosystem is based on the concept of integrated water management. This concept includes, on the one hand, aspects related to the quantity of the resource, its quality and acknowledgement that we are dealing with a resource that is as limited because it is essential. On the other hand, water management nowadays cannot be understood without the introduction of economic, social, behavioural, environmental, territorial, legal and political aspects. All these must be taken into account when it comes to planning water policies in Aragon.

In order to understand the reality in Aragon, it is important to highlight the difficulty that comes with any attempt at planning in water matters, from the perspective of considering it as both a social commodity that must be provided and also as a public service. The first obstacle is the peculiar geographical distribution to be found in Aragon. This can be better understood by studying Table 1, in which it can be seen that Aragon occupies 10% of the surface area of Spain and yet it only accounts for 3% of the Spanish population. The population density in some areas in Teruel is fewer than eight inhabitants/km^2.

As far as the economic activity of the region is concerned, the primary sector accounts for only 12% of the VAB (gross added value), compared with 60% for the services sector, and yet it is by far the largest consumer of water in the autonomous region (around 80%).

Meanwhile, Aragon is blessed with something which is uncommon in the Western world: it has a fair amount of natural water resources and plenty of land, two ideal resources for medium- and long-term development. The territory is home to the hydrological basins of three rivers: Ebro, Júcar and Tajo. The Ebro is far the largest in terms of flow rates and

Table 1. Overall demographic figures for Aragon

	Spain	Aragon
Surface area (km^2)	505,988	47,719 (10%)
Population (millions of inhabitants)	46.1	1.3 (3%)
Density (inhabitants/km^2)	91.2	27.8
Number of municipalities	8,107	731 (1,545 agglomerations)

occupation of land, but this is not to take any importance away from the Júcar which, together with the Tajo (the least significant in the autonomous region) and the Ebro, forms a mass of water covering 281 km^2.

Having said that, what is water used for in Aragon? Surface waters are, obviously, the major source of water (94%) compared with groundwaters. Statistical data show that agriculture uses 77% of water resources, with industrial activities consuming 7%, domestic uses 5%, and 11% reserved for other uses.

Aragon's environmental policies are based on the idea of water as a strategic value for present and future, in which man is the main player in transforming environmental conservation into an engine for social and economical development of the territory, contributing to maintaining environmental heritage as a seal of identity for Aragon. One of the guiding principles of environmental policy is excellence in water management, the aim of which is to make Aragon a leader in sustainable management of water. The key lies in obtaining a balance between offer and demand for the resource.

Role of Local and Regional Governments in Water Management

At present local and regional authorities have an essential role in water management. Although general policies are legislated on a national or even a supranational level, the vast majority of these policies are implemented by local and regional governments.

There are multiple examples of the importance of local and regional governments and of the relevance they are taking in a variety of forums and organizations, such as the World Water Forum in Istanbul 2009 (heads of state, parliamentarians and local and regional governments—consensus document), the Inter-Mediterranean Commission (IMC) of the Conference of Peripheral Maritime Regions (CPMR), the Union for the Mediterranean, regional government networks on water and environmental issues such as ENCORE, NRG4SD, to name just a few.

Participation of local and regional governments in decision-making on water demand management is particularly relevant in aspects such as water treatment, reuse, and modernization of irrigation and supply infrastructures, and in strategies for prevention and action against episodes of drought and flooding, in land regulation and planning matters.

The influence of factors such as demographic growth, migrations towards the big cities or energy dependency lead to social transformations with different territorial peculiarities. Management of water resources at local and regional levels allows a better and more efficient adaptation to these changes.

The application of new models of management becomes mandatory in order to face up new challenges. We subscribe to the concept of shared water management, which is adapted to the characteristics of each territory within the framework of integrated management. In other words, we should aim at a model based on joint integrated planning for all the territories forming the unit or hydrographical demarcation, but a model where management can be carried out by local and regional governments, with adaptations to the circumstances of each territory, as in the case of droughts, floods, quality control, and so on. The above model contributes to achieving sustainable development, which in addition encourages a territorial balance within the region and between regions. Within this model the role of regional governments is irreplaceable, as they are the closest to the citizens and the ones who know their own territory the best.

Water Policies in Aragon

Water policies in the autonomous region are planned and executed from The Water Institute of Aragon (IAA), which is an entity under public law with its own legal personality, within the Ministry of Environment of the Regional Government of Aragon, keeping the condition of public administration of the autonomous region.

The IAA has a number of competences in water matters and water-related infrastructures: planning, execution and management of the infrastructures of the water cycle, and collection of the Sanitation tax (Law 6/2001, of 17 May, on Planning and Participation in Water Management in Aragon; Boletín Oficial de Aragón, 2001). National planning is divided into territories in different hydrographical demarcations, which base their action on the principle of the hydrographical basin unit. This means that competences are shared between a state body and the IAA as a regional body. In the case of Aragon, our political action is aimed at a model of "Shared Management", in which planning should be carried out at state level complying with the principle of basin unity with the participation of all the regions and actors involved, and the plans and programmes being applied by the local and regional governments through appropriate transfer of competences and sufficient financing. In order to achieve a more efficient and agile management, we aim at a maximal transfer of authority towards our regional government when applying the policies that have been agreed previously. Therefore, planning according to the principle of river basin unity at a state level with participation of all regions involved and implementation of those policies agreed by regional governments, through a maximal transfer of authority, who have a better knowledge of the territory.

Water policies in the region are based on sustainable management of water, which is developed in two stages: planning and implementation.

Planning as a Management Tool

The first prerequisite for appropriate planning consists of obtaining information for decision-making. The features defining good information are:

- It must have terminology coding which is known and recognized by everyone.
- Quantity and quality: in terms of quantity, excess and shortage of information are equally damaging when it comes to evaluating data. The quality of the information is more important when it comes to finding out situations and difficulties related to water management.
- It must be comparable and coherent.
- It must be up to date.

It is fundamental to have a common language for all players and all territories, something I have called metaphorically "Hydro-Esperanto", which allows us to share information and experiences in similar conditions of understanding and application.

Similar to the rationale applied in a medical process, once the information has been compiled through adequate "anamnesis", it is essential to perform an appropriate diagnosis of the situation in order to resolve a certain problem. Correct diagnosis requires the evaluation of:

- the current situation of the problem;
- the predicted future and desired scenarios; and
- identification of priorities for decision-making.

Another aspect that determines good planning is the ability to adapt to factors and unpredictable situations, one example of which is the appearance on the world stage (both at a global and local scale) of climate change. On one hand, the feature that best defines the phenomenon of climate change is the uncertainty that generates in the medium- and long-term. In fact, scientists have been forced to set the limits of the degree of uncertainty at two levels: quantitative and qualitative. Quantitatively, we can talk about much/little evidence and different levels of consensus between scientists with regard to the relationship between observed and predicted phenomena. In qualitative terms, scales of confidence in the data observed have been determined and are used to draw up predictive models or models for analysis of variables.

Unlike the above degree of uncertainty, it is certain that climate change is having certain impacts:

- related to agriculture and food safety;
- on human health;
- urban settlements and migratory population movements;
- biodiversity and ecosystems; and
- others: economic, methods of transport, etc.

This involves the need to draw up policies including measures for adaptation and mitigation when faced with these impacts, based on the aforementioned scientific models, which are not completely lacking in uncertainty. The measures for adaptation are designed to reduce the vulnerability of the natural and human systems and the mitigation measures are aimed at reducing contaminating emissions through the introduction of technological changes. And, of course, all this has to be adapted to local and regional variations: some related to climate (precipitations, temperature, evaporation) and others which are not climate-related (changes in the population, food consumption, price of the water, technology).

Implementation as a Management Tool

Implementation is developed through different action plans and, in the case of management of water in Aragon, this is translated into three lines of action:

- availability and security of the resource;
- quality of the resource; and
- consensus through social participation.

The interaction of these three variables leads to sustainable management of the water.

Regarding availability of the resource. Planning this line of action is deployed at three different levels:

- Obtaining an appropriate level of regulation by means of water storage, which allows the water needs of the region to be satisfied with sufficient guarantees.
- Policies for saving, through environmental education and social awareness.
- Responsible management that gives society sufficient infrastructures, which are technologically appropriate and that minimize losses in the water resource. The most representative action in this regard is the so-called "Aragon Water Plan".

The basic principles of the Aragon Water Plan are the following:

- Improvement in water cycle infrastructures in the towns and cities.
- This includes financial, administrative and technical support to the local bodies.
- To increase citizens' quality of life.
- To promote environmental awareness and a commitment to efficient consumption of water.

Construction and maintenance of the infrastructures in the integrated water cycle, in other words, the supply of drinking water and the construction of wastewater treatment networks, is a competence which is legally entrusted to the town councils.

The fact that the population of Aragon is widely dispersed, leading to a large number of small population centres which makes it impossible to have economies of scale to make the necessary investments profitable and, in many cases, the lack of available budget in the town councils, makes it difficult for councils to take on the necessary works alone.

Immigration in the cities, a phenomenon characteristic of the 20th century, had special importance in Aragon in the 1960s, and this reality meant that the infrastructures did not receive all the necessary attention to guarantee they remained in good condition. The ageing of the materials and their use beyond their useful life, the design and antiquated construction thereof, became a problem at the start of the 21st century and constituted a limit to the ability of the town councils to recover their vital equilibrium. This phenomenon becomes more obvious if it is extended to the whole of the 20th century, in which the concentration in the population increased between 1900 and 2008. In 2008, the concentration was 74.4% of inhabitants living in 10% of the most densely populated territory in the surface area of this region compared with 77% recorded at a national level. Zaragoza is the only municipality with a population of more than 100,000 inhabitants, and its population multiplied more than six-fold between 1900 and 2008.

In this historic context, the response had to be the most appropriate one for the social and economic reality of the region. It was the Government of Aragon that, in December 2002, definitively approved the Aragon Water Plan: this included projects by local bodies, which were subsidised by the Aragon Water Institute. These projects included works for supply, sewage treatment and sanitation, improvement of the water quality and the defence of riverbanks. The subsequent development of the Plan has promoted throughout these years a leading role for the local bodies in improving water management, giving them a role to play in solving the problems which affect them.

This is why it is essential, when carrying out the municipal works in the integrated water cycle, to have cooperation between administrations and, particularly, with the regional government, to promote supportive management for structuring the land. This ensures that the necessary projects are carried out for the appropriate supply of water to each and every one of Aragon's towns and villages, making sure that a shortage in the water supply, with sufficient quality and quality, is not an essential limitation for normal rural life and its sustainable development.

On the basis of the above, Law 6/2001, of 17 May, on Planning and Participation in Water Management in Aragon (Boletín Oficial de Aragón, 2001) fully integrates the supply of drinking water and sanitation and treatment of wastewaters in the concerns of the autonomous region; it gives it its own specific regulatory basis, thus dealing with the needs felt and recognized in society, and the irreplaceable role of the autonomous administration for analysing and providing the financial and material investment in the water cycle in the municipalities of Aragon.

Using these regulations, the Aragon Water Institute designs the activity of planning, investment and management in the construction of infrastructures for the supply of drinking water, sanitation and treatment of wastewater, in cooperation with the local administration, by means of direct investments and subsidies. Many of these projects, because of their characteristics and design, receive financing from European Union funds (both the European Regional Development Fund and the Cohesion Funds).

Given the competition for competences and the subsequent development of regulations mentioned above, the activity of planning, investment and management in the construction of infrastructures for the supply of drinking water, sanitation and the treatment of wastewater, in cooperation with the local administration, by means of investments and subsidies, is carried out through a cooperation agreement signed with the town councils. This agreement not only determines the economic conditions and the conditions for working together to construct and maintain the installations, but also promotes specific projects for the spreading and promotion of measures for social awareness and environmental education regarding the efficient use of water. The synergy between the works being carried out and the awareness campaigns amongst the population encourage a proper cultural exchange regarding the use of water.

In those cases where necessary, the town councils cooperate in the agreement by providing the necessary permits to carry out the works, and as many permits and licences, either municipal (works licences, conformity with urban planning) or from other administrations, as necessary for it to be performed (water confederations, National Network of Spanish Railways [RENFE], Ministry for Roads, etc.). There is no doubt that this contributes to a reduction in the administrative deadlines and, consequently, a quicker execution of the works, demonstrating the importance of cooperation among administrations.

The Water Plan has been developed over three programming periods, 2002–2006, 2005–2009 and 2008–2011, and can be summarized by the following approximate figures:

- 1,782 projects.
- €248 million of investment.
- €207 million of subsidies awarded by the IAA.
- 651 municipalities benefited.
- 572,000 inhabitants benefited.

In addition to ensuring demand for human consumption, the improvement in basic services also means a contribution to the correct functioning of the economy in the rural environment and the creation of employment, in places where the sustainable economy can be seen best in the ecological and local production of foods, materials and energy.

All together these projects represent a determined defence of the natural and cultural background of our villages; from a purely environmental point of view, the water saving achieved is important in itself, particularly in the current scenario of climate change, in which the availability of the water resource is on a downward trend, but it also represents a reduction in the pressure on the water resources of rivers, lakes and aquifers from where it is drawn. This aspect of the work of the Water Institute of Aragon together with progress in the treatment of wastewater represents a major contribution to maintaining water resources in a good ecological condition, the major objective set by the European Water Framework Directive.

Regarding quality of the resource. Despite what is usually believed, the major problems we currently face at a global scale are not to do with access to water (i.e. they are not related to the availability of water), but to the quality of the resource. Water quality is the major difficulty that confront different governments, and will require multi-million euro investments in a scenario of world economic crisis which restricts investment.

It should not be forgotten that in order to obtain water of good quality, the most important and least expensive action continues to be prevention of contamination: this is the direction to which a significant part of the projects should be aimed.

Aragon holds a leading position in the treatment of wastewaters amongst countries around the world as far as compliance with the obligation to treat wastewaters is concerned, through the various wastewater treatment programmes currently operating in the autonomous region. To comply with European Directive 91/271/CEE of 21 May 1991 on the treatment of urban wastewaters (European Commission, 1991), the Government of Aragon designed the Aragon Plan for Sanitation and Waste Water Treatment (PASD) as a general planning document for sanitation and wastewater treatment in the autonomous region. Different instruments of implementation integrate the PASD (Figure 1):

- Special Wastewater Treatment Plan (PED).
- Pyrenees Wastewater Treatment Plan.
- Animal Waste Treatment Plan.
- Integrated Wastewater Treatment Plan for Aragon (PIDA).

SPECIAL WASTEWATER TREATMENT PLAN (PED)
Municipalities of > **1000 eq-inhab**
No. of projects: **131 WWTPs**
Pollution load: **588,497 eq-inhab**
Investment: €**1.062 billion**

ANIMAL WASTE TREATMENT PLAN
No. of actions: 4
Investment: €**20 million** during 2008-11

PYRENEES WASTEWATER TREATMENT PLAN
Population nuclei: **292 centres**
No. of projects: **296 WWTPs**
Pollution load: **152,967 eq-inhab**
Investment: €**358 million**

INTEGRATED WASTEWATER TREATMENT PLAN FOR ARAGON (PIDA)
Municipalities of < **1000 eq-inhab**
Pollution load: **234,702 eq-inhab**

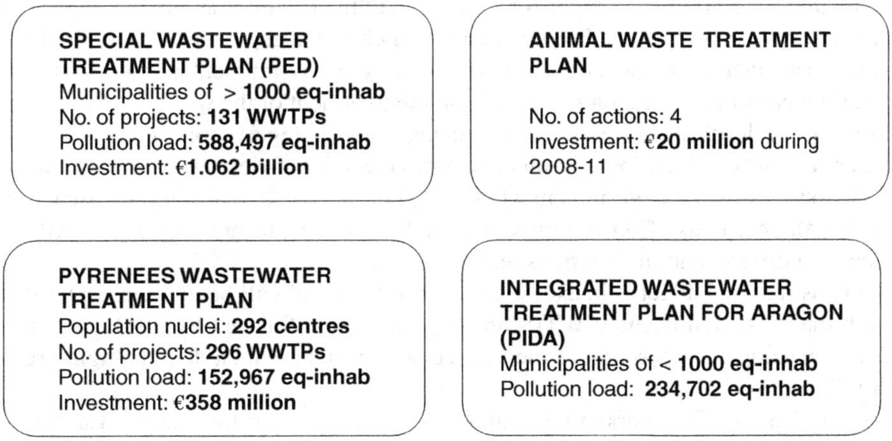

Figure 1. The Aragon Wastewater Treatment Plans.

Special Wastewater Treatment Plan (PED). Implementation of this programme involves treatment of wastewaters in all those municipalities with more than 1000 inhabitants-equivalent which do not already have this service. It is a wide-reaching action, with the problems generated by the wide geographical dispersion of the population centres in Aragon and its depopulation, as well as the orographical difficulties of the terrain (Sodemasa, 2007).

Planning is based on a division of the territory into 13 areas to facilitate the execution of all projects and subsequent operation, with evaluation in all cases of the possible environmental impact. This means a total of 131 wastewater treatment plants (WWTPs) and 40 collectors, which include a contamination load of 588,497 inhabitant-equivalents, and an overall investment of €1.062 billion, including construction and 20-year operation.

The economic viability study is focused on the management of each area through an administrative licensing system lasting for 20 years, thereby introducing an innovative model for public–private cooperation in the construction and operation of wastewater treatment infrastructures.

The PED will enable wastewater treatment for all population centres with more than 1000 inhabitants-equivalent which, together with the infrastructures currently operating, will mean the treatment of approximately 90% of the wastewaters in Aragon.

Pyrenees Wastewater Treatment Plan. The Integrated Water Treatment Plan for the Aragon Pyrenees will allow the region to move towards the objective of complying with European and national regulations, which sets 2015 as the deadline for treating all wastewaters, including the area of the Aragon Pyrenees considered to be of significant environmental value.

Treatment of the rivers in the Pyrenees represents a landmark in the development of the Aragon Statute of Autonomy: it is the first time the Regional Government of Aragon will carry out infrastructures classified as of "general interest of the State" (which means that they must be constructed and financed by the state) through management handed over from the State Ministry of the Environment, Countryside and Marine Habitat to the Government of Aragon, through an agreement to perform these projects in the framework of the National Water Quality Plan: sanitation and water treatment 2008–2015 and the integrated water cycle. The Integrated Water Treatment Plan for the Aragon Pyrenees represents the drive given by the Government of Aragon's Department of the Environment to the improvement of the quality of rivers. Its ambitious nature, which includes all the area at the head of the Pyrenees rivers, represents a commitment to the environment of one of the most outstanding parts of the Aragon region. Complete treatment of the waters in these rivers is focused on the local areas known as Jacetania, Alto Gállego, Sobrarbe and Ribagorza. These areas hold a high ecological value, but from the point of view of design and execution of the works it is also an environment where very complex technical solutions are required, involving an enormous effort in terms of human resources, technology and coordination between the different administrations and companies.

The construction budget is in excess of €128 million and will be undertaken until 2012 (though the most significant WWTPs are expected to be finished by 2011); and if total costs of financing and operation of the licence at 20 years are included, the figure reaches €358 million.

In total, there are 296 works in 62 municipalities of the four local areas (La Jacetania, Alto Gállego, Sobrarbe and La Ribagorza), which will benefit 20,000 neighbours in 292 population centres producing a contamination load of 152,967 inhabitant-equivalents.

It is estimated that the Pyrenees Plan will represent an injection for the Aragon economy of €153 million until 2011. In addition, more than 7,000 job positions, both direct and indirect, will be created throughout the construction period.

Animal Waste Treatment Plan. The projects included in this plan are directed towards the problem of nonpoint pollution. In recent years Aragon has seen a considerable increase in the intensive rearing of pigs, with a great concentration of land. This has led to environmental problems caused by the production of large volumes of manure and animal waste, with the consequent contamination of water resources and soils, and emissions into the atmosphere of ammonia and the production of bad smells. This is why the Aragon's Department of the Environment has launched a series of projects in those areas where the excessive production of animal waste constitutes a serious problem, since the crop fields are no longer capable of absorbing all the animal waste generated in the area, with subsequent contamination of the aquifers. The aim of the animal waste treatment plants is to reduce the nutrient content of the animal waste so that it can be used as a fertilizer product with no risk of diffuse pollution.

The management model for these infrastructures includes the collection of the animal waste, storage, treatment and the constitution of a management centre to direct and supervise all these operations, integrated by town councils and/or local regions (as public partners), together with the pig farmers producing the animal waste (private partners).

This pilot model is being used to solve a serious problem of diffuse pollution which is difficult to control in a strategic sector of Aragon's economy which is showing itself to be very sensitive to the current economic crisis.

Initially, four plants are being constructed, with an investment during the 2008–2011 period of some €20 million.

Integrated Wastewater Treatment Plan for Aragon (PIDA). The so-called PIDA is the last phase in water treatment installations for all the municipalities in Aragon, completing the water treatment in towns and villages with a smaller population (less than 1000 inhabitant-equivalents). It is in its initial stages of programming and execution. Prior studies and field studies have been carried out to identify the possible locations for the WWTPs, the most suitable layouts, the existing sanitation and water treatment infrastructures, and the technology to be applied in each case.

Management Model for the Wastewater Treatment Plans in Aragon

One of the most singular features of the wastewater treatment plans is the management model used to implement them. It is based on a model of public–private participation, through public works licensing contracts, so that investment, construction and the subsequent operation of the WWTPs are shared between the public administration (the IAA) and private companies. This model allows a large number of projects to be undertaken in a short period of time, as it is the private sector that makes the economic investment for the construction of the plants, under the supervision of the public administration which is responsible for ensuring compliance with the regulation governing the treatment of wastewater.

Furthermore, the investment made is not considered a public debt as construction is financed by the licensee companies, therefore not endangering the economic solvency of the autonomous public administration. There is a sharing out between the public partner and the private partner of the set of risks for the project which, traditionally, was completely taken on by the public sector. So that these infrastructures are not considered for the purposes of the European System of Accounts (ESA 95) as an asset of the administration which would be accounted for as a debt, a contractual relationship between the IAA (public administration) and licensee companies has been structured so that the latter take on the risks inherent in this type of contract, following the guidelines set down by EUROSTAT (Statistical Office of the European Communities, which produces figures on the European Union and promotes the harmonization of the statistical methods in the member states. EUROSTAT interprets the SEC-95 regulations on the public deficit and the public debt in their practical application by the member states).

In the case of Aragon, the three risks in the contract are transferred:

- Construction risk: the licensee takes on the increases in cost during the construction of the infrastructures, and will receive no payments until the works are properly completed and come into operation.
- Risk of demand: removing all guaranteed minimum payments and establishing a payment according to the volume of water treated.
- Risk of availability: establishing economic penalties in case of non-compliance with the minimum operating standards.

In this way the private company is responsible for building the infrastructures (which it finances with its own means) and the subsequent operation for a period of 20 years. The licensee companies receive their payments depending on the wastewater flow rate treated following a tariff set out in the contract, thus recovering the investment made in the construction and the operating costs. Participation of the private sector using this model allows the public administration to construct a vast amount of WWTPs within a short period of time, which would have required two or three decades of the total IAA budget totally devoted to WWTP construction through the traditional construction model of management. This, in turn, allows the public administration to free up public funds and improve flexibility and efficiency in the execution of the works.

Financing Model for the Water Treatment Plans in Aragon

Financing has become one of the key aspects for the application of any social policy in all the areas of public investment, especially in times of global economic crisis. In the planning stage of any programme it is vital that it be accompanied by a detailed economic analysis which justifies the model and the origin of the financial resources necessary for start-up (economic viability analysis).

In addition to the public–private participation (PPP) models, it is worth emphasizing the importance, in terms of financing, of the cooperation agreement between the State Ministry of the Environment, Countryside and Marine Habitat and the Autonomous Region of Aragon. This establishes the general scheme for coordination and financing for the execution of projects in the autonomous region of the National Water Quality Plan: sanitation and water treatment 2008–2015 and the integrated water cycle, with a contribution of €169.12 million for the period 2008–2011, which has partly contributed to the financing of the works in the integrated water cycle.

In the case of Aragon, the financing of the wastewater treatment plans is guaranteed with the implementation of the sanitation tax, a tax with an ecological purpose, the product of which affects the financing of activities for the prevention of contamination, sanitation and water treatment considered in Law 6/2001, of 17 May, on Planning and Participation in Water Management in Aragon (Boletín Oficial de Aragón, 2001).

Without going into a detailed description of the management, tariff structure and methods of collecting the Sanitation Tax (laid out in Law 6/2001), which would overstep considerably the objective of this document, I would, however, like to emphasize the three most important principles on which the application of this sanitation tax is based. The taxable event is the production of wastewaters arising from the consumption of water from any source or the discharge thereof, and the three principles are:

- Recovery of costs: the principle of recovery of costs is included in the European Union Framework Water Directive (DMA 2000/60/CE) and consists of the obligation of European Union member states to pass on the costs associated with the execution of water infrastructures to the users thereof.
- Polluter-pays principle: this refers to the responsibility that comes with the production of contamination of any origin on the aquatic medium.
- Solidarity: the sanitation levy is a "solidarity tax",[1] in other words, the tariff applied is identical, regardless of the size of the municipality. It is much more costly per inhabitant to run a WWTP in a village with a small population than in the city of Zaragoza, where a single installation provides a service for a large number of inhabitants (even though this installation is bigger), because of economies of scale. This principle contributes to the territorial balance and social cohesion of Aragon.

Obviously, the economic resources obtained with collection of the sanitation tax are used to pay the expenses involved in the operation of the water treatment plants. This payment is made to the companies that built and now operate the WWTPs in accordance with tariffs which depend on the flow rate treated. Regardless of its collection function, the sanitation tax is also an instrument to make the population more aware of the cost of the water service, encouraging water saving and promoting a more rational use.

Figure 2 illustrates the summary of the management and financing model for the Aragon Wastewater Treatment Plans. The IAA (the public partner) takes on the periodic payments to the licensees of each of the areas into which the territory has been divided for the 20 years of operation. These payments depend on the volume of water treated and will be different in each of the plants. However, the management system is global and the IAA passes on equally to all users of the system the costs borne, thus encouraging balance and territorial solidarity.

The tariff has a two-fold structure. The first section establishes a tariff so that, with the aforementioned volume, sufficient revenues are obtained to cover the costs of carrying out the works, financial costs and fixed operating cost. For the second section a smaller tariff is established so that with a flow rate equal to that expected there are sufficient revenues to give the profitability forecast considered in the viability studies. The total remuneration is the sum of the partial remunerations for all the treatment plants in the contract for the area.

Benefits of the Wastewater Treatment Plans in Aragon

Implementation of the wastewater treatment plans brings a series of benefits covering different areas:

- Environmental: improving the quality of life of the ecosystems and natural habitats.
- Health: improving people's health as a result of a decrease in water-borne diseases.

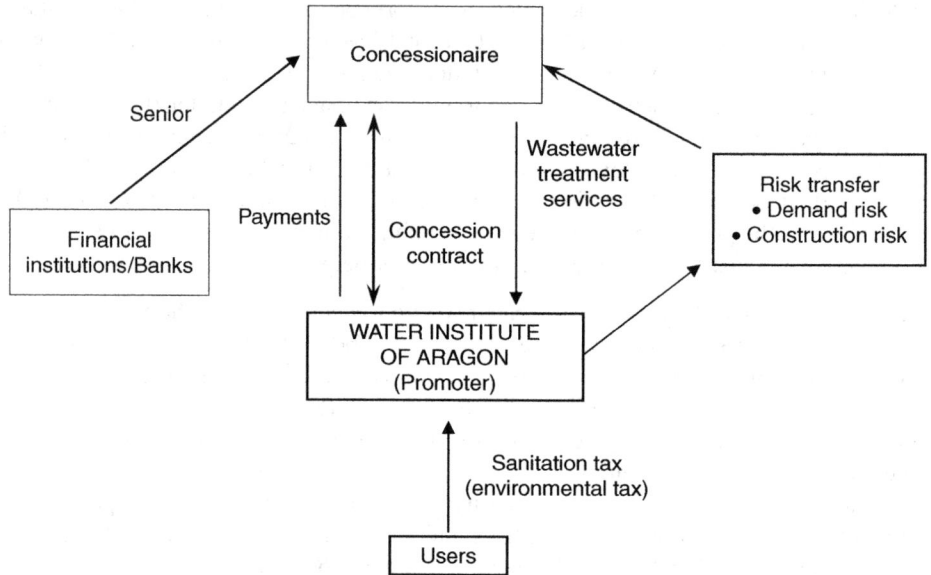

Figure 2. Management model for wastewater treatment plans in Aragon.

- Social: creation of stable and long-lasting employment as companies are set up in the area to operate the water treatment plants for at least 20 years. This contributes to greater social cohesion of the territory.
- Economic: creation of wealth in the territory from the arrival of companies in the water sector and also from other sectors because of the knock-on effect it has on rural tourism, leisure, etc. The companies will want to set up in those places where they have appropriate services, such as quality water, good communications, etc.

In summary, through this PPP model we have managed to transform a major socio-economic and health problem into an opportunity for socio-economic development, which contributes to a better structural and territorial cohesion in Aragon. We not only solve the problem, but also generate a variety of benefits to the local areas.

The presence of a WWTP in every municipality has become part of the territory's set of icons. Formerly, the existence of a church was the most emblematic icon in any municipality from the architectural, social, religious and cultural point of view. Throughout the last century the appearance of the school, the Guardia Civil barracks or the doctor's house/surgery configured the landscape of the towns and villages of Aragon. More recent examples include the building of swimming pools or sports centres and, in my view, the presence of a WWTP will become the next significant icon of social and economic development in many small villages of Aragon.

Regarding consensus in formulating water policies: social participation. The formulation of water policies in Aragon is based on the modern concept of integrated management, which includes as one of its basic pillars social participation. This guarantees the participation of all the actors related to water when it comes to formulating water policies in the region. Social participation is developed using processes of dialogue and socio-political consensus in those fundamental matters which will define water policy in Aragon.

The participatory body *par excellence* on water matters is the so-called Aragon Water Commission (CAA), which has been working since 2003 as a collegiate body for participation of the different players involved with water. The CAA consists of 65 members and 17 groups of stakeholders: social, economic, sectorial and local organizations, neighbourhood movements, experts in water policy, and the university, among others. Yet, despite its wide-ranging membership, it has registered achievements that would have been unthinkable only a few years ago. It has passed judgement on the Aragon Water Infrastructure Plan; drawn up and approved the principles of water policy on the basis of consensus; and reached agreements that unblocked the logjam on controversial projects as reservoirs and water transfers. These successes have generated a scenario without winners or losers in which it will be possible to meet the expected future demand for water, and use it rationally, while respecting environmental values and permitting new developments free of the uncertainty that conflicts generate (Aranda-Martin, 2007; Boné-Pueyo, 2007).

The CAA is a collegiate, participatory body with consultative functions. It is made up of different operating instruments such as the plenary session, committees, working groups and commissions. Thanks to the intense work carried out by the CAA in recent years, historic agreements have been reached on the reservoirs of Biscarrués, Yesa, Cuenca del Matarraña, San Salvador, etc.

The final link in the Management chain is the evaluation of the results obtained after carrying out the policies formulated at the beginning of the process. For this it is important to know the economic data for the projects carried out, the availability of the resource, the quality of the water, and to be prepared to respond when faced with unexpected extreme situations. This allows modification of the policies or certain projects throughout the process.

Tools used in the water management process in Aragon. The process of sustainable and integrated management that is being carried out by the IAA in Aragon requires the use of certain tools which allow the proposed objectives to be met. In our case I would like to highlight three of them because they are the most representative of the political action developed on water issues:

- Governance.
- Transfer of political responsibilities to regional governments.
- Cooperation among public administrations.

Governance is a management philosophy based on the principles of participation, transparency, accountability, coherence and effectiveness.

We have already seen throughout this article some of these elements reflected in the implementation of some programmes. This is the case of the CAA, in which the search for agreements uses the social participation of all the actors involved through dialogue between all the parties. Transparency is an essential quality in any management process and involves making all the information available to all parties to avoid situations of inequality and opacity. Accountability requires the creation of bodies to monitor the efficiency of the processes being developed. With regard to coherence, the information handled in these processes of management and policy formulation must be intelligible to everybody and, in addition, must be applied systematically. Finally, effectiveness must guide each decision within the process of planning and executing the policies formulated (Embid et al., 2007).

Transferring competences, i.e. political responsibilities, is a basic tool of political action for the socio-economic development of the autonomous region. Fortunately, we have a new Statute of Autonomy for Aragon (Organic Law 5/2007, of 20 April, on reform of the Statute of Autonomy for Aragon) which enables a new framework of development of competences from the general state administration to the autonomous region. In this way, from the IAA we are working towards the consolidation of a model of shared management on water matters between the central and autonomous administrations, based on the principle of a basin unit. This means that planning of the projects is carried out within the state (Spain) planning bodies, where all the players involved are represented, including the different autonomous regions of the respective hydrographic demarcations (Ebro, Júcar and Tajo, in the case of Aragon), but where the projects are carried out by the administration closest to the territory and to the citizen, i.e. the autonomous administration. This brings greater flexibility and efficiency in the execution of water policies, as regional administrations have a better knowledge and understanding of local problems and, therefore, are in a better position to offer more adequate solutions.

For the development of this model of shared management, we need to make the transfer of competences from the central administration to the autonomous one in executive matters more flexible; this includes river policing, water licences, authorizations for effluent, actions on drought and floods, etc.

Moreover, the Statute of Autonomy for Aragon enables regional governments to construct and operate hydraulic works belonging to the state through an agreement between the two administrations at the so-called "Bilateral Commission". These commissions have allowed the start-up of the Pyrenees Water Treatment Plan and the execution of the Matarraña Irrigation Reservoirs in the upper basin of the River Tastavins (projects of general state interest, the execution of which was transferred to the autonomous region through the IAA), an historic landmark as it is the first time Aragon will execute hydraulic works of general state interest.

Inter-administrative cooperation, which is very closely related to the previous point, allows priority objectives in water matters to be met, where there are common interests and synergies among different administrations inside and outside the autonomous region. Some examples of this activity are the following:

- General protocol for cooperation between the Ministry of the Environment, Countryside and Marine Habitat (MARM) and the Autonomous Region of Aragon; October 2007.
- Cooperation agreement between the MARM and the Autonomous Region of Aragon as part of the National Water Quality Plan 2007–2015; April 2008.
- Bilateral Aragon–State Commission.
- Agreements with provincial councils.
- Agreements with town councils.

In the framework of cooperation between different bodies of the autonomous region there is a cooperation link with the university for the joint development of teaching, technical and scientific activities. The most important areas of cooperation include climate change, water quality, the legal and regulatory framework on water matters, etc. Some examples are the following agreements currently in force:

- General protocol for cooperation with the University of Zaragoza: OTRI, IUCA.
- Cooperation agreement with the University of San Jorge de Zaragoza.
- Cooperation agreement with the University of Mendoza (Argentina).

Finally, I would like to make an explicit reference to the international activity developed by the IAA. The phenomenon of globalization and the development of information technologies are not unrelated to water-related matters. Water shortage, quality problems, financing difficulties and the application of models for integrated management of water based on governance are problems that have acquired an importance which goes beyond the limits of a region, hydrographical basin and even country, to become a global problem. This means that when it comes to addressing the problems of water, it is important to remember the so-called principle of "think global, act local". In other words, we need to have a global vision of the problems of different territories because they are interrelated and take on a global magnitude. However, to solve global water problems, we need to apply policies and projects at a local level which, when added together, help solve the problems posed.

Modern water management requires knowledge and exchange of experiences between different countries as, in many cases, the problems are similar and the solutions adopted in an area can be adapted to other territories. This aspect is a strategic issue that allows us to show a position of international leadership, for example, in water quality matters, through the implementation of treatment plans for wastewater.

Some examples of our international activity are the following:

- The Environment Department of the Government of Aragon currently holds the Co-Chair of ENCORE and NRGY4SD, two international networks of regional governments developing policies for water, environment and sustainable development.
- European Union: currently in progress are a number of INTERREG Research Projects co-financed by the European Union.
- Organisation for Economic Co-operation and Development (OECD): the water treatment plans for Aragon have been included as case studies for management models applicable to other countries.
- National Water Agency and University of Singapore (Lee Kuan Yew School of Public Policy), with joint projects for reuse of water and the future of water at a global scale.
- Asian Development Bank, Third World Centre for Water Management.

Figure 3 shows the universe in which the IAA carries out its activity and summarizes in graphical terms the need to maintain balances, reach agreements and develop different examples of cooperation within the regulatory framework in force, in line with the concept of governance applied in Aragon.

To summarize, I would like to emphasize that at the IAA we have a strategic vision of water management in which consideration must be given to the territorial, demographic, socio-economic and political aspects that lead to integrated management. To develop management on water matters in Aragon we use governance as a political management tool based on horizontality, a downward and upward approach, and social participation. The political route that will allow us to meet our objectives involves progressive increase of political transfer of competences towards our regional government, within the principle

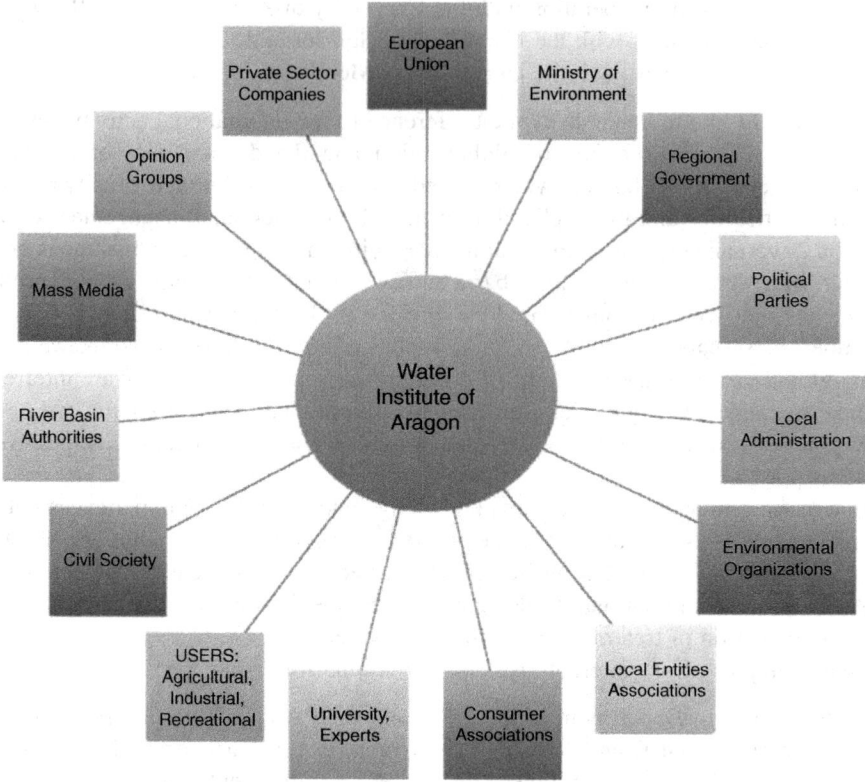

Figure 3. Interrelationships for the Water Institute of Aragon.

of the basin unit, and in accordance with the Statute of Autonomy for Aragon (Organic Law 5/2007, of 20 April, on reform of the Statute of Autonomy for Aragon).

Finally, in order to exercise the aforementioned political action, we need to draw up a new profile for the modern water manager. This profile requires a professional background, which includes the development of a multidisciplinary system of "capacity building". The new manager must have knowledge in subjects as disparate as engineering, chemistry, biology, medicine, politics, sociology, law, and administrative and financial management, amongst others. He/she must promote the use of governance as a political management tool based on horizontality and social participation and must practice what has been referred to as "hydro-diplomacy" between the different players involved, allowing progress to be made through dialogue and social consensus.

Note

1. *Impuesto solidario* in Spanish.

References

Aranda-Martín, J. F. (2007) The Aragon Water Commission: a practical experience, *International Journal of Water Resources Development*, 23(1), pp. 51–61.

Boletín Oficial de Aragón (2001) Law 6/2001, May 17, Planning and Participation in Water Management in Aragon, *Boletín Oficial de Aragón*, no. 64(1 June). [in Spanish].

Boné-Pueyo (2007) Model for social participation in the formulation of water policies, *International Journal of Water Resources Development*, 23(1), pp. 11–20.

Embid, A., Albiac, J. & Tortajada, C. (Guest Eds) (2007) "Water Management in Aragon", 23(1).

European Commission (1991) Council Directive 91/271/EEC of 21 May 1991 concerning urban wastewater treatment (Brussels: European Commission). Available from: http://europa.eu/legislation_summaries/environment/water_protection_management/l28008_en.htm/

Organic Law 5/2007, 20th April, on Reform of the Statute of Autonomy for Aragon.

Sodemasa (2007) Aragon's Action Plan for Wastewater Treatment: a model of environmental management, *International Journal of Water Resources Development*, 23(1), pp. 63–72.

Water Management in the Ebro River Basin: An Approach to the 2010–15 Hydrological Plan

MANUEL OMEDAS-MARGELÍ
Office of Water Planning, Ebro River Basin Confederation, Zaragoza, Spain

ABSTRACT *Water management professionals generally recognize that the management of water in rivers and aquifers is more efficient and sustainable at the river basin level than at the political and administrative levels of regional administrations. The development and consolidation of the river basin authorities has not been without difficulties. Experience has shown that the Spanish river basin confederations, the French water agencies and the US valley authorities have been successful. There have also, however, been failures, many attributable to the difficulties of separating the political power of regions and nation-states. In the Ebro River Basin, integrated water resources management was applied even when it crossed the administrative borders of the Autonomous Communities; otherwise, the water produced by the Ebro would be one-quarter of its current volume. Thanks to the integrated management, the Ebro economic region is supplied by major reservoirs, especially the Ebro reservoir. Its major irrigation systems, the Aragón and Catalonia Canal, Bardenas, the Ebro Delta and others, are projects that were conceived under the principle of integrated water management.*

Introduction

The Ebro River Basin is a depression with an arid climate surrounded by two mountain ranges: the Pyrenees and the Iberian Mountain System. Water management has allowed this sterile territory to develop into a major industrial and food-farming axis that amounts to the fifth of Spain's total agricultural production.

Management of water quality, with its networks of control, sophisticated information systems and management of flash floods and droughts, are all major achievements thanks to the successful governance practices from the Ebro's source to its estuary. Without doubt, the great advantage of integrated management of the Ebro is the absence of conflicts and the integration of users, administrators and civil society in the management of the water. Close to 3,000 user communities manage water in a democratic and participative manner within the confederated organism under one roof: this is one of the major potentials of this form of water management in zones of scarce resources (Figure 1).

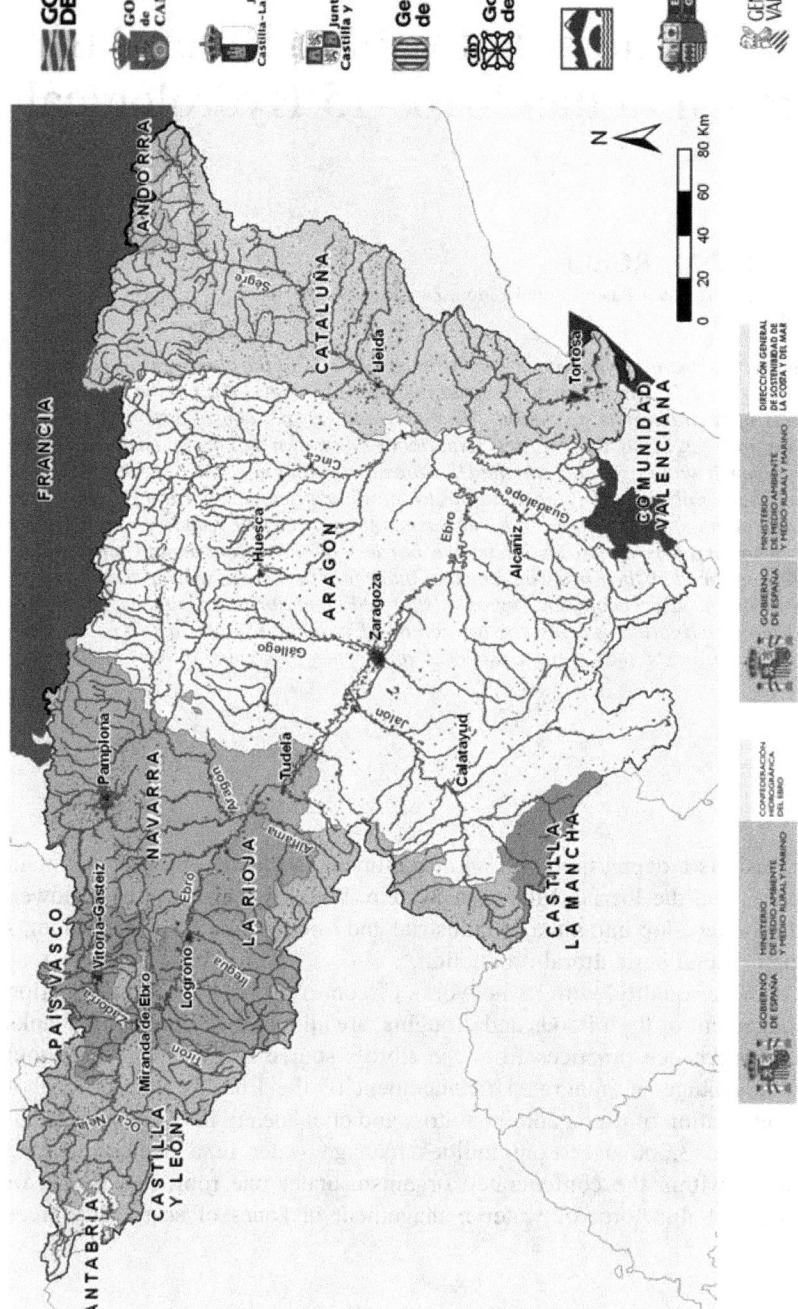

Figure 1. Competent authorities.

Spain: The River Basin Confederations

Spain's experience in the management of water by means of a sole authority within each natural river basin—or groups of small basins—dates to 1926. In that year the Royal Decree-Law created the river basin confederations as autonomous bodies. Article 1 stated that in all the river basins in which the administration finds it appropriate, or in which at least 70% of its agricultural and industrial wealth is affected by the use of its water, it will establish the corresponding Syndicated River Basin Confederation.

The principle of participation of users was established as an intrinsic part of the management bodies and the confederations, among other objectives, were instructed to draw up a coordinated and methodical plan of water use in the basin. As a result, the river basin confederations have operated without interruption since their very beginning. Nor have they restricted themselves to administrative tasks and authorization of particular uses; instead, they have played a major role in the promotion of uses, and they have undertaken works in their respective basins, and stockpiling basic and general water information within their scope of competence.

The Ebro River Basin Confederation was the first basin authority in the world to be constituted, on 5 March 1926. It is highly significant that the Ebro Confederation has lasted for more than 80 years, providing adequate responses to social demands on water management while being independent from the political ups and downs of Spanish politics over the years.

The institutional heritage could not be more positive. The Ebro's waters are distributed equitably among nine Autonomous Communities; there is a sophisticated flood-management system; an efficient system of concessions and conservation of the water resources; more than 150 dams ensure supplies, while more than 900,000 ha of irrigated land account for a large share of Spain's meat production. The use of water for energy purposes accounts for 30% of the nation's primary energy needs. The ecological state of the water bodies is the equivalent, or even better, of other European basins that have greater resources.

Clearly, the Ebro's water management could be improved. However, it is no less certain that the best defence of the Ebro Confederation is that if it did not exist, it would be necessary to create it in order to meet the challenges of the new millennium as expressed by the European Water Framework Directive. The Ebro Confederation is a major organizational water heritage of the basin.

Implementation of the Water Framework Directive has the triple aim of satisfying the demands of water, creating the good ecological condition of the water bodies, and managing droughts and floods. All of these aims have been reinforced by the Spanish River Basin Confederations.

Water management at the river basin level, as established by the Framework Directive, has implied no significant changes to the river basin confederations. This particular model of water management, on a territorial basis and organized by a pyramidal structure of federations, provides the underpinnings of the management of basins that seep through administrative boundaries.

In Spain, the river basin confederations are given the status of public bodies by law, with their own legal personality, different from that of the State. For purposes of administration they are attached to the Ministry of the Environment, Countryside and Marine Habitat.

According to the Water Law,[1] the functions of the basin authorities are:

- To design a hydrological plan of the basin, including updates and reviews.
- To manage and control water resources, including surface flows, beds, banks and margins of the rivers and lakes, and groundwater.
- To manage and control uses of water that are considered of national interest, or those that cross one or more Autonomous Communities.
- To plan, construct and exploit water projects that are paid for by the confederation's own funds and those that are entrusted to the State.
- Those roles that derive from agreements among Autonomous Communities, local corporations and other public and private entities, or those reached with private individuals.

Basin bodies are authorized to implement the following functions:

- To authorize concessions for water. Exceptions are those that relate to water projects and acts of general interest of the State that correspond to the Ministry of the Environment, Countryside and Marine Habitat.
- To inspect and monitor the use of authorizations and concessions.
- To carry out volumetric calculations, hydrological studies, management of floods and control of water quality.
- To study, plan, construct and maintain, as well as to improve water projects included in its own plans, as well as those to which they might be entrusted.
- To define the objectives and programmes on water quality in accordance with hydrological planning.
- To develop plans, programmes and tasks to achieve efficient demand-management practices by means of promoting savings and economic and environmental efficiency of the various uses of water.
- Technical services and consultancy.

The European Water Framework Directive and Decisive Support for the Basin Organizations

The European Water Framework Directive establishes the objectives of environmental quality of rivers as one of the fundamental bases for the achievement of good ecological condition in 2015. This implies a new challenge for the environmental policies of the Autonomous Communities in coordination with the central administration and basin organizations.

The directive establishes the river basin as the unit within which the good ecological condition of the various water bodies should be achieved. The directive does not define which will be the administration responsible for achieving it, for recovering costs or for placing a price on water, but it demands coordination among administrations and among them and society in general.

On the participatory processes for water management, one of the pillars of water policy in the European Union, the basin confederations have had a collegiate, significant representative basis right from their beginnings. These include users' assemblies, overflow commissions, operational boards, water councils and public boards.

Participation of users of water within management can be considered as exemplary since the users themselves take part in the budgets of each sub-basin by means of the exploitation boards. On the other hand, society in general is playing its part by participating in the design of the Hydrological Plan. This active participation is a result of the most holistic concepts of water management in the 21st century, where environmental factors are increasingly important. Finally, the basin bodies, and the Ebro Confederation itself, play the role of a centre of integration among the various interests that surround water management. The Ebro Confederation is, and aims to be, the "river's home" where everyone can have the possibility of presenting his or her proposals and achieve a democratic consensus on solutions.

In the face of the European Union and the fulfilment of the Water Framework Directive, the Hydrological Plan of the Ebro Basin Confederation thus becomes the centre of a network where all the administrations (those of the Autonomous Communities and of the 1,800 municipalities) share a commitment to improve the ecological state of the waters within their respective authorities.

The Aims of the Hydrological Plan

The aims of the plan are as follows:

- To improve of the ecological state of the water bodies: the intention is to achieve good ecological condition and avoid deterioration of the current condition.
- Water, as a factor of sustainable development: satisfaction of current and foreseeable demand, as well as rationalization of use by means of greater efficiency.
- To improve security against flooding, particularly of flash floods.
- To improve the knowledge and administrative skill in water management.

Improvement of the Ecological State of Water Bodies

One of the pillars of the state-of-the-art water management in the 21st century, and of the ability to achieve the aims of the Water Framework Directive on the good condition of water bodies, is the availability of a high-quality information technology platform. The Ebro Confederation's information and control systems are exemplary. The hydrological information system and the water quality information system are comparable with the world's most sophisticated. Even so, investment in maintenance and control of the networks will need to be increased considerably in the coming years.

In recent years, major investments have been made in the network for the control of the state of water bodies (Figure 2).

Currently, the ecological and chemical status of the water bodies is as follows: of the 919 water bodies within the Ebro Basin, 631 (70%) are in good condition, while 277 (30%) fail to meet standards as a result of point and nonpoint source pollution, land use, etc.

Within the context of Spanish and European rivers, the Ebro Basin can be regarded as moderately good, partly due to the low population density of some 36 inhabitants/km^2. The pressure from growth of human settlements is much less than it is in the centre of Europe.

In addition, the ecological state of the water bodies within the Ebro Basin has a dual aspect in common with the population. The 45% of the municipalities in the basin with population density of less than 5 km^2 are almost deserts with no pressure from settlements, therefore representing a major environmental reserve within the European context. The economic axis of the Ebro Valley, by contrast, is densely populated with nonpoint source pollution due to the agro-industrial food industry and point pollution due to the industrial and human waste.

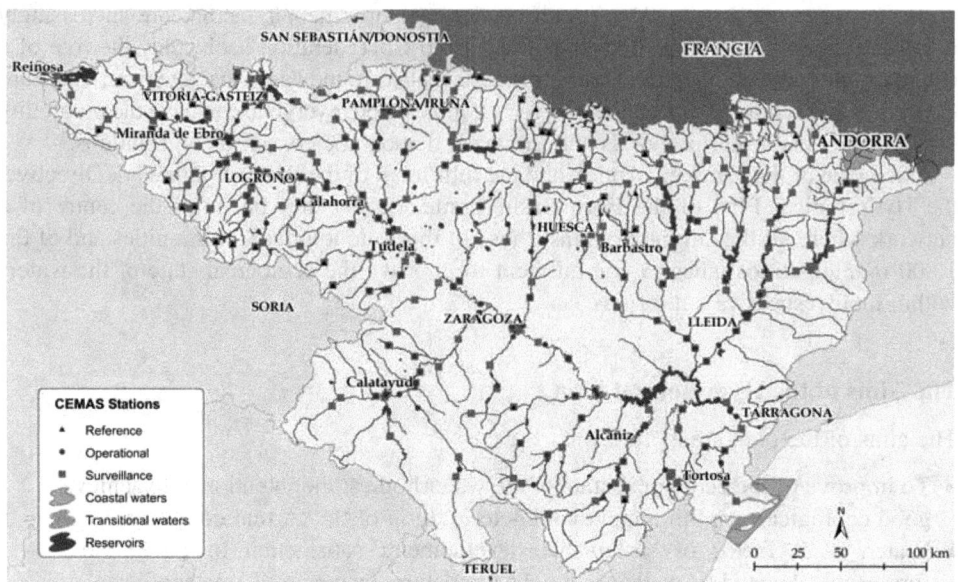

Figure 2. CEMAS network for the control of the state of bodies of surface water.

Table 1 shows the detailed statuses of the various types of water body and the forecast for 2015.

Rivers. Of the 644 water bodies, 478 or 74.2% have good, or very good, ecological condition, while 164 or 25.5% fail to meet that standard.

The target set by the Hydrological Plan for 2015 is that 10.1% of the water bodies are in very good condition and the great majority can become reserves. Taken together, 85.3% of the rivers will achieve a good ecological condition in 2015. Only some 11.8% of the rivers will need the deadline for achievement of good ecological condition to be extended to 2027.

Table 1. Status of the various types of water body and forecast for 2015.

		Assessment up to 2008		Environmental goals for 2015	
		Number of water bodies		Number of water bodies	
		n	%	n	%
Rivers in good condition	Very good condition	–	–	65	10.1
	Good condition	478	74.2	484	75.2
Fails to meet environmental objectives	Fails to achieve good condition	164	25.5	76	11.8
	Less rigorous objective			10	1.6
	Very modified water bodies			7	1.1
	Artificial	2	0.3	2	0.3
Total rivers		644	100	644	100

Just 1.7% of the water bodies will have to meet less ambitious ecological targets because they are rivers that, by their nature, are special cases. The most significant cases are the Elorz and Salado rivers because of their high saline content, the Jalón River in Alhama because of its high content of mineral waters, etc.

Rivers that are very modified represent only 1.1% of the total. These are stretches with a major component of return flows from irrigation, such as the Clamor Amarga stream, the rivers Sio, Cervera, Corp and downstream waters from reservoirs, such as the Guadalope River downstream from the Moros dam. There are, however, downstream stretches from reservoirs that are in good condition and cannot be rated as very modified water bodies. Finally, the Ebro Basin has two river-type water bodies: the Imperial Canal of Aragón and the Jiloca Canal Alto, which have major ecological potential. The Imperial Canal has species of great value, such as Spengler's freshwater mussels.

Achievement of these ecological objectives and, above all, the general improvement of the ecological condition of the water bodies will require a significant investment that has been estimated at €5.915 billion.

The most significant programmes will be the National Water Quality Plan and the Sanitation and Wastewater plans of the Autonomous Communities, as well as the action plan on zero tolerance of effluents, and the modernization of irrigation schemes.

Lakes and reservoirs. By contrast with the water bodies of the rivers, ecological knowledge of the water bodies of reservoirs and lakes represent major uncertainties. In the first place reference terms of different types of lakes need to be updated. Secondly, the biological, physical–chemical, and hydromorphological controls need more time for evaluation and study (Table 2).

The commitments established by the Hydrological Plan for 2015 are those of increasing scientific and technical knowledge, as well as, where possible, preventive measures to avoid nonpoint source pollution in lakes where upstream waters are polluted. The targets for 2015, therefore, represent no change on the current situation.

Table 2. Estimated evaluation on the environmental state and objectives of reservoirs and lakes, pending validation from definitive threshold.

		Assessment in 2008		Environmental targets for 2015	
		Water bodies		Water bodies	
		n	%	n	%
Reservoirs	Good condition	16	29	16	29
	Fails to meet standard	40	71	40	71
Total reservoirs		56	100	56	100
Lakes	Good condition	55	52	55	52
	Fails to meet standard	50	48	50	48
Total lakes		105	100	105	100
Artificial		5	5	5	5

No definitive typification has yet been established for reservoirs. Therefore, data provided in the Hydrological Plan's documentation will be very much improved in the development of the plan. The CEMAS reports, though preliminary, show that about 85% of reservoirs have a good or moderate ecological potential and about 65% are oligotrophic.

In accordance with a Resolution of 10 July 2006, what was then the Office of the Secretary General of Land and Biodiversity declared 16 reservoirs as sensitive areas. Controls are to be increased as well as preventive measures against nonpoint pollution in general and treatment of point sources of contamination. The case of the reservoirs of Mequinenza and Ribarroja was of particular note. These were declared as sensitive, and tertiary treatments were ordered in many of the areas of the basin with population of more than 10,000 inhabitants.

Groundwater. On the state of quality of underground water bodies in the Ebro Basin, all are in good condition, except for the 090.077 Alfamén Micoene aquifer. The recovery of this water body is envisaged as being achieved with supply from surface water of the Imperial Canal and, in the longer-term, from the future Mularroya reservoir. Even so, the authors do not expect the level to be recovered by the 2015 deadline (Table 3).

Preventive measures have been considered for some 11 water bodies with significant exploitation to prevent greater levels of extraction. At the same time, they are subject to regulations on exploitation for a more rational use.

In the qualitative aspects, some 70% of the water bodies are in good condition while 30% are not, nor will they be by 2015. Good farming practices and, above all, the modernization of irrigation will reduce the nonpoint sources of pollution very significantly, but its effects on groundwater will be in the medium-term. As a result, a postponement to 2027 is likely.

Two water bodies, the Urgel alluvial and the Tárrega limestone, face a special problem because of the burden of livestock they bear, and the characteristics of the aquifers. As a result, the aquifers will not be able to achieve good condition within a very long time, and thus the ecological goals are less strict.

Table 3. Groundwater bodies.

Status in terms of quantity	Number of bodies	%
Status in 2008		
Good condition	94	89
Good condition with significant exploitation	10	10
Poor condition	1	1
Objectives		
To be met by 2015	104	99
Deferral to 2021–27	1	1
Good condition	82	78
Poor condition	23	22
Objectives		
To be met by 2015	82	78
Deferral to 2021–27	21	20
Less rigorous objectives	2	2

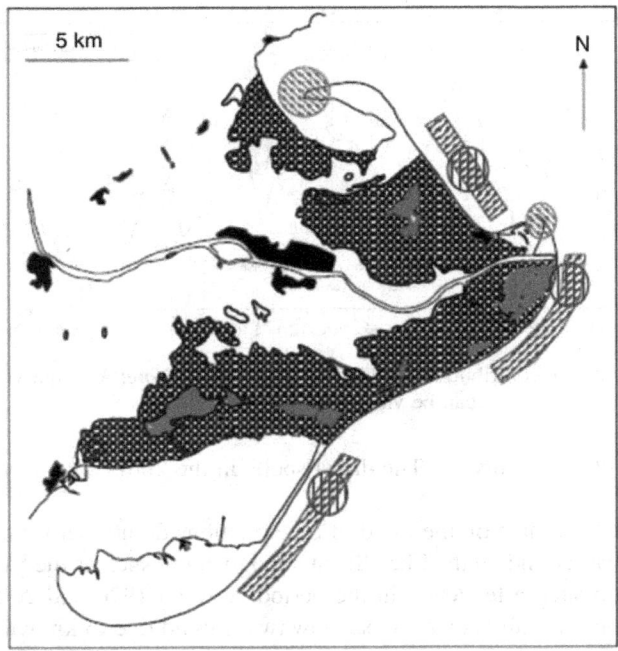

■ Problems associated with relative rise in sea level.
▨ Problems associated with receding shoreline.
⦀ Problems associated with the impact of storms.
⊛ Problems associated with advancing shoreline.

Figure 3. Main areas in the Ebro Delta that have problems related to coastal issues.

Transitional and coastal water bodies. Eight transitional water bodies include the bays of El Fangal, Los Alfaques, the stretch below the Ebro between Tortosa and the river mouth, and the Delta wetlands, as well as three bodies of coastal waters. All present significant pressures of point and nonpoint pollution. Evaluation of the state of these water bodies is complex because of the absence of reference indicators. For the moment, the state of the three coastal bodies and those of the two bays are being evaluated with a view to fulfilling the environmental objectives of the Water Framework Directive. No evaluation is being made of the definitive condition of the remaining transitional water bodies (Figure 3).

Given the environmental characteristics of these water bodies, the forecast for 2015 is that their ecological condition will be good, including parameters that are better than those at present. The most significant measure is the Integral Plan for the Protection of the Delta, a series of measures aimed at maintaining special ecological conditions such as organic accretion, subsidence, regression, etc. Total investment in the plan is €415 million.

The Hydrological Plan as a Factor in Sustainable Development

Existing water

Precipitation. Average precipitation of the Ebro River Basin is 622 mm/year (1920–2002 series). The worst hydrological year was 1949–50 with 452 mm/year and the wettest

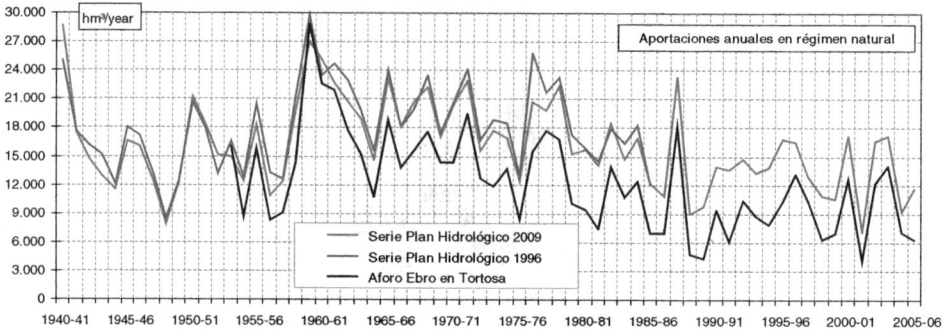

Figure 4. Annual natural contributions to the Ebro River Basin. *Note*: A colour version of this figure can be viewed in the online issue.

was 1958–59 with 809 mm/year. The driest spells in the 20th century were in the 1940s and 1980s.

Statistical studies to date on the trend of precipitations do not permit the conclusion of an overall downward trend on the Ebro Basin, though zones such as the basins of the Segre and Jalón point to such a tendency in the period between 1920 and 2001. On the other hand, erosion from rainfall shows a clear downward trend due to known reasons such as land use. Other reasons, however, remain unknown or inadequately evaluated.

Runoff. The Hydrological Plan of 1996, which includes a series of contributions from the period between 1940–41 and 1985–86, shows a total contribution in the form of a natural increase of 18,217 hm^3/year. The Instruction of Hydrological Planning (Order ARM/2656/2008) requires that the evaluations should be made using the series of water resources that correspond to the periods 1940/41–2005/06 and 1980/81–2005/06, with the latter period as the basis of the assignation and reserve of resources. The Instruction of Hydrological Planning also proposes that, from time to time when the evaluations of the climate scenario prepared by the Ministry of the Environment, Countryside and Marine Habitat are not available, a percentage of the global reduction of natural contributions should be applied at 5%.

The provisional estimates of natural runoff in the Ebro River Basin for the following periods are included in Table 4. On the runoff not used to meet demand, measured at the Ebro station at Tortosa, the years of lowest volume were 1988/89, 1989/90 and 2001/02 (4,756, 4,283 and 4,128 hm^3, respectively). Figure 4 presents the volumes registered since 1950.

Water consumption and ecological volume. In the current situation, consumption amounts to 34% of the total, taking into account a series of runoffs in recent years, as well as climate change.

The two extreme scenarios for 2027, one with moderate demand and the other with all the requests for demand registered with the nine water councils of the river basin, place consumption of water within the range 37–50% over natural contributions.

The utilization of the water resources of the basin will be more than the current level, despite the reductions for the efficient use of water and, in addition, most of this consumption will be about the equivalent of half of the existing amount.

Table 4. Provisional estimates of natural contributions in the Ebro River Basin.

Period	Maximum	Minimum	Mean	Percentile 1%	Percentile 5%	Percentile 10%	Percentile 25%	Percentile 50%
1940/41–2005/06	28,679	7,047	16,195	7,576	9,306	10,852	12,539	15,906
1980/81–2005/06	23,305	7,047	13,869 (14,500)	7,515	8,983	9,441	11,075	13,868

Table 5. Status of implementation considered in the 1998 Hydrological Plan (RD.1664/1998).

	1998 Plan (hm³)	Implemented 1998–2010	To be implemented 2010	Planned	Not planned	Unviable or not wanted
Aragón	2,736	215.8	942.0	639.3		626.3
Cantabria						
Castilla León	101	1.5			101.2	138.2
Cataluña	488	402.0	80.0	62.4	1.8	
Rioja	90	4.5	54.8	10.5	6.8	29.5
Navarra	522	418.0	7.2	32		119.7
País Vasco	15	2.0	0.9			19.5
Total Ebro Basin (hm³)	3,952	1,043.8	1,084.9	746.0	108.0	933.2
Number of reservoirs	64	14	13	20	5	36

The ecological volume of the current 1998 Hydrological Plan estimates the current stretch of the Ebro at 3,238 hm^3 (100 m^3/s + 83 hm^3 ecological volume of the rice fields). This volume represents 23% of the existing water and 73% of the current consumption of water in the basin.

The river basin's hydrological plan will set a regime of variable volumes monthly. With the data now available, more volumes than those of the existing plan could be assured in the Delta. However, the volumes demanded by the ACA could not be guaranteed because they are very much higher.

The development of the irrigation systems of the Ebro Valley does not determine the environmental flow of the delta. This is because the Mequinenza Reservoir allows for regulation of the flow and reduces the impacts of upstream withdrawals.

Regulated water. Within the Ebro River Basin 109 main reservoirs (of more than 1 hm^3), with a total capacity of 7,580 hm^3, are currently in operation. Of these, 40% have been built for purposes of regulation for consumptive uses, and 60% were built for hydroelectricity. The capacity for runoff amounts to 52% of the average contribution and capacity, while the capacity for consumptive uses amounts to 21% of the average contribution (during the period from 1980–81 to 2005–06). There are 850 weirs, some 10,000 ponds and 35,000 groundwater wells. When the infrastructure is all in use or under contract, the 1998 Hydrological Plan (RD.1664/1998) presupposes a capacity of 3,949 hm^3 (Table 5).

A total of 27 reservoirs with a capacity of 2,128.71 hm^3 have been built or are under construction under the 1998 Plan. This amounts to 54% of the volume that was originally contemplated. Twenty-five reservoirs with a total of 854.11 hm^3 are planned or under preliminary plans, amounting to 22% of the reservoir capacity foreseen by the 1998 Plan.

The current 2010 Hydrological Plan scraps 36 reservoirs that would have been built in the 1998 Plan (RD.1664/1998), but have been ruled to be economically, social or environmentally unviable. These reservoirs would have amounted to a capacity of 933 hm^3. The Autonomous Community of Aragón reserves the right to propose other projects to replace the reservoirs that were scrapped within its authority.

Uses of water

Urban supply. The Ebro River Basin has 3.05 million inhabitants in addition to those who live in the areas where water has been transferred to the zones of the Basque Country and Cantabria, as well as the internal basins of Catalonia. In total the Ebro Basin provides water supplies for some 5 million inhabitants. The demand for water for these uses amounts to 490 hm^3/year.

The major systems of a general nature have fully assured supplies, with the exception of the Zadorra system that supplies Vitoria and Greater Bilbao. The system has suffered crises from time to time, though the most severe restrictions were incurred between 1980 and 1990. Shortages have also arisen in the Camp de Tarragona region of Catalonia because of strong seasonal demand in summertime.

On the other hand, the minor nuclei of the basin are vulnerable to droughts, given their dependence from small springs and aquifers of considerable weakness. As a result, 120 nuclei suffered severe restrictions and 60 had to be supplied with water trucks during the 2005 drought.

In the qualitative aspect, of the 146 sampling points for surface water for the supply of more than 500 persons, the results obtained in 2005 (an especially dry year) showed that only 16% of the population supplied within the Ebro Basin or from it did so from water of A3 quality or of less than A3. This situation, however, will change when the new supply to Zaragoza and its area of influence comes into operation. Without counting that particular zone, the percentage would have fallen in 2005 to 3% of the inhabitants (Figure 5).

In recent years major efforts have been made in investments in high-capacity water supplies in which more than half the population has gained from improvements in regulation and transport of potable supplies, as well as collectors and systems of treatment of wastewater. Currently, 89% of the population has access to secondary treatment of wastewater.

Forecasts for the future and measures in the plan. In the qualitative aspect of the water supplies to Zaragoza and its environs, the joint supply from the Navarra Canal and some other individual sources will have eliminated the supplies from the Ebro between Miranda and Mequinenza. There the water is not fully up to potable standard.

On the other hand, a tendency continues for the provision of joint supplies in order to improve the high-capacity water supplies to Zaragoza and its environs from Yesa to the communities of Lleida and Pamplona, to Huesca from Montearagon, to the lower Jiloca from Lechago, to the municipalities of the Oja river, Las Garrigas, the lower Ebro, etc.

In terms of sanitation and wastewater treatment, the State and the Autonomous Communities comply with European Directive 91/271/EEC by means of the National Plan of Sanitation and Wastewater Treatment for 2007–15. This plan includes measures for sanitation and purification of a large number of urban zones of more than 2,000 inhabitants that lack water-treatment facilities, as well as the measures established by the new declaration of sensitive zones and zones of fewer than 2,000 inhabitants.

The Resolution of 10 July 2006 declared the reservoirs of the Bajo Ebro, Mequineza and Ribarroja as Sensitive Zones. As a result, the population centres of Utebo (130,000 inhabitants), Zaragoza-Almozara and La Cartuja (1.14 million), Ejea de los Caballeros (62,200), Río Huerva (62,200) and Lérida (190,000), among others, should eliminate the phosphorous from their effluents within seven years.

Agrifood uses. The food-farming complex (crops, stock-rearing and food industry) constitutes the second leading productive axis of the Ebro Valley after metallurgy and transport. It is, however, a sector that is of fundamental importance in terms of land use of the rural communities of the river basin.

Farming in the basin is one of two very different types. The periphery, which forms part of the Pyrenees and the Iberian System, is very weak in productive terms, with much abandoned farmland. Meanwhile, the centre of the Ebro Valley produces one-fifth of all of Spain's agricultural production, and a major renaissance of farming is underway.

The meat-producing sector of the Ebro Valley (cereals plus feed plus fodder) accounts for 32% of Spain's production, while the output of fruit accounts for more than 60% at the national level.

Land under irrigation with concession rights amounts to 965,698 ha in the basin, though the amount of irrigation in effective terms amounts to some 700,000 ha. Demand is rated at 7,339 hm^3/year, though the volume of water supplied is less than that.

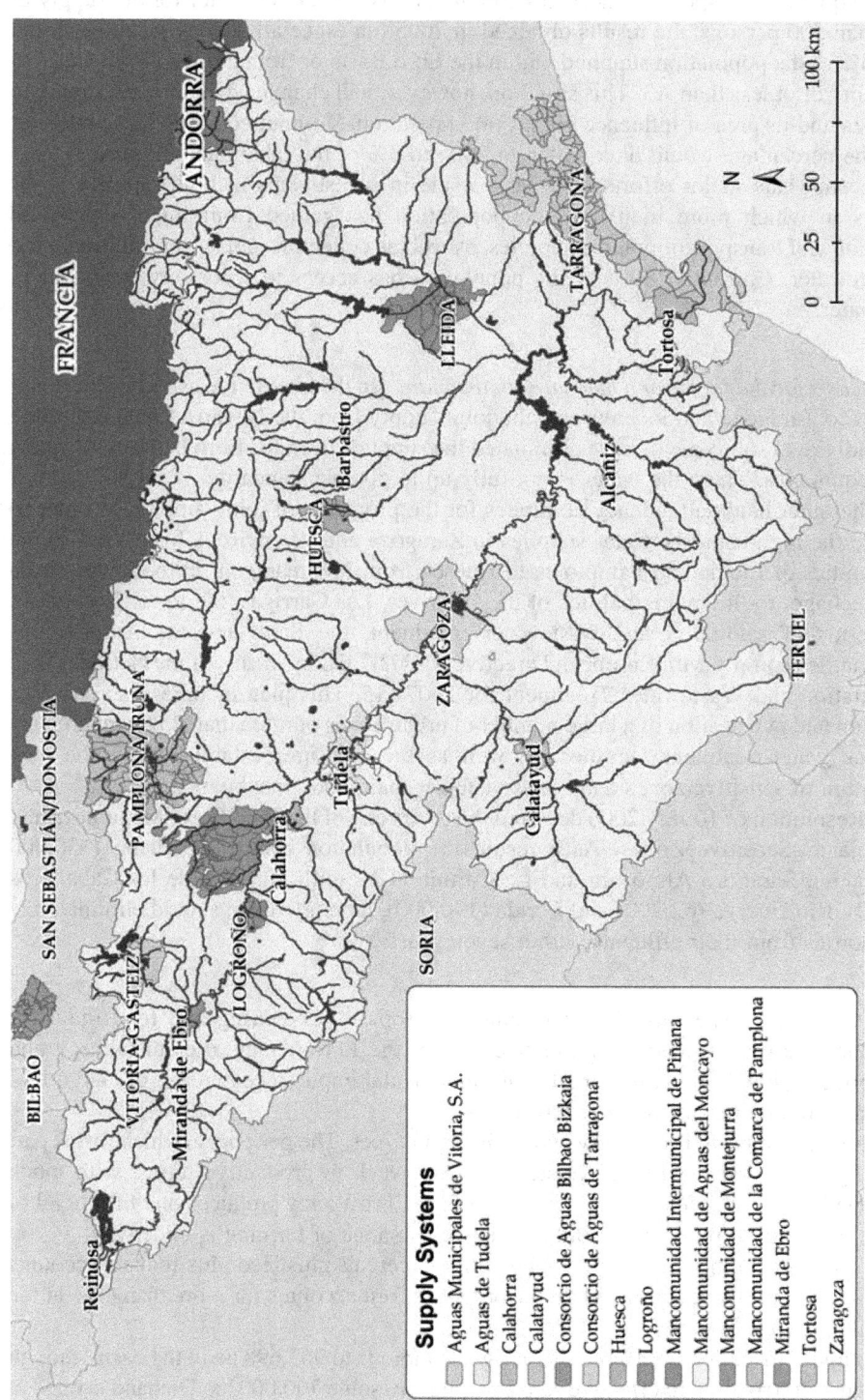

Figure 5. Supply systems for more than 20,000 inhabitants.

The Plan estimates the deficit at 950 hm^3/year. There are two main causes: a lack of water resources, an issue of more relevance on the right bank, which is also expected to suffer the impact of climate change with greater intensity; and a deficit of regulation and transport on the left bank, especially the lack of regulation (Figure 6).

According to a 2006 study of the quality of surface water of irrigation return flows in the Ebro Basin, the annual amount of inorganic nitrogen (N) exported by the Ebro in Tortosa was 25,907 metric tons/year, the equivalent of a weighted median concentration of 2.50 mg, of which 88% is in the form of nitrate (NO_3) and the rest almost all in the form of ammonium (NH_4). This is moderate in terms of other European rivers that have a much higher component of nutrients for crops.

On territorial distribution, nonpoint pollution from food farming is concentrated on the central third of the Ebro Valley. In nutrients, the most significant points of pollution are principally those that gather the irrigation return flows, as on the Arba River at the point where they arrive from the Bardenas irrigation system, at Clamor Amarga where the returns arrive from the Aragón and Catalonia system, or at the Alcanadre and Flumen rivers where the return flows arrive from the Alto Aragón system. In the nets that control Lists One, Two and Special and Priority Substances, high levels have been found in the Alcanadre River in Ontiñena, Clamor Amarga in Zaidín, Arba de Luesia in Tauste, and Segre in Serós.

Nonpoint pollution from the food-farming complex, though important, is concentrated in very localized river stretches, so helping its control and the lessening of its impact.

Groundwater evaluations have pointed to the presence of nonpoint pollution by nitrates and pesticides. Nitrates were measured at 157 points in 2007, with 30 zones affected or at risk (concentrates of nitrates higher than 50 or 25 mg/l, respectively). Pesticides were detected at 27 points of the 584 samples analysed between 2003 and 2007, of which only 10 had contents that were higher than the established limit. The zones that were most affected coincide with those described as vulnerable in the following map. The two zones with the highest levels of pollution are the Urgel terraces and alluvial system of the Ebro axis.

One positive aspect of the irrigation systems of the Ebro Valley is that some 400,000 ha of irrigation lie in impermeable tertiary terrain. As a result, there is no adverse impact on the groundwater (Figure 7).

Forecasts and measures in the plan. The Ebro Valley Foundation, consisting of savings banks, business organizations and universities, as well as the members of the Ebro Federation of irrigation systems and representatives of the Departments of Agriculture, held a meeting in Tudela on 30 November 2005. They concluded that the Ebro Valley is competitive within Europe and can continue to be so. One of the fundamental challenges lies in producing quality raw materials from high-technology irrigation systems.

The recent analysis of its development shows sustained annual growth of 2.6%, showing that the trend of the food-farming complex of the Ebro (crops plus livestock plus food industry) continues to be one of the pillars of the valley's productive system, together with the metallurgical and transport sectors.

Quantitative aspects. The 1996 Plan envisages a horizon of irrigated land covering up to 1.315 million ha, of which 125,000 ha will support irrigation. The National Irrigation

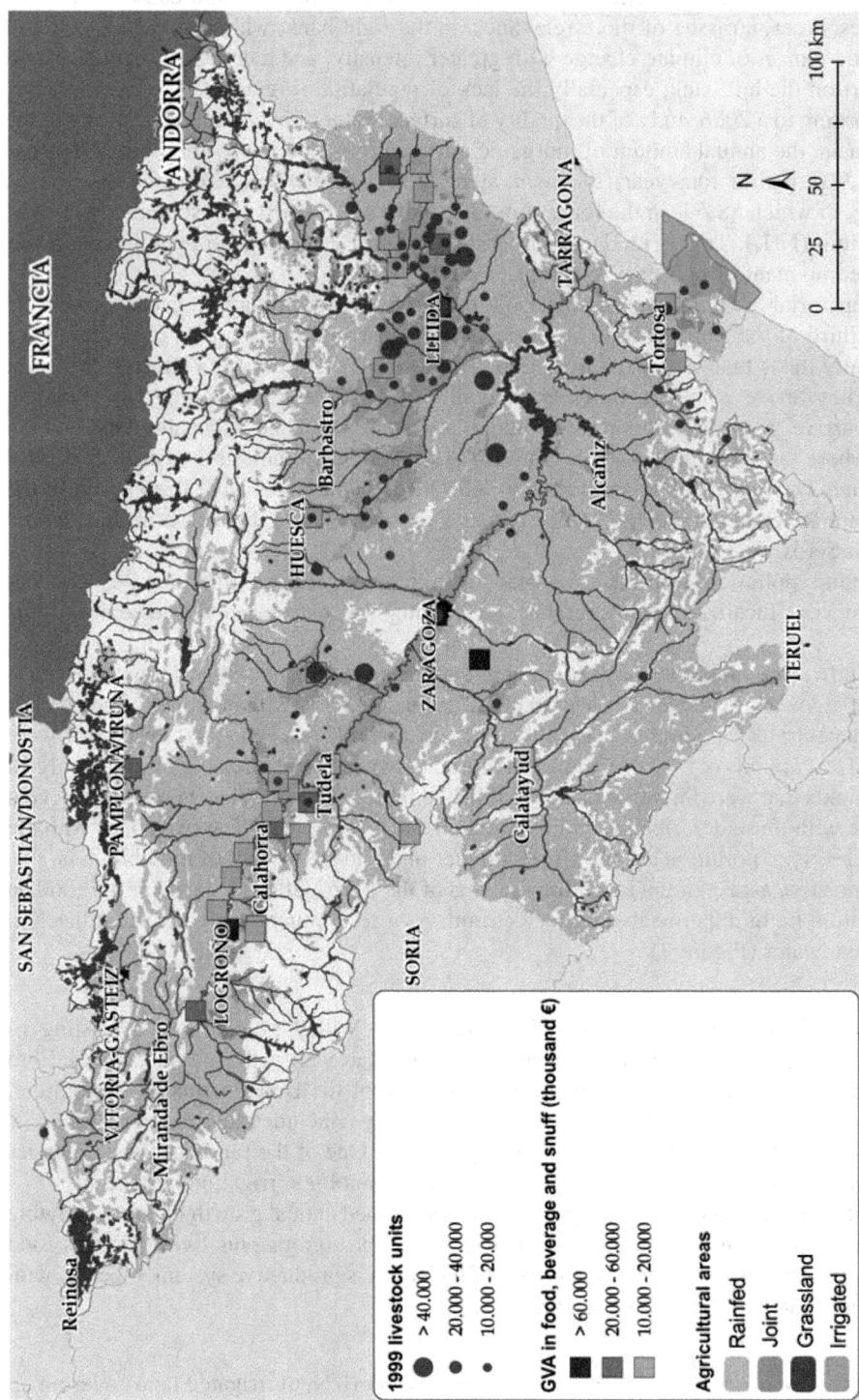

Figure 6. Food-farming complex of the Ebro Valley.

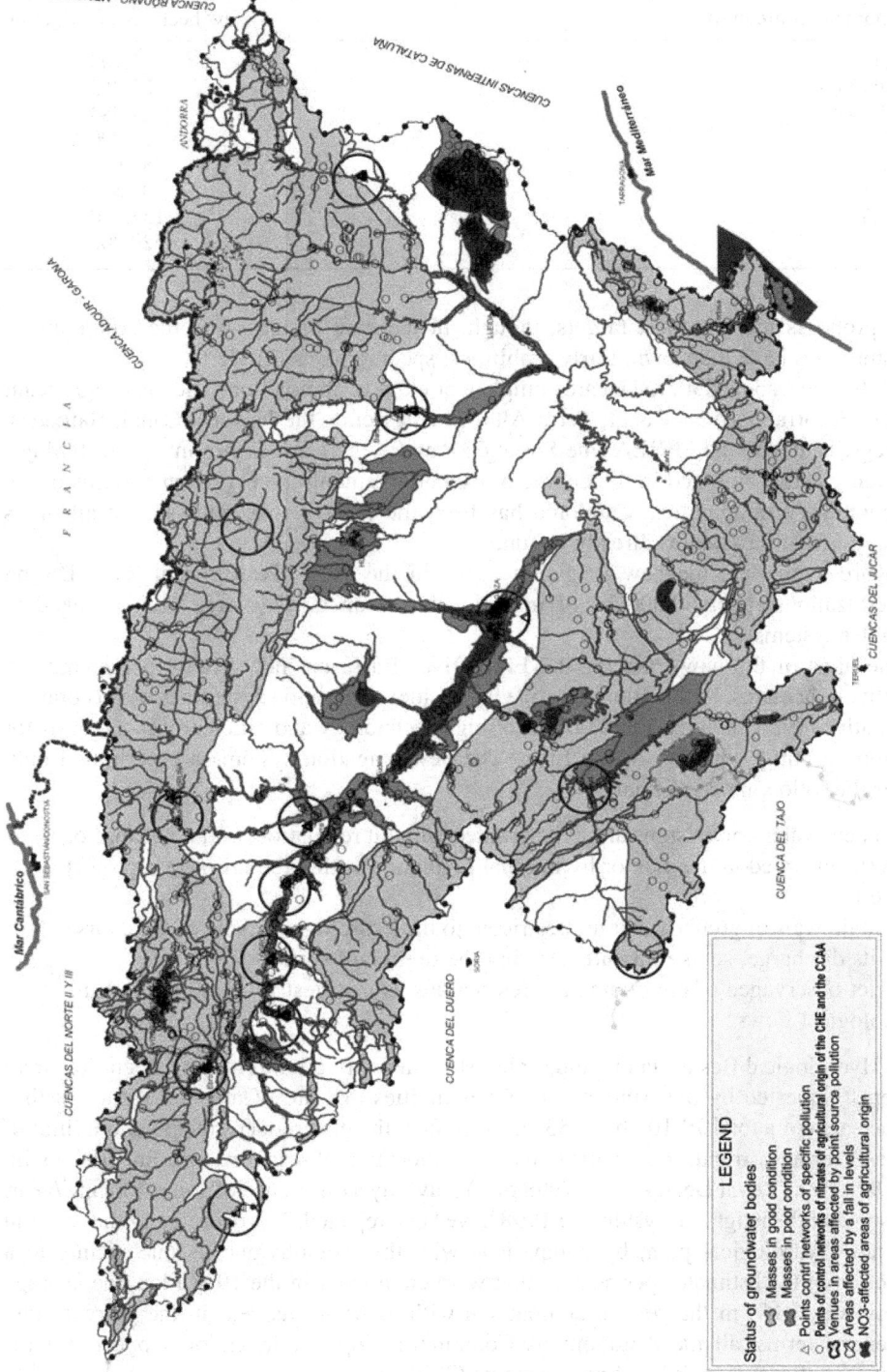

Figure 7. Status of groundwater bodies, 2008.

Table 6. Irrigated land (ha).

Autonomous community	New hectares of irrigation
Cantabria	2,700
Castilla León	23,765
País Vasco	37,784
La Rioja	17,900
Navarra	82,635
Aragón	207,534
Cataluña	157,590
Total	529,908

Plan proposes more modest targets, though, in the case of the Ebro, the Autonomous Communities have their own, fairly ambitious, specific plans (Table 6).

Of this area, some 290,000 ha are complete or close to completion. The most significant are the Segarra-Garrigas Canal, Terra Alta, Cherta-Cenia, the Navarra Canal, Bardenas, Monegrosand, and the PEBEA (the Strategic Plan for the Lower Ebro in Aragón). Major differences have emerged in the degree of execution among the irrigation systems of the Autonomous Communities. Catalonia has been the most active, with almost all of its planned irrigation systems already begun.

Figure 8 presents the new irrigation zones in the 1996 Hydrological Plan. During modernization, about 60,000 ha of irrigation based on gravity have been changed to sprinkler systems.

The aims of the new Plan for the Ebro River Basin are not simply to increase the quantity of hectares. Rather, the objective is to achieve irrigation systems that can compete internationally, with at least 800,000 ha of high-technology and efficiency as the basis for the food-farming complex of the Ebro. The new irrigation systems are permitted only under the following conditions:

- No new water concessions are to be granted without regulation, either because of works included in the plan or by internal regulation within the utilization project itself.
- Withdrawals of groundwater are restricted to the hydrological cycle of the course of its discharge, so as to avoid affecting the regime of surface water.
- Strict observance of environmental restrictions as manifest in the regime of the ecological flows.

This Hydrological Basin Plan includes almost no new irrigation projects, except for small systems requested by the Autonomous Communities that are of a social nature, with a surface area of a total 70,101 ha in 53 cases. In fact, the plan reduces the potential limit of requests for new irrigation systems in terms of allocation of resources in comparison with the 1998 Plan (Royal Decree 1664/1998 of 24 July), by a difference of some 400 hm^3/year, and some of the irrigation systems of 1998 have been rejected. The decrease of the water in this new hydrological plan, by comparison with the previous one, is due mainly to a reduction of the estimates per hectare that were envisaged. In the 1998 Plan, the average supply was 6,444 m^3/ha/year in comparison with 4,761 m^3/ha/year in the current plan. In general terms, all the Autonomous Communities opt for irrigation-support systems rather than those of traditional new systems (Table 7).

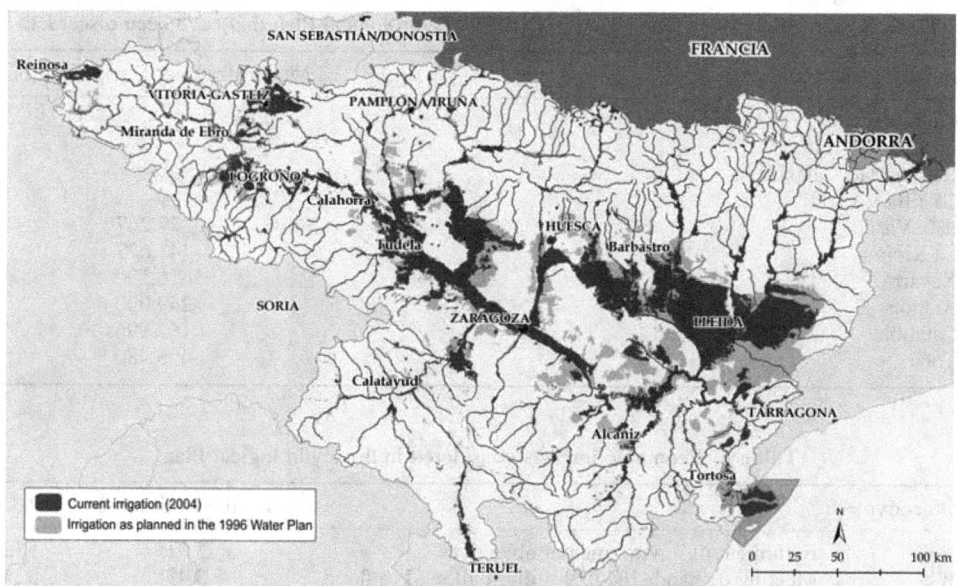

Figure 8. Current and planned irrigation in the Hydrological Plan, 1996.

Table 8 shows the irrigation systems proposed in the 1998 Hydrological Plan that have been discarded by the basin's Autonomous Communities in the 2010 plan. Modernization of the irrigation systems, and greater efficiency in their management and reuse, together combine to a reduction in water consumption.

Major projects for mains water are also included in the new Hydrological Plan in order to improve efficiency in water use, as well as internal regulations in all the major irrigation systems of the left bank: the management and modernization of the lower Aragón, the lower Galicia, lower Gallego, lower Cinca, irrigation systems that depend on the Val and Montearagon, etc. In addition, a major improvement is proposed in the efficiency of the recovery of the irrigation returns of Bardenas and the Upper Aragón systems as well as the Aragón and Catalonia Canal. Meters are also to be introduced to measure the volumes of the canals.

Table 7. Irrigation projects.

Autonomous community	Hectares removed from the 1998 Hydrological Plan	Hectares of the 2010 Plan not included in 1998
Aragón	28,465	13,596
Cantabria	1,323	–
Castilla-León	15,565	4,323
Cataluña	5,405	37,106
La Rioja	1,247	10,374
Navarra	19,184	511
País Vasco	24,595	–
Total	95,784	65,910

Table 8. Irrigation systems proposed in the 1998 Hydrological Plan that have been discarded.

Autonomous community	Modernization of irrigation (ha)
Cantabria	–
Castilla-La Mancha	–
Comunidad Valenciana	–
Castilla y León	–
País Vasco	27,267
La Rioja	41,627
Navarra	16,590
Aragón	249,000
Cataluña	163,996
Total	498,480

Table 9. Economic impacts considered in the Hydrological Plan.

Objective		Total (€, millions)	%
A	Fulfilment of environmental objectives	5,915	57.3
B	Meeting demands (€440.9 million allotted to flood prevention)	3,498	33.9
C	Extreme incidents (€440.9 million included in the part of regulation allotted to flood prevention)	903	8.8
	Total Hydrological Plan for the Ebro Basin	10,316	100.0

A fundamental aspect of the water management of the irrigation systems in the Ebro Basin will be the new National Strategy of Sustainability of Irrigation Systems, which is intended to be implemented by 2015 and will affect both the infrastructure of the major irrigation systems as well as the type of irrigation itself. Under the terms of this strategy, the modernization of irrigation systems proposed by the Autonomous Communities (Table 9) are expected to have major financial support.

Qualitative aspects. The modernization of irrigation is expected to provide very significant improvements in the volume of pollution exported by the systems. According to data supplied by the Aragón Centre for Research into Farming, the modernization of Monegros is expected to reduce by 30% the volume of nutrients exported, some 8% of pesticides and some 8% of salts.

The same research centre has studied the evolution of nitrates at several points in the Ebro Basin. Although the situation is not alarming, annual increases have been significant. These include at Arba in Gallur (0.91 mg/l/year; 2.7%), Bayas in Miranda (0.71 mg/l/year; 7.2%), Tirónin Cuzcurrita (0.66 mg/l/year; 4.3%), Ega in Andosilla (0.56 mg/l/year; 4.2%), Cinca in Fraga (0.38 mg/l/year; 4.6%), etc. For that very reason, the modernization of irrigation systems and the use of water from the returns of the systems will imply fundamental measures to reduce nonpoint pollution.

Several measures to reduce nonpoint pollution form part of the Hydrological Plan. These include the following:

- Environmental checks of the irrigation systems and their returns.
- Preventive measures in designated vulnerable zones as demanded by European Directive 91/676/EEC (the Nitrates Directive).
- Introduction of appropriately designed and managed animal waste banks in farming regions.
- Improvements in the control of the elimination of stock-rearing waste in areas where crops are grown.
- Introduction of animal waste treatment and composting plants in zones where the available area for farming is less than those that were necessary for the pilot projects.

Energy uses. The Ebro Basin produces 32% of Spain's nuclear energy, as well as 21% of its hydropower and 11% of its thermal energy production. The river basin has three conventional thermal plants, four combined-cycle, two nuclear plants and a hydroelectric park that consists of 360 plants.

Hydroelectricity uses some 38,000 hm^3/year of water that produces some 9,400 GWh/year, with an installed capacity of 4,000 MW. Taking account of the volume normally used, the unit production is some 0.5 kWh/m^3.

Demand for water for cooling of thermal plants, which have 5,430 MW of capacity, is some 3,100 hm^3/year, and most of it is required for the nuclear reactors of Santa María de Garoña (Burgos) and Ascó (Tarragona) that have a combined capacity of 2,521 MW. The combined-cycle plants at Arrúbal, Castejón, Castelnou and Escatrón are located at various points all along the Ebro and have a larger capacity than the traditional thermal plants at Teruel, Escucha and Escatrón. Demand for these thermal plants comes to some 30 hm^3/year (Figure 9).

The most significant impacts on water is that the use of the hydroelectric plans affect the volume of water in 990.5 km of the river and of the flooding of dams by 314.7 km. Problems exist in the river's continuity due to barrier effects at some 260 points.

On the other hand, the availability of hydroelectric reservoirs provides an assurance of the river's minimal volumes, which are fundamental to the maintenance of ecosystems. Downstream demand of the reservoirs of the Mequinenza and Ribarroja complex are fundamental to the maintenance of the Ebro Delta, given the current intensity of human activities. In addition, the 9,400 GWh/year of hydroelectricity produced saves about US$494 million/year on Spain's trade balance, while avoiding the emission to the atmosphere of 5.26 million metric tons/year of carbon dioxide, 60,000/year of sulphur dioxide and 11,190/year of nitrogen oxide.

Forecasts for the future and measures in the plan. Since the 1996 Hydrological Plan was passed, the following concessions have been approved (Table 10). Many of these plants have not been built, nor are they likely to be within the plan's timeframe.

The plan's forecasts were the fruit of the participation processes of the companies of the sector, Spain's electricity network and other departments of state industry, and of the Autonomous Communities. Water is now a major factor in the development of renewable energy, especially wind. For that reasons alone, growth of some 2,000 MW of power is expected.

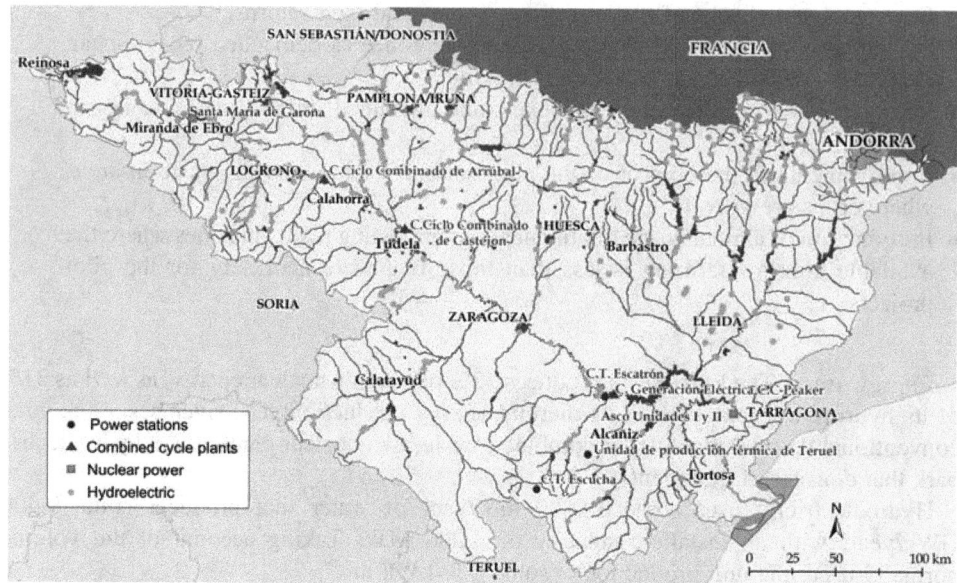

Figure 9. Power plants in the Ebro River Basin.

As far as the increase of production is concerned, the forecast is that the regulatory infrastructure not yet constructed will remain as such and the existing growth in power is to be increased. The combined-cycle plants are expected to increase and thus also the installed capacity, but their use can be affected because of the composition of the energy mix. The installation of solar energy plants could be of some importance, but their development is linked to issues of energy efficiency and pricing policies. The biofuels plants that have been constructed, or are under construction, can transform more than 300,000 metric tons/year, but much of the raw material is imported. Within the perspective of the plan, some 75,000 ha of irrigated land are expected to produce crops for energy purposes.

On the management of existing energy uses, the following are foreseen:

- Areas around several hydroelectric reservoirs are particularly sensitive. These include Sobrón, Mequinenza, Ribarroja, Flix, Ullívarri-Urrúnaga, etc. Follow-up will be required on the effectiveness of the measures taken.

Table 10. Power plants: capacity and estimated volumes of water.

Type	Number	Capacity (Mw)	Estimated volume (hm^3/year)
Biofuels	4		0.766
Combined cycle	17	13,877	160.228
Cogeneration	2		44.623
Hydroelectric	123	958	2,304.702
Mechanical	1		
Renewable refrigeration	1		0.616
Solar	4	200	3.479
Dual fuel	2	604	
Total	152	15,035	

- The trophic state of the reservoirs must be followed up, especially those that are most significant in terms of their eutrophic state, such as those of Mequinenza and Ribarroja.
- An exhaustive study of the Ebro's water quality must be continued at Ascó.
- A plan to improve coordination among users of hydroelectricity, rafting and irrigation.
- A study to harmonize wind power with reverse pumping.
- Adjustment of the withdrawals from the plants in order to meet environmental requirements, especially in the Santa María de Garoña plant.
- Achieve flexibility in the concessional volumes, seeking agreement with users on improvements in management that need to be introduced.
- Voluntary agreements within the framework of the investment and/or compensation programs.
- Extension of concessions in exchange for reductions in volume.
- Provide turbination of the ecological flows as a compensatory mechanism.

Industrial uses. Industry is of major importance in the Ebro Basin. The sector's gross value added is €11.278 billion/year, or 28% gross value added in the whole river basin. Industry also provides jobs to 317,000 people, some 25.5% of the active population. The principal activities are metallurgy, the auto industry and food, all close to water. The leading industrial centre is Zaragoza, followed by Vitoria, Pamplona, Logroño and Lérida.

Demand by manufacturing industry in the Ebro River Basin runs to some 460 hm^3/year, of which 260 hm^3 corresponds to industries that are not connected to the municipal networks. In addition, some 45 hm^3 of water from the river basin are also transferred to Greater Bilbao and the Camp de Tarragona region.

For the river basin as a whole, industrial discharges have been estimated for 2001 in chemical oxygen demand (COD) 9,171–11,367 metric tons, CFD = 2,826 mt, suspended solids = 1,422–3,740 mt, total nitrogen = 260–996 mt, phosphorus = 77–189 mt, and heavy metals = 21 mt (Figure 10).

Forecasts for the future and measures in the plan. Industrial activity is expected to continue its 3–4% annual growth, though there will be variations within the various subsectors. The measures foreseen are as follows:

- Definition of the criteria for the authorization of effluents (especially, pollution of an urban or industrial nature).
- Studies for the reduction of detailed emissions of dangerous substances.
- Studies by sector of effluents to the recipient medium and proposal for plans to reduce pollution.
- Measures aimed at control of effluents (control checkpoints and frequencies of overall sampling in the network for control of effluents).
- Treatment of the major focal effluent points in the Ebro Basin.
- Promotion of the creation of effluent associations.
- Support for industries to help them carry out the clean-up of effluents demanded by law and which will be more urgent when the Water Framework Directive comes into force.

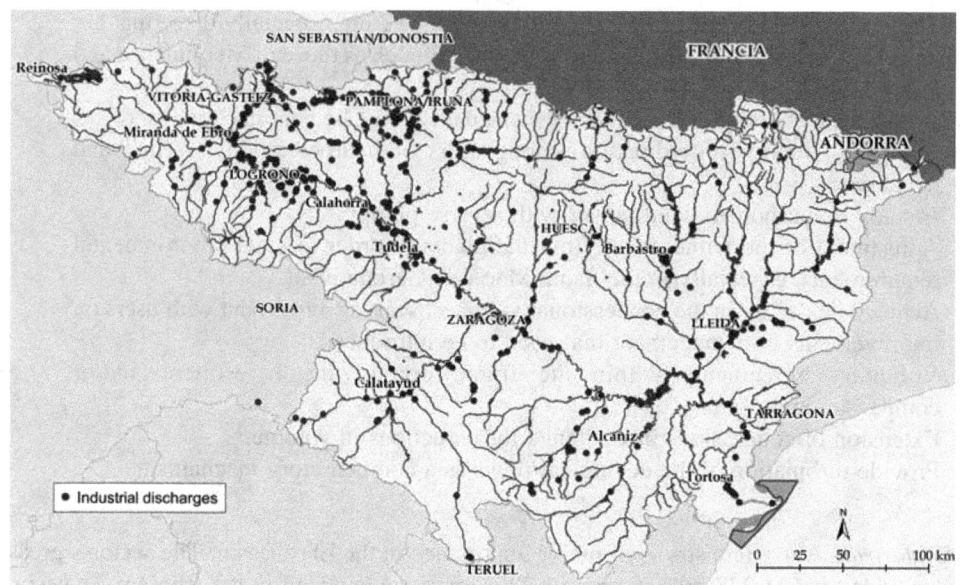

Figure 10. Industrial effluents in the Ebro River Basin.

Recreational uses. Tourism is not very developed in the Ebro River Basin. Nor does it represent any significant pressure on the water ecosystems. Within the overall demand in the basin, tourism represents less than 300 hm^3/year, which is an insignificant amount.

The most significant uses are the production of artificial snow, now introduced to almost all the basin's ski stations, and the irrigation of 21 golf courses (Figure 11). Recreational users are as shown in Table 11.

Forecasts for the future and measures within the plan. The trends within the Ebro River Basin in recent years show a 4% sustained increase in the official numbers of golfers, 4.2% growth in winter sports and a double-digit increase in adventure sports related to water. On the other hand, the numbers of licences for anglers, as well as for boating on reservoirs and rivers, appear to be stabilizing.

Measures to encourage recreational use of water have been many of the 2,000 requests raised in the participation processes involving society in general. The most significant such measures since the plan came into effect include the recreational dams of Itoiz and Rialp, the plan for the renewal of the La Loteta reservoir, the environmental renewal of the urban stretch of the Ebro in Logroño, and the recreational measures of the Integrated Plan for the Protection of the Ebro Delta (PIPDE), etc.

The promotion of recreational activities and scientific tourism related to water are increasingly being requested by society. In coming years they are one of the niche measures to be undertaken.

On management issues, some of the challenges are the fulfilment of the Resolution of 15 May 2007 to combat the zebra mussel, the angling plan and the democratic election of recreational users into the River Basin Water Council.

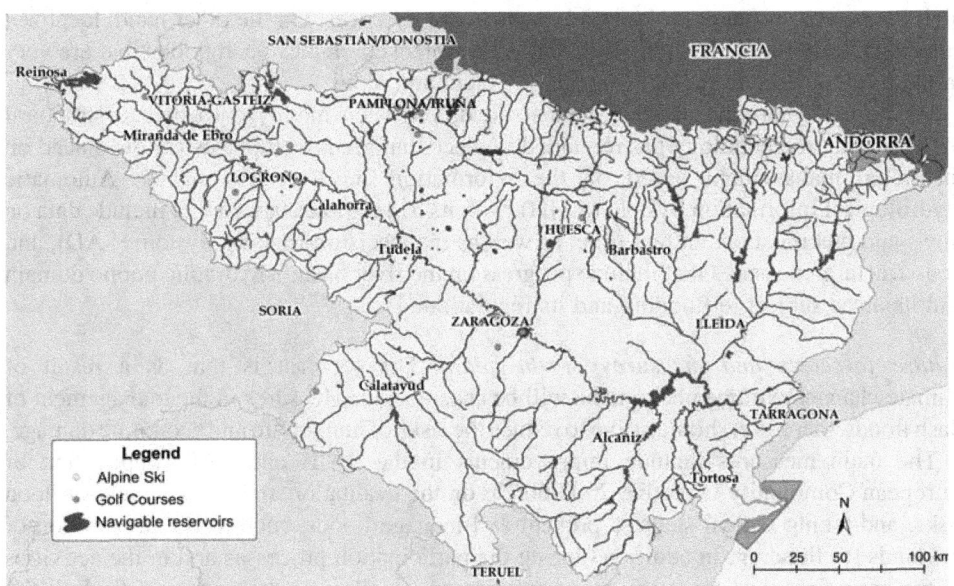

Figure 11. Recreational activities in the Ebro River Basin.

Other uses. Other uses include fish-farming, the management of arid areas and forestry.

Aquaculture. The Ebro River Basin has 51 inland aquaculture farms mostly for the production of rainbow trout for human consumption. The Ebro Delta has 13 fish farms, most for shellfish. The basin also has 20 aquaculture farms which are not used or are abandoned, as well as 10 that are to be developed. Fish-farming requirements for water come to about 1,000 hm^3/year.

Aggregates extraction. The extraction of aggregates authorized by the Ebro Confederation as an average over 12 years is 698,000 m^3/year, though the amount varies widely from year to year. The trend shows a reduction in extraction for environmental reasons.

Forestry within the hydraulic public domain. This activity is not very well developed in the Ebro River Basin and it has tended to decline.

Flood Management

Within the Ebro River Basin floods are naturally due to two types of meteorological situations. On the one hand, persistent rain over wide areas is made worse by unusually

Table 11. Recreational use.

Use	Number of users
Skiing	3,000,000
Boating and sailing	1,700,000
Adventure sports	100,000
Angling	125,000

high temperatures that provoke rapid thaws in the Pyrenees. On the other hand, localized convective rainfall of short duration and major intensity creates flash floods that are very limited in territorial terms but extremely violent and rapid.

For several years the defence against floods has been mainly entrusted to structural barriers. That methodology has remained in place, but greater emphasis is now placed on integrated management based on the information and forecasts of the Automatic Hydrological Information System (SAIH), with its 514 remote stations that include data on flows and precipitations in real time, as well as the Decision-Making System (SAD), and non-structural measures that include progress on the river basin's hydraulic public domain and its areas subject to flooding and its regulation.

Future forecasts and measures of the plan. The forecast is that as a result of climate change, extreme phenomena will be accentuated. Measures in the management of flash floods, therefore, should strive to reduce the risks of human life and economic damage.

The main measures include improvements in the SAIH and SAD, application of European Community Directive 2007/60/CE on the evaluation and management of flood risks, and taking radical steps to prevent and manage floods, such as the preparation of lowlands for flooding. In addition, during the participation processes and in the activities of the basin organization, many requests are made to alleviate the impact of flash floods through management measures.

Management of floods is expected to account for about 15% of the investment in the River Basin Plan.

Management of Droughts

At an overall level, the Ebro Basin has no problems with water supply in the main inhabited areas. All are connected to major irrigation systems and drought simply implies a switch in priorities. However, shortages have occurred, especially in the transfer of the Greater Bilbao region. Small villages are also vulnerable at times to drought.

Droughts, however, can mean severe economic problems. The economic analyses of the 2004–05 drought estimated the losses to farmers of €540 million, and the total loss of production of hydroelectricity estimated at €98 million.

Forecasts for the future and measures in the Plan. The Special Action Plan for Situations of Alert and Contingency for Drought in the Ebro Basin forms part of the Hydrological Plan of the Ebro River Basin. It establishes the thresholds and criteria in order for Exploitation Boards to classify droughts as Normal, Pre-Alert, Alert and Emergency, with a series of measures to be taken for each case in each zone. In addition, many measures of a structural nature ensure supplies to urban areas as well as improving guarantees for other uses, and avoiding the environmental risks posed by droughts.

Investments

The Investment Scenario

Given that the plan is currently being drawn up, the amounts of investment that are necessary have yet to be established in precise terms. However, from the plans, programmes and measures now under way, and those that can be foreseen, a sketch can be provided of the needs of the plan. The plans and programmes included in the Hydrological Plan are, among others, the following:

Fulfilment of environmental objectives:

- National Plan for Quality and Sanitation and Wastewater Planning.
- Measures from the National Strategy for Restoration of Rivers.
- Action plan for checks on water supply and connections.
- Action plan for zero tolerance on effluents.
- Plan for the reuse of urban effluents and irrigation returns.
- Plan for agro-environmental measures in irrigation systems.
- Plan for the protection of groundwater.
- Plan for the modernization of irrigation through environmental priorities.
- Introduction of ecological flow systems in priority stretches.
- Programme for reviews of concessions.
- Improvement and development for control networks.
- Integral environmental projects in order to achieve good conditions in the stretches foreseen by the plan.
- Plan for an improvement in the quality of water yet to reach potable standards.
- Integral plan for the protection of the Ebro Delta.
- Action plan for non-native species.
- Treatment of contaminated sediments.
- Plan for environmental and voluntary education.
- Plan for long-term continuity of rivers.
- Hydrological forestry measures.
- Adaptation plan for climate change.
- Programme for studies and projects.

Attention to demands:

- Plans for consolidation and extension of irrigation systems.
- Interconnectivity of exploitation systems.
- Construction of infrastructure for regulation purposes.
- New ways of generating energy in the existing infrastructure.
- Management of groundwater.
- Territorial development of areas affected by waterworks.
- Plan for the development of recreational uses.
- Plan to establish the value of water heritage.
- Plan for the conservation, maintenance and security of waterworks.
- Plan for the modernization and development of pipelines and related infrastructure.
- Programme for studies and projects.

Extreme weather phenomena (floods, given that measures for the prevention of drought can be considered within the previous section):

- LINDE programme.
- Maintenance and improvements to the SAIH–SAD system.
- Cartography of zones subject to flooding.
- Programme to clean up rivers.

- Defence mechanisms in urban stretches and at critical points.
- Measures for the recovery of river space.
- Management of flash-floods in the middle Ebro Valley.
- Infrastructure plan to regulate floods.
- Plan for flash-flood abatement infrastructure.

The provisions of the Hydrological Plan foresee annual investment of some €1.719 billion, which will be funded by the public sector and the rest by the private sector.

Public investment in this Hydrological Plan supposes a 14% increase on the amount that appears on the government's 2009 Budget and in the budgets of the Autonomous Communities and local administrations.

Governance in the River Basin Plan

Our pledge to the European Union, as expressed in the Framework Directive, is to establish budgetary and timeframe commitments among the administrations in a coordinated fashion so as to ensure the River Basin Plan's investments in the environment. In that way, joint responsibility will be truly exercised in order to achieve the aims of the plan.

Competent Administrations

Only a very small part of the territories are shared with France and Andorra, both within the Ebro as within the French river basins of Adur-Garona and Ródano-Mediterranean. International coordination for the development of plans and the application of measures are achieved through a series of coordination meetings held in the framework of the Toulouse Treaty and several already existing international commissions.

In the Spanish part of the Ebro River Basin, areas of responsibility concur with those of the national authorities, the Autonomous Communities and local corporations (Figure 12 and Table 12).

The complex institutional structure for water management in the Ebro River Basin requires the establishment of a permanent forum. There, measures can be adopted democratically and in a participative fashion in terms of responsibilities to the European Union on the commitments acquired to achieve the good condition of water bodies.

On the other hand, decentralized water management by communities of users, and the presence of society in general in decision-making, is vital to the existence of the permanent forum.

The Ebro River Basin Confederation, faced by the river basin's new Hydrological Plan, emerges as a major element of organizational heritage. If this confederated organ did not exist, it would be necessary to create it, with all the consequent sacrifice in political and administrative terms.

Table 12. Institutions involved in the water management of the Ebro River Basin.

Administration	Head offices
Central government	8
Nine Autonomous Communities	27
Provincial governments	17 provinces
Local councils	1,724 towns and villages

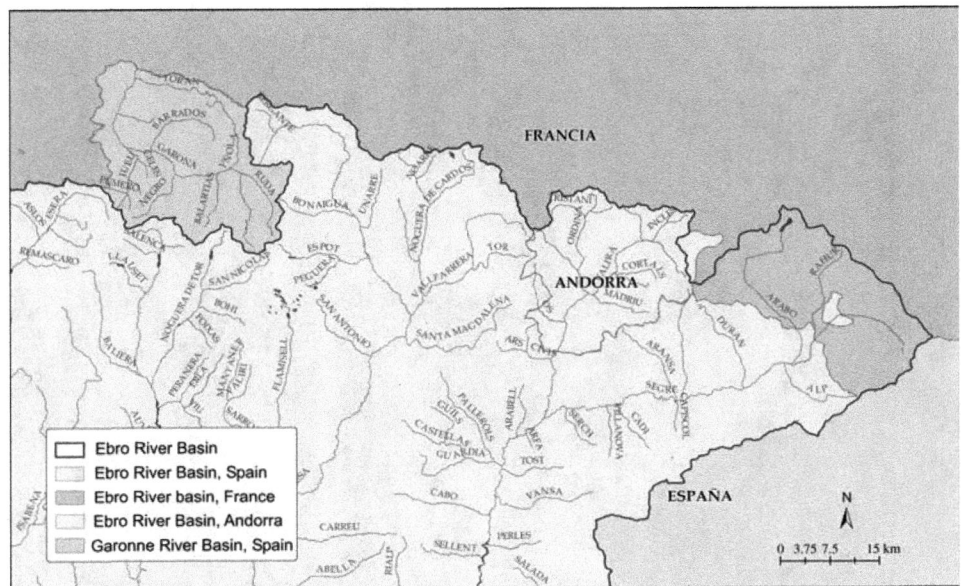

Figure 12. International sections of the Ebro River Basin.

The new Confederation has a very complex task to develop. A job that makes it necessary to achieve the potential of the major challenges set by the Water Framework Directive and the decentralized nature of the administration of the state that we Spanish have adopted.

Note

1. Royal Legislative Decree 1/2001, 20 July 2001.

Water Quality in Zaragoza

JAVIER CELMA
Environment and Sustainability Agency, Zaragoza City Council, Zaragoza, Spain

ABSTRACT   *The severe droughts of the 1990s made clear that the Strategic Plan of Zaragoza and the Action Plan of Local Agenda 21 in terms of water management were not appropriate to satisfy the development needs of its economy and the future demands of a growing population. In response, the city redefined its water management model, from one of continuous exploitation of resources to the search for solutions to reduce consumption. The city's efforts included a comprehensive programme of stakeholder involvement, rehabilitation of drinking water treatment and distribution infrastructure, improvement of water quality, and reform of the billing system. After 12 years, the city has managed to deduce its total consumption by nearly 30% and improve the quality of its water very significantly.*

Introduction

Zaragoza lies in the Ebro Depression, some 130 km from the most southerly glaciers in Europe and in the middle of a semi-desert, the most northerly north of Africa. The coasts of the Mediterranean to the east and the Cantabrian to the west are each about 300 km from the city.

With an arid continental climate, the city has precipitations of less than 350 mm. Seasonal variations are very marked between summer and winter (maximum temperatures of 43°C contrast with minimums of $-10°C$). Winds are frequent and evapotranspiration is high. The vegetation is determined by droughts in summer and low temperatures in winter.

The population of Zaragoza has grown steadily in recent years to about 700,000 inhabitants, rising to 850,000 in the outlying areas. The city is one of the most important economic centres of Spain. Prominent sectors include the car industry, paper and pulp, and logistics. The city is also a major commercial centre for all of north-eastern Spain.

The need for water, in terms of both quantity and quality, has been a historical concern to the city in its everyday business. Debates on the sources of supply, their quality and the necessary infrastructure have created a breeding ground of major thinkers on water issues (Ramon de Pignatelli, Joaquín Costa, Lorenzo Pardo and others). Pardo was the creator of Spain's Hydrographic Confederations, of which the Ebro was the first, at the beginning of last century. The formula of integrated basin management that was then established is now the basis of the European Union's Water Framework Directive.

Sources of Water Supply to Zaragoza

The central element of Zaragoza's water supply is the Imperial Canal of Aragón, constructed in 1784. The canal's waters come from the Ebro River, at what is known as the Pignatelli connection, some 100 km upstream from the city. The system is complemented by the La Almozara pumping station in Zaragoza. La Almozara collects water directly from the Ebro to the potabilization station when the canal has supply problems.

This infrastructure allowed Zaragoza to solve its problems of water quantity and quality throughout many episodes of plague and cholera, as well as supply difficulties. The irrigation zone was amplified to what now is the city's market garden. Water has also been a means of transport and the driving force of industrial development in Zaragoza and its surroundings.

The quality of the water supply is not of the best due to the high dissolved saline content that forms part of the Ebro's natural make-up. The level of salinity varies along the course of the year; it is highest in the dry summer weather when the flow of the river is reduced, as seen in Figure 1.

The major increase in irrigation areas upstream of Zaragoza is resulting in an increase in nonpoint pollution, related to the presence of nitrates. The levels do not, however, pose a danger to the consequent potabilization of the water.

The loss of the Ebro's volume flow as it passes through Zaragoza is also significant. A historical average of 7,304.7 hm^3 that was registered since 1913 has dropped to 5,730 hm^3 in the last 20 years and in the last four years to 4,907 hm^3. This reduction in volume has mainly been caused by the increase in irrigation, the reduction of hillside erosion as a result of reforestation, and an increase in evapotranspiration.

Problems of quality in the water supply, uncertainty of climate change and the need to take adaptive measures against climate change as well as to ensure enough water supply resulted in a twin-pronged strategy in Zaragoza:

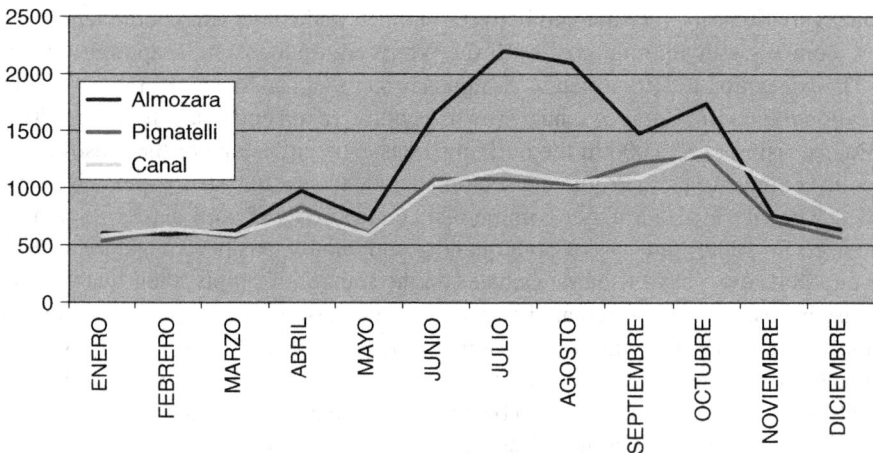

Figure 1. Changes in conductivity in the Ebro River throughout the year. *Source*: Ebro River Hydrographic Confederation.

- Water was to be brought from the Yesa reservoir in the Pyrenees.
- The city's waterworks had to be adapted to meet the new supply requirements.

Water in Zaragoza: A Strategy

In 1994, work was launched on the Strategic Plan for Zaragoza and its Area of Influence, taking into consideration the international background of the 1992 Rio Summit and the Fifth European Union Environment Programme on sustainable development.

Throughout the whole process, protection of the environment and the sustainable management of resources have been important elements of the work on the strategic plan. As a result of these principles, measures on water policies for Zaragoza have formed part of the framework of the local version of Agenda 21. The basic aim was to reduce the consumption of water without affecting the supply of the basic service to the citizens and to improve the quality of the supply. To this end, the aim set for 2010 was the reduction of the city's total water consumption to 65 hm^3. In order to achieve these aims, several measures were taken. The following are the main ones:

- Water to be brought from the Pyrenees.
- Quality to be improved to meet the use for which water is required.
- Leaks and other losses to be reduced in storage and distribution networks.
- Promotion by public bodies and centres of the water-saving technologies.
- Criteria of environmental accountability to be introduced for all municipal services.
- Standards to be set for efficiency and sustainable use of water.
- Adjustments in the policy of tariffs.
- Information and publicity campaigns.
- Programmes for cooperation with non-governmental organizations.
- International projection.

These measures form part of the Plan for the Improvement of the Management and Quality of Water Supply, the new policy on tariffs, as well as citizens' participation forums. In addition, they include the Municipal Regulations on Efficiency and Integral Management of Water, campaigns to raise awareness, the European LIFE and SWITCH programmes, and cooperation with non-governmental organizations.

Plan to Improve the Management and Quality of the Water Supply: Diagnoses and Objectives

In 2001 Zaragoza City Council drew up a Plan for the Improvement of Management and Quality of Water with the aim of updating the municipal facilities and networks. Full use was to be made of the quality of water supplied by the new system from the Yesa reservoir in the Pyrenees.

In 1979 consumption peaked at 106.39 hm^3, or 507 l/day/person. The priority in those years was the need to treat wastewater, but the first measures were also taken to rationalize consumption. These included suppression of the discharge chambers of the sewers, extension of consumption controls and use of phreatic flows in the irrigation of the greenbelt. At the same time there was an important increase in the price of water, which was partly the result of the costs of its treatment. All this contributed to a very notable reduction in the level of consumption of water in the city; by 2001 it had dropped to 79.39 hm^3 or 349 l/person/day.

This Plan derived from an analysis of the conditions in which water was supplied. The conclusion was that much of the facilities and the distribution network were in very poor condition because of the lack of modernization and renovation. In addition, consumption was not sufficiently metered or otherwise checked, and even less was billed. All of this together meant that a considerable percentage of the total volume of water was unaccounted for.

On the basis of this diagnosis, a series of measures stretching over the seven years between 2002 and 2009 came to a cost of €84.59 million. They comprised seven major areas. The principal measures that formed part of the Plan were as follows:

- Improvement in water quality. The main measure consisted of the construction of a group of intermediate facilities to strengthen chlorination in order to achieve the most uniform chlorine level possible in the network.
- Improvement in quality control. Equipment was acquired to improve the control of water quality for the treatment plant itself as well as for the distribution network.
- Improvement of facilities. Measures of very varying types were taken to improve the treatment plant as well as the storage and pumping facilities. The main project consists in the complete remodelling of the deposits that form the main reserves of drinking water of the city, and which date from the beginning of the 20th century. The perimeter walls and the control mechanisms were also renovated.
- Renovation of the distribution network. This is the most important item in terms of investment, accounting for 65% of the total. This was a direct consequence of the European Union's demand to reduce the level of losses very significantly. The aim, during the plan, was to increase greatly the speed of renovation and rehabilitation of the pipes of the distribution network until it increased to more than 30 km a year (Table 1).
- Management of consumption. Measures were stepped up to renew the municipal stock of water meters. Meters were also installed on municipal buildings and green spaces that lacked them. The aim was to achieve effective control of total consumption.

Table 1. Reduction in the number of pipeline leaks.

Year	Own responsibility	Responsibility of third parties	Total
1996	525	139	664
1997	600	86	686
1998	614	123	737
1999	597	149	746
2000	448	190	638
2001	502	202	704
2002	487	138	625
2003	440	89	529
2004	424	96	520
2005	427	95	522
2006	305	91	396
2007	207	112	382
2008	262	69	331

Source: Zaragoza City Council. Department of the Integrated Water Cycle.

- Modification of private facilities. Several improvements were made both to the supply points and the domestic deposits. These include the removal of water tanks and improvement of main connections.
- Use of technology. Measures of various types included the promotion of agreements to carry out research and raise awareness. Other measures included the approval of materials and products, standardization of indicators, adjustment of quality standards and improvements in services to users.

As the most visible result of the achievement of the plan, a very significant outcome was obtained in terms of the city's total consumption. When the plan was devised, total consumption was 80 hm^3/year; by the end of the plan it was 65 hm^3/year.

Water Quality

The main problems in the quality of the water supply were its high conductivity and organic material which treatment results in the formation of trihalomethanes.

Evolution of the Conductivity in the River Ebro

Conductivity is one of the most critical parameters in terms of water quality. This parameter is directly related to the quantity of ions dissolved in the water. The existing ions in major concentrations are responsible for the high degree of conductivity in the waters: sulphate, chlorine, calcium, magnesium and sodium.

High conductivity renders the water suitable for some uses, and it also sometimes exceeds the sulphate content, one of the parameters of quality. In industrial terms the water needs to be softened and demineralized in order to reduce deposits in circuits and the equipment for industrial processes.

Table 2 presents the parameters as they were in 2008, according to the tests carried out by the Ebro Hydrographic Confederation at Almozara, in the area that flows into the city of Zaragoza when the Imperial Canal of Aragón is not available or in maintenance (one month a year). The values associated with conductivity vary from one year to another, in accordance with the rainy season. In addition, they also vary between each hydrological year as a function of rainfall variability and the flows resulting from both seasonal thaws and discharges from reservoirs.

Table 3 compares the volumes of water (hm^3/year) and conductivity in 2002 and 2008 for the Ebro as it passes through Zaragoza. An explanation for the difference in conductivity between the two years is clearly related to the volume of water in each hydrological year. In 2002 the volume was the lowest since 1912, the first year that records were established by the Ebro Hydrographic Confederation.

Water Supply from Yesa Reservoir

In 2008 the pipes from the main supply in the Sora–Loteta and Loteta–Zaragoza stretches came on-stream from the Yesa reservoir, including the Casablanca deposits in Zaragoza (Figure 2).

In the summer of 2009, an agreement was established for the provisional supply of water from the Yesa–Bardenas system to Zaragoza. The accord allows for the availability

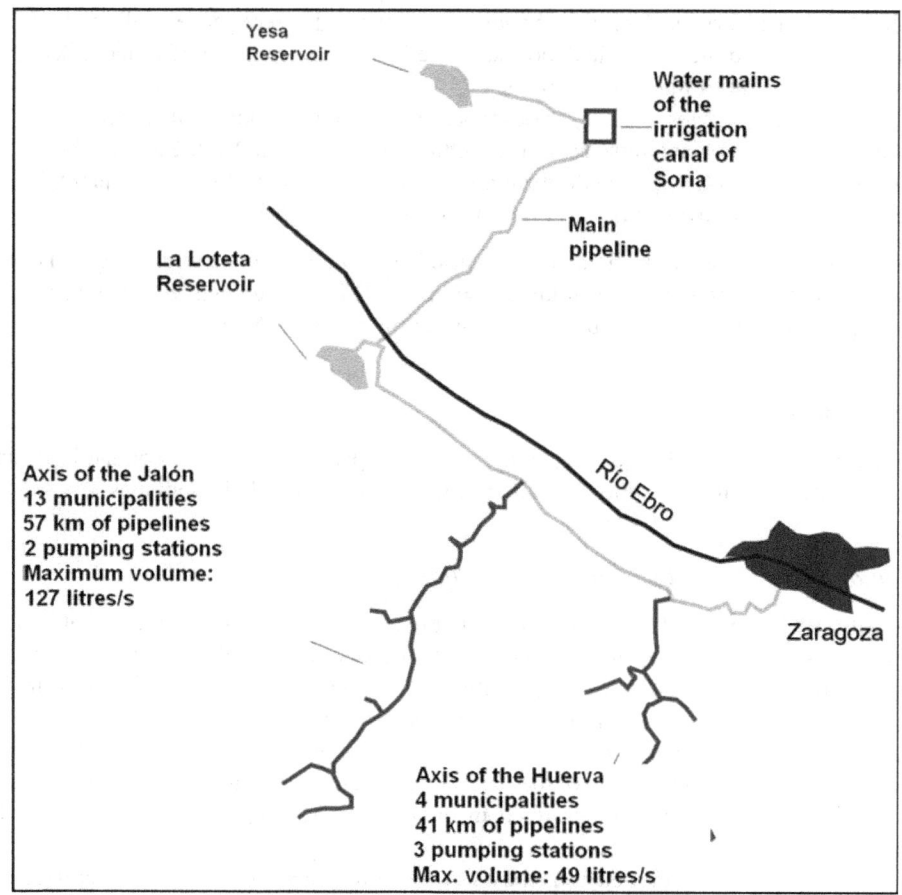

Figure 2. Plan of the main water supply to Zaragoza and other locations. *Source*: Agency for Environment and Sustainability, Zaragoza.

Table 2. Analysis of water quality supplied at Almozara.

Almozara	Conductivity	Sulphate	Chloride	Calcium	Sodium	Magnesium
January 2008	932		135.2			
February 2008	1,291		203.4			
March 2008	1,580	291.4	251.6	149.3	184.6	35.4
April 2008	447	60.7	41.4	80.2	28.7	8
May 2008	1,161		184.9			
June 2008	619		88.6			
July 2008	1,370	257.1	210.8	128.1	119.6	21.8
August 2008	1,570					
September 2008	1,546		228.8			
October 2008	1,271	194.6	176.8	114.6	120.4	22.1
November 2008	1,143		152.1			
December 2008	948		111.2			

Source: Agency for the Environment and Sustainability, based on data from the Ebro Hydrographic Confederation.

Table 3. Comparison of flow volumes and conductivity at Almozara.

	2002	2008
Volume flow (hm^3/year)[a]	2,282	5,625
Mean conductivity (μs/cm)	1,597	1,156

Note: [a]October 2001 to September 2002; and October 2007 to September 2008.
Source: Agency for Environment and Sustainability, based on data from the Ebro Hydrographic Confederation.

of 1 m^3/s, which amounts to 50% of the city's consumption, and allows improvement in the quality of water from the intake of the Imperial Canal. This is a limited and partial supply because it remains subject to the availability of the water in Yesa until the necessary works are completed.

Quality of Water from the Imperial Canal of Aragón and Yesa Reservoir

Table 4 presents the average values of several parameters for the Imperial Canal of Aragón and Yesa reservoir. In general terms the values of all parameters of the water from Yesa reservoir are much lower than those of the Imperial Canal of Aragón. In addition, the

Table 4. Comparison of the quality of the water supplied by the Canal and Yesa.

Parameter	Units	Yesa reservoir	Imperial Canal of Aragón
Conductivity	μs/cm	276	950
pH		8.29	8.1
Alkalinity	Mg/l CO$_3$Ca	156.2	181
Harness	°HF	16.8	31.7
Sodium	mg/l	8.365	95.1
Potassium	mg/l	0.912	2.99
Calcium	mg/l	43.55	94
Magnesium	mg/l	8.21	20
Fluorides	mg/l	0.183	0.204
Chlorides	mg/l	9.27	131
Nitrates	mg/l	4.35	11.6
Sulphates	mg/l	19.37	145.2
Bromide	μg/l	12.13	102.6
Total organic carbon	mg/l	1.38	3.83
Oxidizability	mg/l oxygen	2	4.1
Boron	mg/l	0.024	0.04
Manganese	μg/l	0.71	17.6
Iron	μg/l	70	95
Nickel	μg/l	0.2	1.1
Copper	μg/l	0	1.2
Arsenic	μg/l	0.1	0.1
Selenium	μg/l	0.1	0.1
Cadmium	μg/l	0	0
Antimony	μg/l	0	1.5
Mercury	μg/l	0	0.001
Lead	μg/l	0.1	1.78
Uranium	μg/l	0.1	1.6

Source: Zaragoza City Council, Integrated Water Cycle Office.

fluctuations in quality of the canal are much greater, above all in terms of parameters such as turbidity, which is not included in the analyses.

Water from Yesa reservoir into Zaragoza's network implies an improvement in the quality of supply as a result of a lower content in salts and micro-pollutants. Table 5 shows the most important parameters. A visible improvement in water quality can be noted in the distribution network, if one compares the average data from 2007 with that of 2009.

Evolution of the Sources of Water Supply to Zaragoza

Historically, the supply of water has come from the Imperial Canal of Aragón, using the direct catchment of the Ebro when the canal is undergoing maintenance. Table 6 shows the sources of the water supply to Zaragoza in 2008 and 2009.

The source of the water supply is changing radically. At the moment, the Yesa reservoir provides about 50%, but is expected to reach 100% by the end of 2010.

Improvements in the Quality of the Water Supplied

The quality of the water supply is controlled by Zaragoza City Council's Municipal Institute of Public Health. The institute meets the RD 140/2003 standard that sets the sanitary standards for the water quality of human consumption. Under this standard, the municipalities have to ensure that water supplied at each point of delivery to the consumer is apt for human consumption (Table 7).

The critical parameters in the water supplied are the sulphate ions and the trihalomethanes. The content of sulphates was duly exceeded as a result of the continuous and natural dissolution of the chalk. Particularly high levels are registered when the volumes of water are very low.

Table 5. Improvement in water quality from the Yesa supply.

Parameter	Units	Imperial Canal of Aragón	Yesa	Reduction (%)
Conductivity	μs/cm	950	276	71
Hardness	°HF	31.7	16.8	47
Chlorides	mg/l	131	9.27	93
Nitrates	mg/l	11.6	4.35	63
Sulphates	mg/l	145.2	19.37	87
Total organic carbon	mg/l	3.83	1.38	64

Source: Zaragoza City Council Integrated Water Cycle Office.

Table 6. Sources of water supply in Zaragoza.

| Source | 2008 | | 2009 | |
	m^3	%	m^3	%
Imperial Canal	54,723,727	89.58	44,611,469	74.47
Ebro River	4,873,883	7.98	1,634,125	2.73
Yesa Reservoir	1,491,191	2.44	13,657,050	22.80
Total	61,088,801	100.00	59,902,644	100.00

Source: Zaragoza City Council Integrated Water Cycle Office.

Table 7. Improvements in water supply to Zaragoza.

	Units	2007	2009	Reduction (%)	Normal limit
Trihalomethanes	μg/l	96.2	43.6	55	100
Nitrate	mg/l	16.2	7.0	56	50
COT	mg/l	2.03	1.07	47	–
Chloride	mg/l	182.5	118.2	35	250
Conductivity	mS/cm	1,185.8	755.1	36	2,500
Sodium	mg/l	129.7	81.2	37	200
Sulphate	mg/l	195.5	106.8	55	250
Hardness	mg/l	371.3	236.6	36	–

Source: Zaragoza City Council.

The trihalomethanes are not a natural component of water. Rather, they are formed in the process of adding chlorine to make water potable. Their concentration increases with the existence of certain organic compounds, mainly of natural origin, and when additional chlorine is used to treat water of low quality. The permissible value of trihalomethanes established by the standards was reduced from 150 to 100 μg/l in 2009. However, the level was frequently surpassed when chlorine was added to the water from the canal.

In terms of nitrates, although we may be well away from levels of risk, its content can be considered to be high, mainly as a result of nonpoint pollution from farming activities.

All three parameters, sulphate, trihalomethanes and nitrates, will be sharply reduced by the supply from Yesa reservoir, especially the content of trihalomethanes. The improvement in the quality of water, together with new chlorination stations at various points in the network, will mean lower doses of chlorine and therefore there will be less of a risk of the formation of trihalomethanes.

Tariffs

The European Union's Water Framework Directive establishes the principle of cost recovery of services related to water supply of good quality, treated with the best technology available.

Principles

In Zaragoza tariffs are applied in accordance with the following principles:

- Sufficiency. The expected income will have to recover the inherent costs of the services of water supply, sewerage, purification, control and management in each fiscal period. Any new expense incurred, or any savings, in any particular year must be included in the tariffs of the following year.
- Equity. The prices applied must be equal to the benefits conferred. Similarly, prices have to be different when the benefit obtained is not the same. To this end, the price will be in accordance with the use of the supply and the quantity of water consumed, but measures may be taken to avoid the fact that the build-up of consumption in a specific point of supply reflects the resulting price. Finally, universal access has to be ensured for basic consumption at affordable prices where tariffs have to take into consideration the economic capacity of the users.

- Effectiveness. Pricing has encouraged rational behaviour by consumers when savings have been encouraged, and excessive consumption has been penalized. Where tariffs have two elements, one portion includes the fixed element (costs plus the rate), while the variable proportion depends on consumption. This ensures progression in terms of billing of the amount consumed.
- Transparency and economy. The tariffs are as straightforward and clear as possible. The costs of their application, for both the City Council and the users, must be as low as possible. It is also necessary to show a profit in relation to the income obtained as well as the improvement achieved in terms of efficiency and equity. Users need to have all the information available in order to know and control their consumption and the costs it involves. In addition, mechanisms will have to be established as to make it easy for users to take part in decision-making of tariffs and also on revising them.
- Sustainability: The tariffs for the services of supply, sewerage and wastewater treatment have to apply the principle of "the polluter pays". This means that pollution can be a consequence of both excessive consumption and of a precise amount of contaminants in the effluent. In all cases priority must be given to positive incentives for users that take preventive action on their consumption and waste of water.

Tariff Structure

In Zaragoza most houses are part of a system of 320,000 water meters, representing a ratio of 2.03 inhabitants per meter. Similarly, consumers without a meter (with supply contracts) are penalized in the form of a lump sum. These represent a total of about 600 accounts.

The structure of the municipal tariffs has the effect of redistributing charges to accommodate special cases such as the following:

- Citizens of lower income (pensioners, the unemployed, very large families) are favoured.
- In order to promote the efficient use of water, 10% reductions are applied to those who have reduced their consumption from the previous year (more than 33,000 of the 320,000 existing accounts in the last year have made savings of 1 hm^3 in water consumption).
- Families are favoured. The first 6 m^3/month have a reduced rate of more than 50%. This tariff applies to some 30% of consumption.
- Penalties are imposed for excessive consumption. The price per cubic metre of the third band is almost five times as much as in the first.
- The reduction of pollution in wastewater from industrial activities is encouraged. The tariff for sanitation is proportional to the pollution in effluents.

Domestic Consumption and Price

Follow-up of consumption is achieved by means of various codes on the meters in accordance with the different types of consumption: domestic, commercial, industrial, irrigation, etc. (Table 8).

In general, it can be said that there is one meter for every two persons, and only one user in 500 has no meter because of technical or economic difficulties of his or her facilities.

Table 8. Domestic consumption and tariffs in 2008.

Type of meter	Number of meters	Total (%)	Consumption per account (m^3/year)
Domestic	295,080	91.09	91.1
Commercial	25,664	7.92	332.3
Industrial consumption and similar uses	1,357	0.42	1345
Irrigation, refrigeration, fire prevention	1,164	0.36	237
None (lump-sum charge)	663	0.20	219.5
Total number of accounts	323,928		

Source: Zaragoza City Council, Finance Department.

The percentages of meters are very similar to those in the year 2000, when there were 91.5% of domestic users (91.09% at present), 8.2% were commercial and industrial users (now 8.34%), and 0.3% were public uses (now 0.36%).

In 2000, there were 291,137 accounts, representing an increase of 11%.

Table 9 compares the percentage of consumption and accounts among the main users, with data available in 2008.

Lump-sum Accounts

At the end of 2001 there were 1,405 accounts included in the lump-sum arrangements in Zaragoza. Of these, 1,204 were domestic and 201 were commercial. By the end of 2008 there were 662 lump-sum accounts still remaining, of which 575 were domestic and 87 commercial.

Price of Water for Users

The price of water includes a fixed term for the provision of the service, and a variable cost in accordance with the amount registered on the meter. Two different items, supply and sanitation, are included. This is because, in certain cases, the water supplied is not municipal; in other cases, wastewater is not removed by the municipal collectors. Prices vary depending on whether consumption is domestic or non-domestic (Tables 10 and 11).

The average consumption of a domestic account in 2008 was 91 m^3/year. Table 12 shows the total price of water for this type of user, with 2010 tariffs. The total charge for a domestic account for water supply and sanitation would be €95.69/year, with a cost of €1.05/m^3 or €0.26/day.

Table 9. Percentage of consumption and accounts in 2008.

	Consumption (m^3)	Consumption by sector (%)	Accounts per sector (%)
Domestic consumption (m^3/year)	27,025,136	71.77	91.09
Commercial consumption (m^3/year)	8,529,657	22.65	7.92
Industrial consumption (m^3/year)	1,825,498	4.85	0.42
Others (m^3)	276,595	0.73	0.56
Total		100	

Source: Zaragoza City Council, Finance Department.

Table 10. Prices for domestic consumption.

	Supply	Sanitation	Overall
Fixed term	€0.077/day	€0.05/day	€0.127/day
Block 1 (0–200 l/home/day)	€0.210/m^3	€0.215/m^3	€0.425/m^3
Block 2 (>200–616 l/home/day)	€0.503/m^3	€0.515/m^3	€1.018/m^3
Block 3 (>616 l/home/day)	€1.258/m^3	€1.288/m^3	€2.546/m^3

Source: Zaragoza City Council, Finance Department.

Table 11. Prices for non-domestic consumption.

	Supply	Sanitation	Overall
Fixed term	Varies with calibration of the meter		
Block 1 (0–200 l/account/day)	€0.503/m^3	€0.515/m^3	€1.018/m^3
Block 2 (>200–616 l/account/day)	€0.503/m^3	€0.515/m^3	€1.018/m^3
Block 3 (>616 l/account/day)	€1.383/m^3	€1.416/m^3	€2.799/m^3

Source: Zaragoza City Council. Finance Department.

Table 12 Annual costs of domestic accounts.

	Overall	Annual cost of 91 m^3/year
Fixed term	€0.127/day	€46.35 (no limit)
Block 1 (0–73 m^3/year)	€0.425/m^3	€31.02 (for the 73 m^3)
Block 2 (>73–225 m^3/year)	€1.018/m^3	€18.32 (for 18 m^3 consumed in the block)
Block 3 (>225 m^3/year)	€2.546/m^3	€0 (all consumption is charged)

Source: Zaragoza City Council, Finance Department.

Table 13. Annual cost of industrial accounts.

	Overall	Annual cost of 1,345 m^3/year
Fixed term (40 mm meter)	€2.4716/day	€902.13 (whatever the consumption)
Block 1 (0–73 m^3/year)	€1.018/m^3	€74.31 (for the first 73 m^3)
Block 2 (>73–225 m^3/year)	€1.018/m^3	€154.74 (for 152 m^3 consumed within the block)
Block 3 (>225 m^3/year)	€2.799/m^3	€3,134.88 (for 1120 m^3 consumed within the block)

Source: Zaragoza City Council, Finance Department.

The average cost of an industrial account in 2008 was 1,345 m^3/year. The total price of water for this type of user, assuming the meter has a calibration of 40 mm, and with 2010 tariffs, would be as shown in Table 13. The total cost of an industrial account for 1,345 m^3/year of water supply and sanitation would be €4,266.06/year, with a cost of €1.05/m^3 or €11.69/day.

Income Structure (2008 Figures)

Income is produced for the supply and sanitation water-related services for both domestic and non-domestic uses. The structure for 2008 is shown in Table 14.

Table 14. Income structure for 2008 (€, thousands).

	Supply	Sanitation	Total
Total income	26,232	20,956	47,188
Income from domestic consumption	17,039	13,432	30,471
Income from non-domestic consumption	9,193	7,524	16,717

Source: Zaragoza City Council, Finance Department.

Domestic Consumption

The income for domestic consumption in Zaragoza is presented in Table 15.

Table 16 presents forecasts of the real costs and income for Zaragoza City Council.

Results and New Challenges

What is proposed here is the management of water resources, taking into consideration that there is a limit for them. We also work towards the concept of more for less, which is to achieve more with fewer resources.

Table 15. Income from domestic consumption.

	Fixed quota (€,thousands)	%	Variable quota (€,thousands)	%
Domestic supply	9,299	55	7,740	45
Domestic sanitation	6,075	45	7,357	55
Non-domestic supply	1,816	20	7,376	80
Non-domestic sanitation	1,165	15	6,359	85
Supply income	11,515	43	15,116	57
Sanitation income	7,240	35	13,716	65
Total	18,355	39	28,832	61

Source: Zaragoza City Council, Finance Department.

Table 16. Costs and income forecast for 2007–2010.

Costs and income (€)				2006	2007	2008	2009	2010
Direct costs				43.06	44.92	47.57	50.52	49.47
	Personal costs			11.27	11.89	12.41	13.66	13.99
	Operational costs			31.79	33.02	35.16	36.86	35.48
		External services		7.94	7.75	8.79	8.73	9.53
		Water treatment		16.68	18.03	18.9	20.49	18.36
		Water cost		2.51	3.46	3.57	3.69	3.7
		Energy cost		1.49	1.12	1.08	1.1	1.1
		Merchandise		2.32	1.77	1.92	1.91	1.8
		Rates		0.7	0.74	0.75	0.8	0.77
		Other costs		0.15	0.16	0.16	0.14	0.22
Indirect costs				4.81	5.34	5.89	6.02	7.55
Amortization				6.12	7.03	7.58	8.63	9.07
Total cost				54	57.29	61.04	65.17	66.09
Income				0	50.8	50.3	53.55	57.48

Source: Zaragoza City Council, Finance Department.

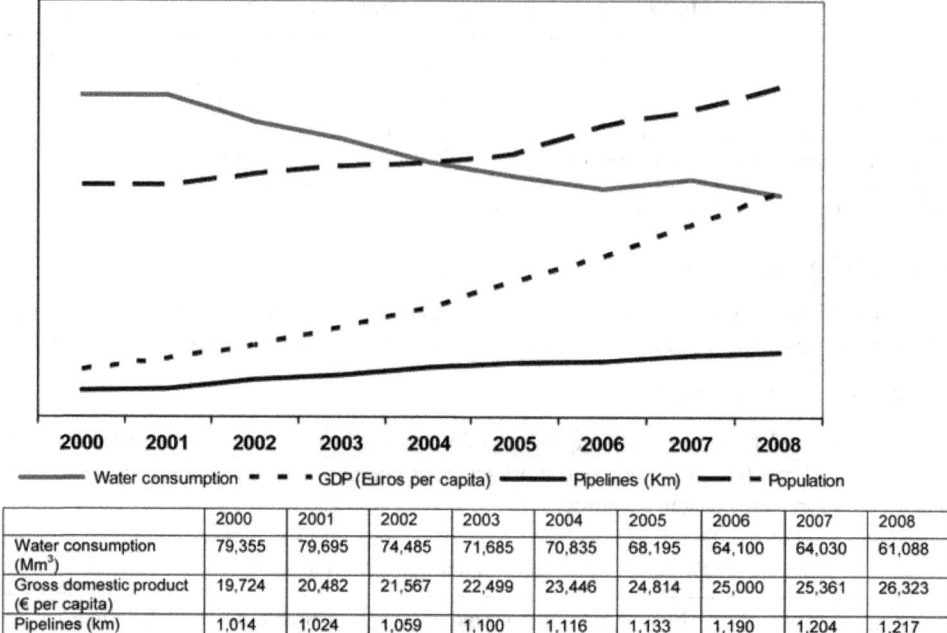

	2000	2001	2002	2003	2004	2005	2006	2007	2008
Water consumption (Mm^3)	79,355	79,695	74,485	71,685	70,835	68,195	64,100	64,030	61,088
Gross domestic product (€ per capita)	19,724	20,482	21,567	22,499	23,446	24,814	25,000	25,361	26,323
Pipelines (km)	1,014	1,024	1,059	1,100	1,116	1,133	1,190	1,204	1,217
Population (n)	613,433	618,432	622,602	626,081	630,942	650,592	659,282	688,412	693,086

Figure 3. Gross domestic product, number of inhabitants, water consumption and growth of Zaragoza. *Source*: Zaragoza City Council, Environment and Sustainability Agency.

Figure 3 shows increases in income (gross domestic product) per capita, the number of inhabitants and the length of water pipes as an indicator of the physical growth of the city. Meanwhile, the overall consumption of water has fallen very significantly in the city.

Currently, Zaragoza has the lowest ratio of domestic consumption per person/day in all of Spain: 106 litres. In 2006, total consumption of 64 hm^3 met the target set for 2010 of a maximum of 65 hm^3. In December 2008, consumption reached 61 hm^3 and the most recent figures reached less than 60 hm^3/year by the end of 2009.

For 2015, we are working on new targets that will set the city's total consumption at 58 hm^3/year and domestic consumption of less than 90 l/inhabitant/day. Figures 4 and 5 show the supply of water as well as the domestic consumption from 2000.

Within the framework of Local Agenda 21, a new plan of action will help to meet the targets set within a framework that will further strengthen savings. This will include a municipal by-law which is to be introduced on savings and efficiency in the use of water, and the development of programmes based on the creation of cluster of companies that use water efficiently.

New Actions

Apart from providing continuity along previous lines, a second phase of the Infrastructure Improvement Plan (2008–2013) includes the investment of €37 million and aims to tackle the following issues:

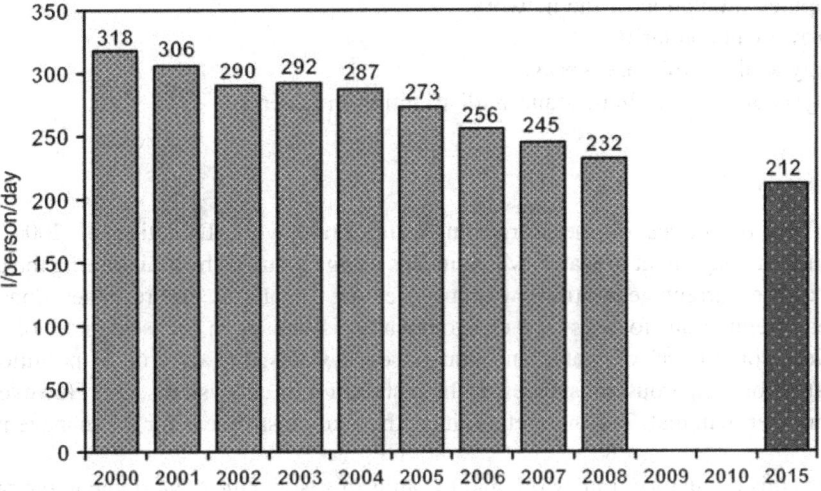

Figure 4. Water supplied to Zaragoza City. *Source*: Zaragoza City Council, Environment and Sustainability Agency.

- Creation of storm tanks.
- Improvements in the sewerage network.
- Improvements in the Almozara treatment plant.
- Networks divided by sector.
- Flood prevention for the Ebro.

On the other hand and, in accordance with the final results of the Switch Programme, progress is to be made on:

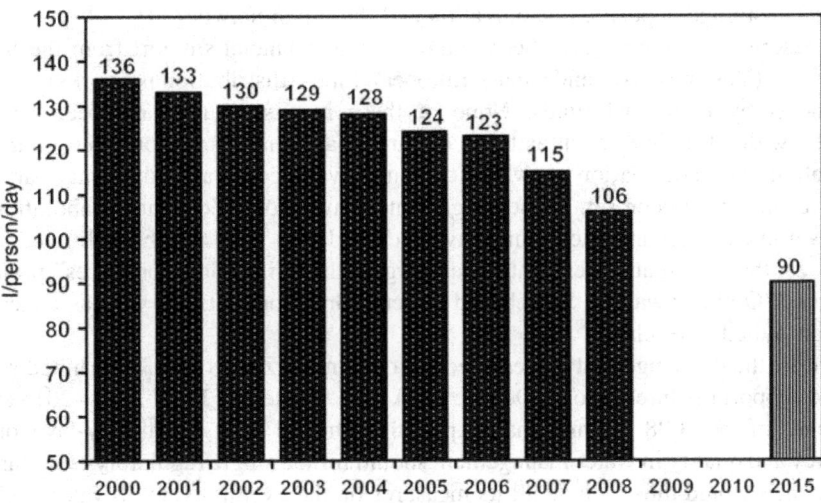

Figure 5. Domestic consumption in Zaragoza. *Source*: Zaragoza City Council, Environment and Sustainability Agency.

- Computer simulation of the network.
- Improvements on tariffs.
- Energy analysis of waterworks.
- Analysis of losses in homes and in distribution in general.

Conclusions

One of the objectives of the European Water Framework Directive of 2000 is the sustainable management of water, which means using resources by maintaining the quality of life of the current generations without affecting negatively future generations. This European regulation focuses on environmental issues as a consequence of "good ecological conditions" of water and aquatic ecosystems, the control of pollution, the elimination of dangerous substances or the restoration of ecosystems, etc. However, it is much broader than that. Taken together, it can be a key instrument for the management of water.

Within this regulatory framework and the aims of sustainable development, the rational use of water must be sought with criteria of efficiency and savings, as well as ensuring supplies to citizens in terms of quantity and quality. This implies reducing the consumption of water, recycling and reusing it as much as possible while polluting it as little as possible as it goes on to the stage of treatment. Then it can be returned in an acceptable condition to its natural state with minimal impact to ecosystems.

With its new regulations, Zaragoza City Council aims to maintain the innovative spirit of its previous by-laws. These set the limits of the quality of effluents and maintain progressive penalties for high consumption as well as pollution of wastewater to the municipal sewerage.

Since the year 2000, a huge amount of work has gone on to improve the overall cycle of water in Zaragoza. This has included looking for the safest sources of water in terms of both quantity and quality, as, for example, the water of the Pyrenees through the Yesa reservoir and on to the canal from Sora to Zaragoza. Improvements have also been achieved in storage deposits, potabilization and the urban network.

All these water projects have been achieved with financial support from the Spanish government (Yesa reservoir and canal), European funds (distribution networks) and major investments by municipal funds. None of these results would have been achieved, however, without policies such as those on tariffs, and mainly without a programme on information and participation on which citizens have become a fundamental part in the objectives that have been set. These programmes have promoted joint collaboration with non-governmental organizations and have included, but have not been limited, to the "Zaragoza, the City that Saves Water Campaign", the "Fifty Good practices" project on water use, "Optimisawater", Switch and Green Homes projects, as well as educational proposals aimed at teachers.

Faced by the challenge that represented the opening in Zaragoza of the United Nations office to support the International Decade for Action "Water for Life" 2005–2015 and the experience of the 2008 International Expo,[1] the aim has been that the by-laws on eco-efficiency and quality in water management should form a single regulatory text. Zaragoza City Council would thus include all its measures on the overall cycle of water, including aspects such as ensuring the supply and quality of water, its efficient use, the encouragement of sustainable habits, and citizens' rights to be informed.

The regulations consider new urban developments that establish systems of recovery of rainwater, homes with water-saving equipment and systems for the reuse of water from showers (with savings of 40%). Public gardens are to be developed that use plants that save water, and intelligent irrigation systems based on the experience of the Parque Oliver, where 68% savings have been established in irrigation.

The new rules were approved by the local government board in February 2007. Currently they are pending approval to be remitted as information for the public.

Note

1. See http://www.zaragoza.es/TribunadelAgua/CajaAzul/.

Bibliography

Agenda 21 (2009) *Update of the Sustainability indicators of Zaragoza of the Environment and Sustainability Agency for 2009* (Zaragoza: Zaragoza City Council). Available from: http://www.zaragoza.es/contenidos/medioambiente/Cuaderno17.pdf/. [in Spanish].

Bueno-Bernal, V. & Celma-Celma, J. (2002) *Audits of the Management and Use of Water in Zaragoza* (Zaragoza: Environment and Sustainability Office, Zaragoza City Council).

Bueno-Bernal, V. & Celma-Celma, J. (2010) *Audits of the Management and Use of Water in Zaragoza*. Draft (Zaragoza: Environment and Sustainability Office, Zaragoza City Council).

Ebro Hydrographic Confederation (n.d.) Available from: http://www.chebro.es/

Entralgo, J. R. (2002) *Plan to Improve the Management and Quality of Water Supply* (Zaragoza: Zaragoza City Council). Available from: http://www.zaragoza.es/cont/paginas//grandesproyectos/pdf/gestion.pdf/. [in Spanish].

Entralgo, J. R. (2009) *Plan to Improve Management and Quality of Water Supply. Results of its Implementation* (Zaragoza: Zaragoza City Council). [in Spanish].

European Commission (2000) Water Framework Directive, 2000/60/CE, 23 October 2000 (Brussels: European Commission). Available from: http://eur-lex.europa.eu/LexUriServ/LexUriServ.do?uri=CELEX:32000L0060:ES:NOT/

Garcia-Lucea, J. (2010) *Data on Water Consumption and Related Taxes* (Zaragoza: Zaragoza City Council).

National Institute of Statistics (INE) (2010) *Environmental Statistics on Water. Questioners on Water Supply and its Treatment for 2009* (Madrid: Department of Infrastructure, INE).

Zaragoza City Council (2010) *Fiscal Regulation 24.25 of the City Council of Zaragoza on the Taxes of Water Supply and Wastewater Treatment Services in 2010* (Zaragoza: Zaragoza City Council). Available from: http://www.zaragoza.es/contenidos/normativa/ordenanzas-fiscales/24-25.pdf/. [in Spanish].

Zaragoza City Council (various years) *Reports on Water Supply and Sanitation for 2007, 2008 and 2009* (Zaragoza: Department of Economics, Zaragoza City Council).

Water Quality Management in China: The Case of the Huai River Basin

JUN XIA*, YONG-YONG ZHANG*, CHESHENG ZHAN* & AI ZHONG YE**
*Key Laboratory of Water Cycle and Related Land Surface Processes, Chinese Academy of Sciences, Beijing, China; **Beijing Normal University, Beijing, China

ABSTRACT *This paper addresses the importance of water quality management and the impacts of water pollution control and water development projects. The case study of the Huai River Basin is an example of the major challenges on water quality management that China is facing, and why water quality management will play a key role on its sustainable use and management. Three urgent issues for the Huai River Basin are identified: water and ecosystem interactions on the river system due to the impacts of increasing pollution and water development projects; comprehensive assessment on impact of dams and sluices on changes of river flow regimes, water quality and ecosystems; and improvement of water quality, and the restoration of river ecosystems through state-of-the-art environmental monitoring and integrated water management practices.*

Water Quality Management in China

Water quality management is a key issue throughout the world. China is a developing country with a variety of climates and increasing stress from its population and economic development (Xia & Chen, 2001; Xia *et al.*, 2007, Xia & Zhang, 2008; Shen & Yang, 2001), which in addition is affected by floods, droughts and water environment-related issues. In terms of water shortage, it was estimated by the Ministry of Water Resources (MWR) (2010) that 59% of the total area of China faces water stress, that it has an impact on 60% of the population at the national level, in 67% of the cities and on 72% of the gross domestic product produced from these impacted areas. The annual water shortage reaches 100–110 BCM with annual economic losses almost US$30 billion.

In terms of floods, the major disasters occurred in the Yangtze Rivers in 1998. Just this sole event caused at least US$22 billion-worth of damage. The Huai River, analysed in this paper, is also one of the most vulnerable zones in China. In 2007, floods seriously affected the basin: a maximum rainfall for 7 days reached 304 mm, and maximum rainfall for a single day reached 138 mm, resulting in serious natural disasters in the basin.

In terms of water resources per capita, the Hai River Basin has only 305 m^3, that is, only one-seventh of the national average and 1/24th of the world average. Water crises have occurred in, for example, 1972, 1999 and 2000 in Beijing and other big cities in Northern China due to severe droughts. Water shortages have also resulted in ecosystem degradation, such as some rivers drying up, wetland degradation, over-extraction of the groundwater and decreasing outflow from the Hai River Basin to the Bohai Sea (Xia *et al.*, 2007; Xia & Zhang, 2008).

Another major problem in China is that of water-related environmental issues. It is clear that China is facing huge challenges to do with water quality due to rapid social–economic developments. Poor natural water quality in some areas reaches 25% of surface water and 35% of groundwater. In some rivers the total pollution load is beyond the carrying capacity of the water resources, since it has accounted for 40–50% of runoff. Unsafe drinking water-related problems in rural areas are also urgent issues since they affect almost 0.3 billion people. Thus, water quality management has become a priority issue for water security (Xia *et al.*, 2009; MWR, 2010; Cao & Wang, 1999; World Bank, 2006).

To tackle water quality-related constraints at the national level, the MWR, the Ministry of Environment Protection (MEP), and other water-related institutions have carried out extensive work on environmental monitoring, assessment and control of water pollution since the nation entered the reform path in 1979, particularly since industrial growth began in the mid-1980s. China has established a water quality classification system based on the use of water and setting targets for protection, following Environmental Quality Standard GB 3838-2002 (China State Environmental Protection Administration, 2002). This water quality standard has five grades:

- Grade I, mainly applicable to the source of water bodies and national nature reserves.
- Grade II, mainly applicable to a class A water source-protection area for centralized drinking water supply, sanctuaries for rare species of fish and spawning grounds for fish and shrimps.
- Grade III, mainly applicable to a class B water source-protection area for centralized drinking water supply, sanctuaries for common species of fish and swimming zones.
- Grade IV, mainly applicable to water bodies for general industrial water supply and recreational waters in which there is no direct human contact with the water.
- Grade V, mainly applicable to water bodies for agricultural water supply and for general landscape requirements.
- Grade V+, essentially without any use.

Based upon this classification, water quality is being monitored on a major systematic basis in China.

For instance, the MWR, in collaboration with the MEP and other government sectors, established task forces for land and water quality assessment in 2000, taking into consideration the related river systems, lakes and reservoirs, and head water areas for drinking water (Zhou & Wang, 2005). For this comprehensive job, a huge amount of representative monitoring information and a data set on water quality in China were collected, which included the water chemical characteristics from over 2,442 monitoring stations and river water quality from 5,952 monitoring stations covering the major rivers; water quality from 237 lakes and 813 reservoirs; drinking water quality from 1,073 head water areas, etc. The changing trend of water quality in the last 30 or more years was also analysed based on 846 representative monitoring stations. Assessment and analysis showed that most water mineralization is of a standard sufficient to satisfy domestic, industrial and agricultural demands; however, both point and nonpoint pollution have increased in China. Based on statistics, good water quality (i.e., water quality within Grades I–III of water quality standards) reaches 66% in all the assessed rivers; bad-quality water (i.e., water quality within Grades IV and V) reaches 18%. The lowest-quality water (i.e., water quality over Grade V) is about 16%. Polluted water in river systems can be characterized by a combination of organic concentrations (biological oxygen demand and chemical oxygen demand (COD)) and nutrients (ammonium nitrate concentrations), particularly in the tributaries of river systems in North China such as the Hai and Huai rivers, etc. However, high organic concentrations, particularly high COD loads, are a critical factor in poor water quality. Lake eutrophication in China is also a very serious problem. It significantly threatens water supply security and fisher products and reduces its recreational uses.

The qualifying rate at the head water levels for drinking water in China is 75.3%. The population without access to safe drinking water is 0.323 billion people, representing a major limiting factor for regional sustainable development. The trend analysis shows that even though specific pollutants can be controlled for, total surface water quality is still declining. It also appears that increased agricultural activities and higher urbanization are resulting in significantly higher ammonia nitrate concentrations.

The efficient control of the trends in water pollution in China is still a huge challenge, which makes it a priority to achieve water quality management in the country. Over the last 30 years or more China has established an extensive water pollution-control system with a large set of institutions, a variety of legislation and policy instruments, and comprehensive investment plans that were largely shaped within the traditional five-year plan preparation process. In the earlier phases water pollution control was characterized mostly by command-and-control instruments (industrial permit systems, simultaneous control programmes) and partly by economic instruments (such as pollution levy fees and discharge permits). In recent years there has been a gradual increase in voluntary approaches (such as environmental management systems like ISO 14000 and cleaner production) and public disclosure (World Bank, 2006).

Along with China's social and economic development, water policy has changed significantly. The government has recognized the need for more effective management of its water resources. Water resource management issues have been given a high priority in the 2006–2010 and 2011–2015 five-year plans. Among the many Water Resources Ministry (WRM) issues that the government must address, priorities include (1) ensuring water availability in water-scarce areas; (2) providing a clean supply of water in order to reduce the heavy burden of water-related diseases; (3) treating wastewater, with a focus on the impact on receiving water bodies; (4) rehabilitating the heavily polluted water bodies that are critical for local communities; and (5) protecting drinking water sources (World Bank, 2006).

Water policy in China is shifting from water quantity management to water quality management. Major actions highlight the need for an overall strategic plan for water quality management that establishes a long-term vision and realistic targets for five-year plans over the next 20–30 years. The Ministry of the Environment has also implemented new modifications in the Comprehensive Water Resources Programme in the river basins since 2005. The Ministry of Environment Protection and the Ministry of Science and Technology (MOST) have carried out the Key Water Environmental Project (KWEP) since 2008 which considers all issues related to China's water pollution based on scientific and new technology studies and institution innovation on water quality management.

New policy instruments include effluent standards, environment impact assessment (EIA) requirements, discharge permits, fee and levy systems, ISO 14000, and total load control. The 2002 Water Law (re-draft) in China addressed river basin management directly, specifying that "the state shall adopt a 'combined division responsibility' (CDR) system of river basin management in conjunction with jurisdictional management". It indicated that river basin management should be set up by the water administration department under the State Council, and it set out the following river basin management functions: (1) planning; (2) protection of water resources, water areas and projects (including pollutant discharge loading and sewage facilities); (3) water resources allocation; and (4) water dispute resolution. China is thus facing the dilemma of reducing water pollution problems and, at the same time, continuing with its social and economic development on which water quality management plays a fundamental role, in addition to big challenges such as floods and droughts.

Water quality management is thus related to integrated water quality and quantity management and also to water projects operation. The Huai River Basin (HRB) is a very complex case in terms of water quality management. It is presented here as an example of the major challenges in water quality management in China and why water quality management will play a key role in water sustainable use and management. This paper also addresses how water quality management is impacted by wastewater control and the operation of water projects.

Background of the Huai River Basin

The HRB ($30°55'-36°36'$N, $111°55'-121°25'$E) is one of the top seven river basins in China. It is located between the Yangtze River Basin and the Yellow River Basin (Figure 1). It flows through the five provinces of Hubei, Henan, Anhui, Shandong and Jiangsu. It is the most densely inhabited river basin and the main grain-producing area of China. In 2005, the total population and grain yield accounted 13.1% and 16.1%, respectively, of the national total. Its average population density is approximately five times of the national average.

Although annual mean precipitation and water resources of the basin are 888 mm and 83.5 billion m^3, respectively, water resources per capita and unit area are less than one-fifth of the national average. Moreover, because 50–80% of annual precipitation is concentrated during the rainy season (June–September), the basin faces both flood and drought problems. Before the establishment of the People's Republic of China in 1949, the average flood and drought disasters occurred 94 and 59 times more per century, respectively. The basin was thus named the "disastrous river basin" in China.

Due to demands on flood control and water supply in the HRB, more than 5,700 reservoirs and 5,000 sluices have been constructed in most of the main streams and tributaries. This is a positive development since dams and sluices in the HRB provide a useful engineering solution for flood control, irrigation, water supply, etc. On the other hand, the water project's impact on water quality and the ecosystem are also considered as a new and urgent issue for the sustainable development of the HRB. One warning was the river pollution event in 1994 during a period of flooding in Huai main stream. Due to heavy rainstorms upstream, the water level in the reservoirs was alarmingly high and gates had to be opened to discharge the excess water. This event allowed 0.2 billion m^3 of polluted water accumulated in sluices during the non-flood period to flow downstream of

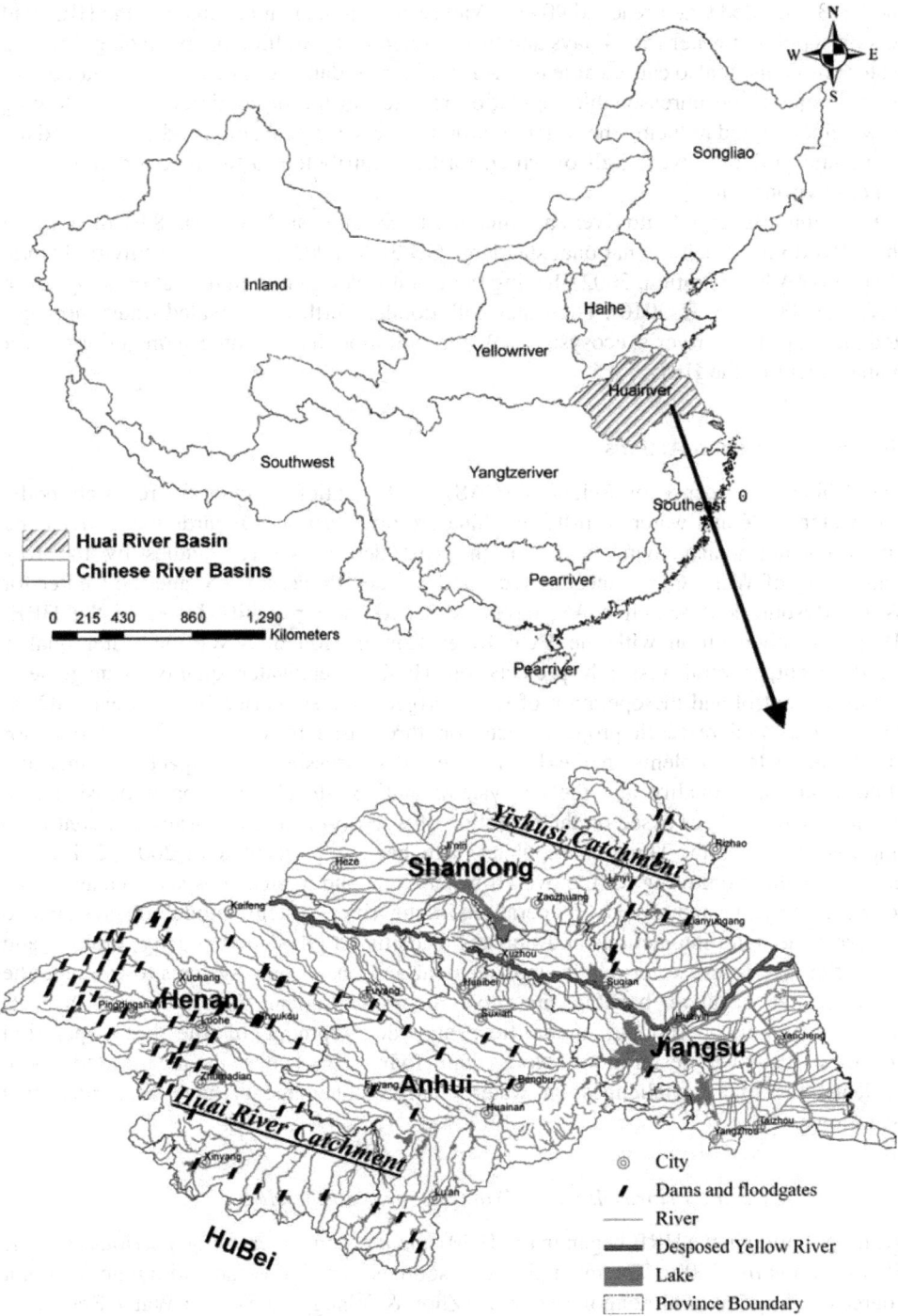

Figure 1. Location of the Huai River Basin.

the HRB. Polluted water reached 90 km. Waterworks in the main channel of the HRB had to stop supplying water for 54 days and this resulted in 1.5 million people facing drinking water problems. It also caused at least RMB2 billion of damage. The excessive number of water projects and unreasonable regulation resulted in the discontinuous flow, slowing flow velocities and reducing the water environment-carrying capacity, and so triggered the accidental pollution events, all of which further contributed to the deterioration of the water environment.

Pollution discharged into rivers has increased year on year. More than 83% of rivers in the HRB do not reach the national standard (GB 3838-2002; China State Environmental Protection Administration, 2002), having the worst water quality in the nation's top seven basins in 2005 (MWR, 2010). The water pollution has further aggravated water shortages and destroyed the river's ecosystem. Water pollution has become a major issue for management of the HRB.

Research Plans and Actions

The Chinese Academy of Sciences (CAS) is the national scientific research body. The water cycle and water security in China are priority issues regarding resources and environmental management fields. In the past decade several studies by the Key Laboratory of Water Cycle and Related Land Surface Process, CAS, and the Center for Water Resources Research, CAS, have focused on water-quality issues in the HRB. Thanks to cooperation with the Huai River Commission of MWR on water quality management, several research projects on Huai River water quality management, pollution control and the operation of water projects are supported by CAS and MOST. Major goals and research projects focus on three aspects: (1) identifying the major causes of water problems in the Huai River; (2) assessing the impact of dams and sluices on water quality and the ecosystem; and (3) developing comprehensive best management practices based on the operation of water projects and wastewater control to improve river health. Three-phase plans have been implemented in 2007–2011, i.e., environmental monitoring for the overall water cycle and related ecosystem changes due to water projects and reservoir operation in the river system; modelling systems to integrate the main interactions and impacts of dams on physico-chemistry, biology and other characteristics; water quality management and operational systems to evaluate the positive and negative benefits on economic, sociological and ecological aspects under different operational modalities; and development of the best operation schemes for river restoration. More than four years after having started these projects, the major challenges on water quality managements have been identified as follows.

Water Quality is the Priority Issue on Water Security in the HRB

Water pollution in the HRB began in the 1970s and it became increasingly serious after the 1980s (Xia *et al.*, 2009). The main pollutant sources were industrial and municipal point sources and agricultural nonpoint sources (Zhou & Wang, 2005). The Water Resources Protection Bureau of the Huai River Commission has implemented as routine a water quality monitoring system in the whole basin since the 1980s with the support of the local governments in Henan, Anhui, Jiangsu and Shandong provinces. Monitoring activities

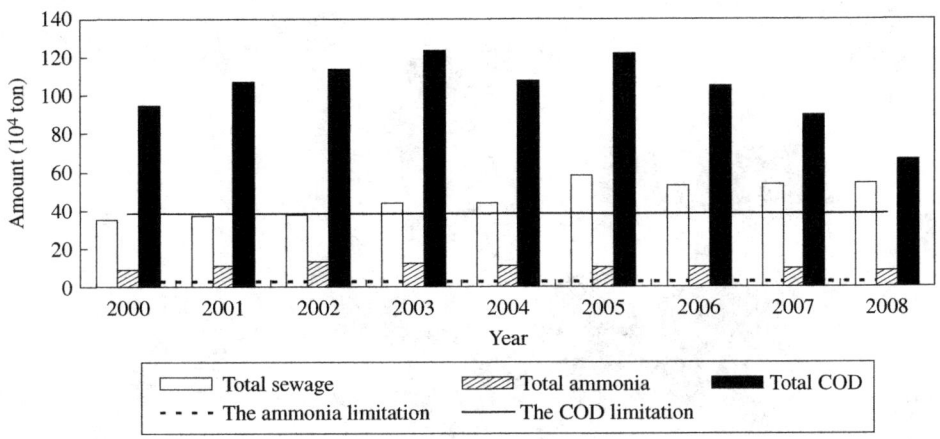

Figure 2. Pollutant discharge amount in the Huai River Basin from 2000 to 2008.

include monthly sampling of the polluting load in the sewage outlets into rivers, in rivers of 220 cities; and water quality in 153 areas of different water resources and 51 provincial boundary cross-sections. Water quality monitoring indexes included water temperature, turbidity, pH, dissolved oxygen, organic pollutants, nutrients, metal material, etc. (He & Wang, 2001; Xia et al., 2009).

According to the water resources bulletins from 2000 to 2008, the main pollutants in the HRB were NH_3-N and COD, and the total pollutant amount in the rivers has always been above the emission limitation (COD = 38.2 tons, NH_3-N = 2.66 tons). In 2008, the amounts were 0.74 and 2.05 times more than the limits, respectively (Figures 2 and 3). The industrial and municipal point sources are the main pollution sources. In addition, the nonpoint sources of pollution were also considered because HRB is the major grain-producing area. Its contribution was nearly 30% of the total amount of pollutants going into the river in 2000. The main sources are the excessive use of pesticides and chemical fertilizers, livestock and poultry farming, and straw decay, all of which are washed into the rivers and thus increase the load of COD, total nitrogen and total phosphorus.

The excessive pollution load has caused most rivers to become seriously polluted. The Huai River Commission has insisted on assessing water quality status in the whole basin every year based on the national standard (GB 3838-1983, 2002, etc.; China State Environmental Protection Administration, 1983, 2002). It has also published a water resources bulletin since the 1990s. The rivers that were in or below Class V (unsafe water) were always nearly 50% of the total samples assessed, and the rivers in Classes I–III (suitable for drinking water) were less than 40%. Furthermore, the percentage of non-safe water was greater than 60%, and the water suitable for drinking use was only nearly 20% between 1994 and 1998. (Wang et al., 2005) (Figure 4).

The monitoring of provincial boundary cross-sections has shown that the water quality status of these cross-sections was even worse than in other areas from 1994 to 2005. Thus, although great efforts have been made to improve the environmental quality of the river basin, the goal of "Huai River water is clean" has not been achieved so far. The situation of water pollution is extremely critical.

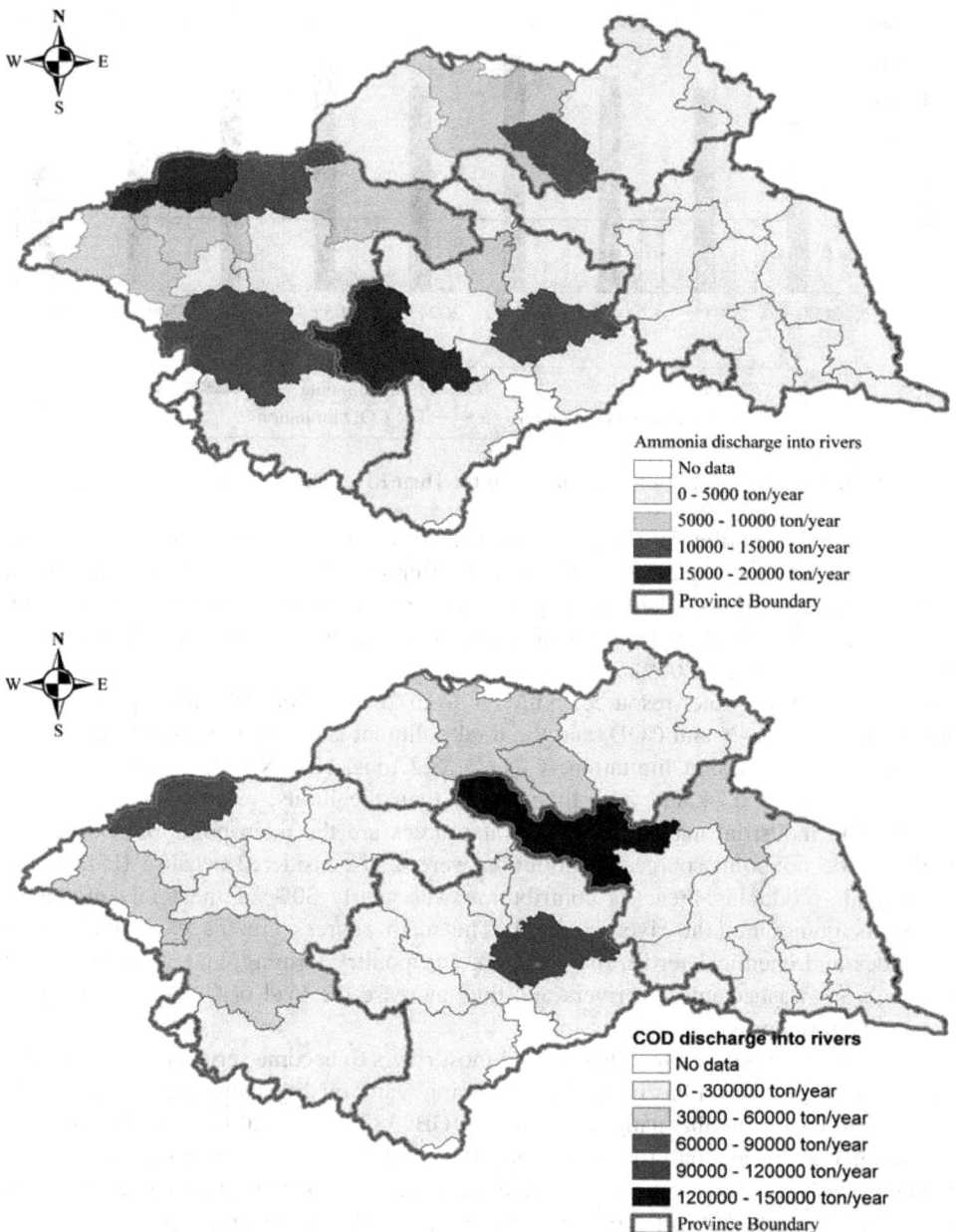

Figure 3. NH$_3$-N and COD sources distribution in the Huai River Basin in 2000.

Control of wastewater pollution sources is thus essential for water quality management in the HRB. Total maximal daily load (TMDL) and its load content in the basin was suggested to have been identified by the National Key Water Project (2009–2019) where water quality is the priority issue from a water security viewpoint in the HRB.

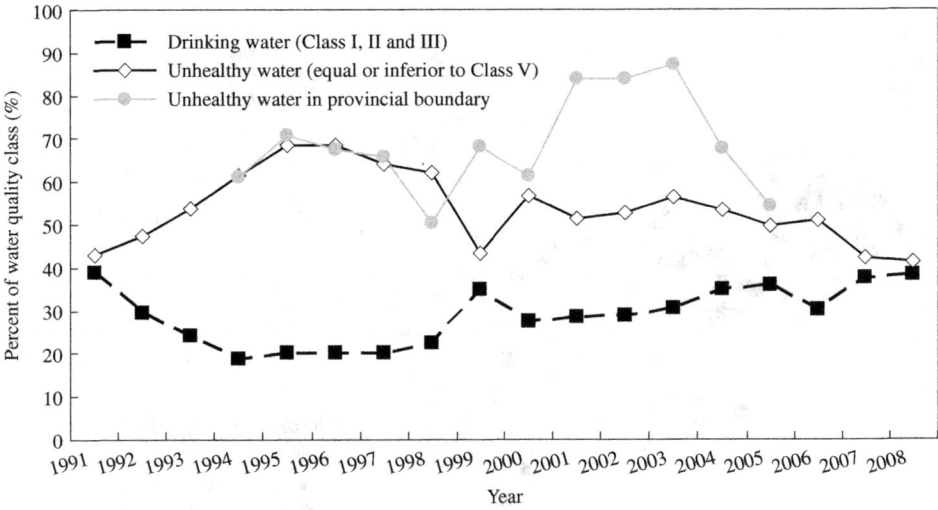

Figure 4. Percentage of river water quality classes from 1991 to 2008 in the Huai River Basin.

New Challenges on Integrated Water Quality and Quantity Management

The Huai River is a strongly regulated river due to a large number of water projects for flood control and water supply. The traditional regulation of water projects has focused mainly on its use (e.g., flood control, electricity generation, water supply, irrigation and aquaculture), but it has not taken into consideration environmental issues. However, so many water pollution sources have posed a huge risk of pollution for the river. Thus, new challenges will focus on how to integrate water quality and quantity management as well as the operation of projects in the HRB.

Some research on the impact of water projects on the river flow regime, the environment and ecology in this basin was carried out by CAS (Xia et al., 2008). The research focused on the 87 main dams and sluices, and the typical year was 1999 ($p = 95\%$, an extremely dry year) when the impacts of dams and sluices were significant (Figure 5). The results showed that compared with the pre-dams' scenario, the annual runoff at the outlet of the river basin in the post-dams' scenario decreased by 13%, the downstream carrying capacity of the water environment decreased by 16%, the contribution of a single dam on water quality in 1999 was less than 30%, and the average contribution was nearly 10%. Furthermore, the contribution of the dam groups in the upper and middle stream of the HRB was less than 38% (Xia et al., 2008; Wang & Xia, 2010; Zhang et al., 2010). The impact of water projects on water quality cannot be disregarded.

Water projects in river systems have thus the potential to increase water pollution, but only when there is no appropriate control of polluting sources. Thus, it is a very important issue that China integrates water quality and water quantity management, and the operation of water projects. This issue should be emphasized in the overall operation of river systems.

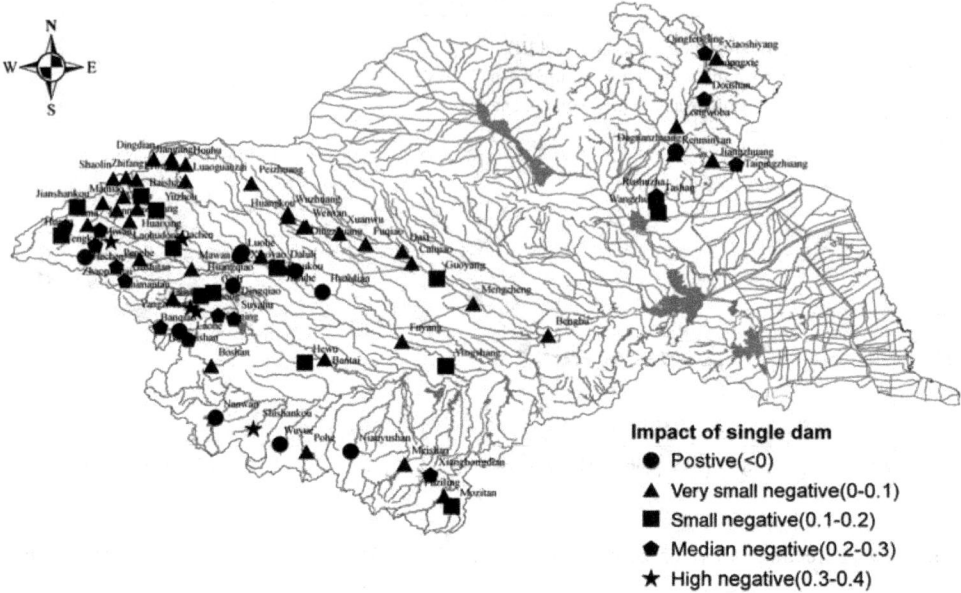

Figure 5. Contribution of 87 single dams and sluices on water quality in 1999 ($p = 95\%$, extreme dry year) in the Huai River Basin.

Challenge on Institutional Innovation on River Management Mechanisms

The State Water Resources Ministry, Environmental Protection Ministry and the local provincial governments are the management authorities in terms of river basins in China. Therefore, all these organizations have collaborated in the environmental management of the HRB directly, including the Water Conservancy Bureaus of Anhui, Henan, Jiangsu and Shandong provinces, the Environmental Protection Bureaus of the four provinces and the Huai River Commission.

The Huai River Commission was considered the main administrative department in the Water Law of P. R. China. However, it could not be a true leader in terms of environmental management because of the limitations it faced in terms of law enforcement. Every department that participated in this study worked on its own, within different jurisdictions. In addition, contradictions and a lack of effective communication and cooperation occurred among different departments, especially between the environmental protection and the water resources departments.

There were also some blind spots regarding law enforcement, such as in the provincial boundaries. At the same time, law enforcement existed in the different sectors, such as the water conservancy departments and environmental protection departments for a considerable time. Both the Water Resources Ministry (WRM) and the Environmental Protection Ministry (EPM) had their own monitoring system and there was a great gap between these two data sources. For example, the COD load discharged into the rivers, as published by the WRM, was 94.7×10^4 tons in 2000 and the load was 123.2×10^4 tons in 2003, which means an increase of 30%. However, the data published by the EPM were 81.2×10^4 tons and 71.2×10^4 tons in 2000 and 2003, respectively. This means that the load in 2003 decreased by 12.3%. The management departments were confused and could not make reasonable decisions because of this discrepancy in the information.

Furthermore, economic development and environmental protection have been traditionally regarded as opposite issues. Gross domestic product growth was a key indicator for evaluating the achievements of government officials in their positions, while the indicator related to environmental quality was not considered at all. The implementation of Environmental Protection Law was obstructed by local protectionism. The harmonious development of the socio-economy and environmental protection in different regions was always emphasized, but was difficult to realize.

Conclusions and Recommendations

Water quality management in China is facing enormous challenges due to pressures from its population and economic growth. Along with China's social and economic development, water policy has gone through significant changes. The government has recognized the need for more effective management of its water resources. In fact, water resource management issues are given a high priority in the 2006–2010 and 2011–2015 five-year plans.

In the case of the HRB, water pollution has resulted from both the impacts of wastewater load increase and the development of water projects. It is a very good case study of how to handle effectively environmental impact assessments and water quality management-related issues.

The control of pollution sources in the HRB will be the most essential issue for water quality management. TMDL and its load allocation in the basin area could be used to identify major polluted areas as well as sources by the National Key Water Project (2009–2019).

Water projects in river systems also contribute to the increased risk of water pollution when there is no appropriate control of polluting sources. Therefore, integrated water quality and water quantity management, as well as the operation of projects, have become very important issues in China.

Several suggestions regarding integrated water quality and water quantity management as well as for operation of projects are suggested below.

Strengthen Integrated Monitoring and Unify Different Data Sources

The current monitoring processes of water quality are not continuous nor systematic because of a shortage of funding. The frequency of routine monitoring was only two or three times per month and real-time monitoring was very difficult also because of low investments (He & Wang, 2001). It was common that some small but highly polluting workshops or factories would close their outlets during the monitoring process to avoid being fined, only to open them again later. In addition, routine monitoring was of no use for tracking spatial and temporal variation of water quality along the rivers whenever there was an accident which involved pollutants. Therefore, the monitoring plan should be strengthened and new techniques included into routine monitoring plans such as real-time monitoring systems, auto-monitoring systems or remote sensing (RS) monitoring techniques. Moreover, the different data sources from WRM and EPM should be unified.

Strengthen Pollution Source Control and Adjust Industrial Structure

The control of wastewater load is still a significant issue for Huai River Basin. The COD and NH_3-N amounts discharged into rivers were 0.74 and 2.05 times more than the limits set in 2008 (Wang *et al.*, 2005). Thus, pollutant emission is still the major cause of water pollution in the HRB. The construction of sewage treatment facilities (pipe networks, treatment plants, etc.) should be further speeded up and the improvement of industries should be strengthened. Strict control and discharge permit management for the total pollutant amount should be carried out. Some highly polluting enterprises (paper-making, tanning, brewing, chemical industry, etc.) should construct sewage treatment facilities as soon as possible, or control the pollutant load by modifying their infrastructure, including closing, combining or transforming some facilities. In addition, agricultural nonpoint source pollution control should also be strengthened.

Strengthen Integrated Water Quality Management and Research at the Basin Scale

Water pollution is the biggest water problem in the HRB. Although improvements have been carried out for nearly two decades, the achievements have not been remarkable. One of the main reasons is that all the previous works were confined to a particular region or one water problem, but they ignored the influence of different regions or other water problems. Following the advanced experience of river basin management, the best practices should integrate all water related-managements issues (e.g. flood control, pollution control, water supply, ecological restoration, regulation of water projects, etc.) at a basin scale. The basin water cycle (hydrological and socio-economic) should be the foundation for the integrated management which should also explore the interactions of these water problems in the different regions. For integrated water quality management, some primary works should be completed in the HRB, for example:

- Identify the spatial and temporal distribution of pollutant sources.
- Identify the types of pollutant materials (organic pollutants, persistent organic pollutants (POPs)) and their contribution to water pollution.
- Develop integrated water quality warning and prediction systems based on the water cycle model by using RS and geographical information system (GIS) techniques, and quantify the risk probability of water pollutant events.

Water quantity and quality regulations of water projects should be made jointly. The regulation of water projects should give considerable attention to the environment and the ecology, and not just to flood control and water supply in the HRB. A joint water quality–quantity operation of water projects is an important approach to improve the environment (Chen *et al.*, 2005; Zhang *et al.*, 2010). In highly polluted rivers (Shaying, Guo, Hongru, etc.), operations should keep a small discharge rate in order to make sure that water storage is not affected by any form of upstream pollution during the non-flood season. In the flood season, operations should control floods and extend the discharge time of polluted water by using the storage capacity of the dam groups. Moreover, the dams in highly polluted streams should dispatch with the dams in lesser polluted streams to lighten the damage of polluted water, especially in the event of pollution accidents.

Strengthen Legal Management and Public Participation

In August 1995 the State Council enacted China's first basin-wide water pollution control laws and regulations: "Interim Regulations on the Prevention and Control of Water Pollution in the Huai River Basin".

The State Council also applied successive programmes for the prevention and control of water pollution and the 9th and 10th five-year plans of the HRB in 1996 and 2003, respectively (Zhou & Wang, 2005). Nearly 2,300 water management laws and regulations have been issued. The excessive management departments and laws made decentralization and the legal system imperfect.

Environmental protection management should insist on the principle of integrated legislation, integrated planing, integrated supervision, division of labour with individual responsibility and responsibility of the government. The implementation of all planning and activities should follow the laws and regulations. The Huai River Commission should become the implementing agency for integrated water quality management in the HRB; and the environmental protection responsibilities of the authorities at the local level, the environmental protection departments and water resources departments at all levels should be clearly defined by law, especially in the Interim Regulations. The accomplishment of the water pollutant protection objectives should be an indicator to evaluate the performance of local governments and officials. All local governments should be required to sign a responsibility pledge committing to fulfil the protection, reducing the target set up in the 11th five-year plan.

Finally, multi-department cooperation and consultation systems and public participation processes should be developed. Environmental protection and improvement is not just the mission of management departments. All residents should also play active roles in this long, complicated and difficult task.

Acknowledgements

This study was supported by the National Key Water Project (Grant Number 2009ZX07210-006) and MWR project (Grant Number 200801001).

References

Cao, D. & Wang, J. (1999) *Economics for China's Industry Pollution Control* (Beijing: China Environmental Science Press).

Chen, X. S., Jia, L. & Shen, Z. S. (2005) Joint water pollution defense to reduce water pollution damage in the Huai River, *China Water Resources*, 6, pp. 21–24.

China State Environmental Protection Administration (1983) *Environmental Quality Standards for Surface Water*. GB 3838-1983 (Beijing: China Environmental Science Press). [in Chinese]

China State Environmental Protection Administration (2002) *Environmental Quality Standards for Surface Water*. GB 3838-2002 (Beijing: China Environmental Science Press). [in Chinese]

He, D. J. & Wang, Z. H. (2001) River discharge outlet supervision for Huaihe River Basin, *Water Resources Protection*, 3, pp. 12–13.

Ministry of Water Resources (MWR) (2010) *Water Resource Assessment for China* (Beijing: China Water and Power Press).

Shen, Z. S. & Yang, Z. (2001) Some cognitions about the investigation and disposal of water pollution accidents in the Huai River Basin, *Harnessing the Huai River*, 6, pp. 29–30.

Wang, B., Chen, X. S., Jiang, Y. S. *et al.* (2005) *Research on Pollutant Carrying Capacity and the Control Limitation in the Huai River Basin* (Beijing: Ministry of Water Resources of China).

Wang, G. S. & Xia, J. (2010) Improvement of SWAT2000 modelling to assess the impact of dams and sluices on streamflow in the Huai River basin of China, *Hydrology Processes*, 24, pp. 1455–1471.

World Bank (2006) *China Water Quality Management: Policy and Institutional Considerations, Environment and Social Development: East Asia and Pacific Region*. Discussion Paper (Washington, DC: The World Bank).

Xia, J. & Chen, Y. D. (2001) Water problems and opportunities in hydrological Sciences in China, *Hydrological Science Journal*, 46(6), pp. 907–921.

Xia, J. & Zhang, Y. (2008) Water security in north China and countermeasure to climate change and human activity, *Physics and Chemistry of the Earth*, 33(5), pp. 359–363.

Xia, J., Chen, X. S., Zuo, Q. T., Jiang, Y. S. & Wan, Y. (2009) *The Integrated Carrying Capacity of Water Environment and the Control Countermeasure* (Beijing: China Science Press). [in Chinese]

Xia, J., Zhang, L., Liu, C. & Yu, J. J. (2007) Towards better water security in North China, *Journal Water Resources Management*, 21, pp. 233–247.

Xia, J., Zhang, Y. Y. & Wang, G. S. (2008) Assessment of dam impacts on river flow regimes and water quality: a case study of the Huai River Basin in P.R. China, *Journal of Chongqing University (English edn)*, 7(4), pp. 260–274.

Zhang, Y. Y., Xia, J., Liang, T. & Shao, Q. X. (2010) Impact of water projects on River Flood Regime and Water Quality in the Huai River Basin, *Water Resources Management*, 24, pp. 889–908.

Zhou, Z. Q. & Wang, F. (2005) Causes for water pollution in Huai river basin and prevention measures, *China Water Resources*, 22, pp. 23–25.

Water Quality Management in Egypt

SAFWAT ABDEL-DAYEM
Arab Water Council, Cairo, Egypt

ABSTRACT *One of the greatest water-related challenges facing Egypt is the pollution of its surface and ground water resources from agricultural, domestic and industrial sources. The cost of environmental degradation due to water quality deterioration is relatively high with serious health and quality-of-life consequences. The closed water system of the country makes it more vulnerable to quality deterioration in a northward direction. The water quality of Lake Naser upstream of the High Aswan Dam and the main stem of the River Nile from Aswan to Cairo is good and traces of pollutants, if any, are far below the levels set in the quality standards set by Law 48. However, water quality in the irrigation and drainage canals deteriorates downstream and reaches alarming levels in the Delta. Monitoring water quality of the Nile system (Lake Naser, the main Nile and its branches, irrigation canals, drains and groundwater aquifers) started as early as the 1980s. The complexity of water quality management required the development of other mechanisms including policies, institutional and governance arrangements, infrastructure for monitoring and analytic laboratories, awareness and skilled human resources. This paper describes the different aspects of water quality management in Egypt and the current state as it stands by the end of the first decade of the 21st century. It also presents the methodology used in turning several monitoring programmes managed by different institutions into one national integrated system. It argues that water quality management is multifaceted and while progress along one aspect could be significant, other aspects could be lacking due to multiple reasons, the high cost involved in pollution reduction at the source is not the least.*

Introduction

Fresh water availability is a prime sustainable development concern in Egypt. The water management situation is fragile mainly due to water scarcity and less-than-efficient water management. As the population grows and the economy expands while holding the amount of fresh water constant, the problem of water availability intensifies. The scarcity of fresh water resources and protecting them from pollution and wasteful uses is an issue that concerns all Egyptians, and threatens the sustainability of the development of Egypt.

The total area of Egypt is approximately 1 million km^2 divided administratively into 29 governorates. Egypt suffers the extreme arid weather characterizing the great desert land of North Africa. Most of the land is barren desert and only about 4% is inhabited. With water being the basis of economic and social development, Egyptian life since the dawn of history has developed on the narrow strip of land along the banks of the Nile River and in its Delta (Figure 1). The majority of the Egyptian population, now counted as

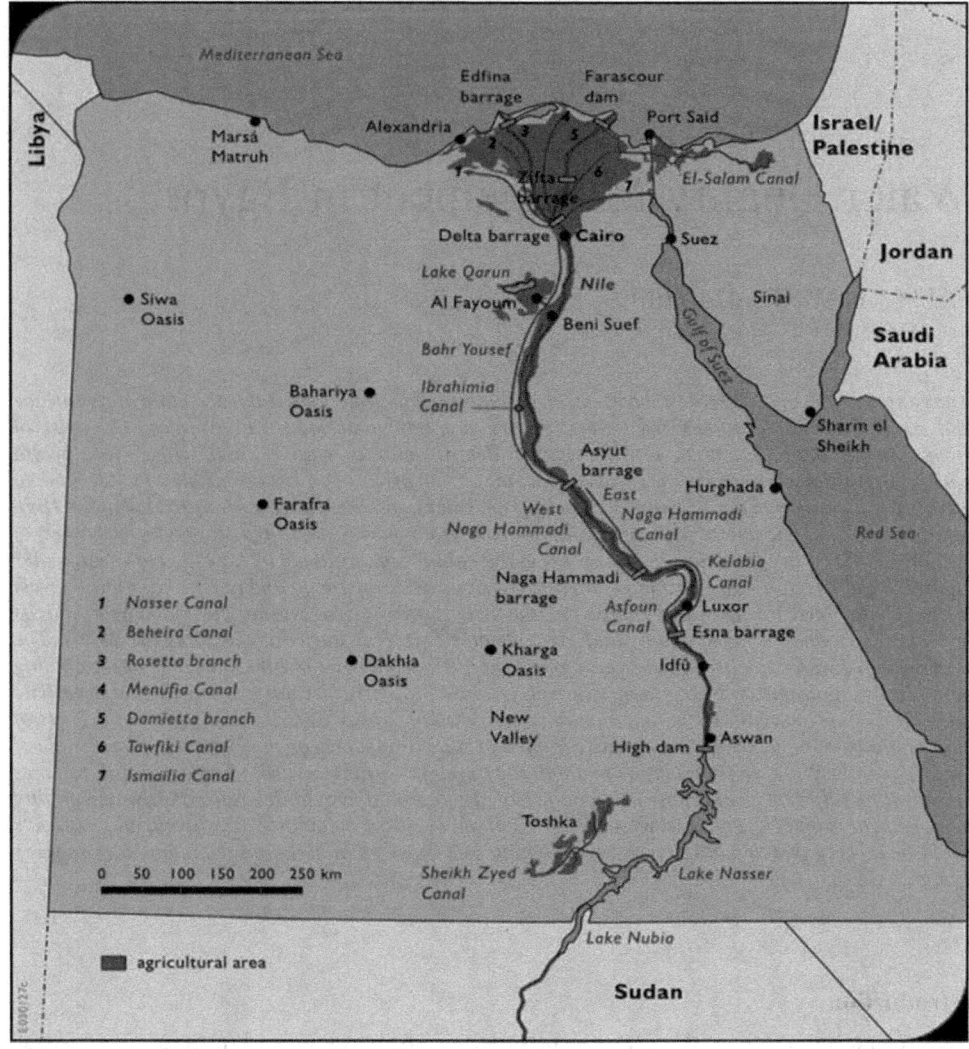

Figure 1. Map of Egypt.

approximately 82 million people living in cities, towns and villages, is concentrated in this piece of land. Cairo, the capital, has a population exceeding 15 million. The urban population represents about 43% of the total, while the rest is mainly rural depending on irrigated agriculture for their living. Industries are located near big cities, particularly Cairo and Alexandria. Tourism is a major activity and spans over Egypt's historic sites and the coasts of Sinai and the Red Sea. The growing population, at an annual rate of 2%, and rising standards are increasing pressure on natural resources, with water being the most critical resource. The outcome is not only a falling per-capita water share, but also increasing trends of pollution.

The decline in water quality has, in part, been caused by problems related to the fast growth of urbanization in the Nile Delta and Valley region, insufficient and inefficient municipal and industrial wastewater treatment facilities, disposal and reuse of agricultural drainage water, poor or non-existent solid waste management, and weak pollution-control and abatement programmes. The cost of environmental degradation, mainly due to water-quality deterioration, has emerged as a development issue (Arab Water Council (AWC), 2009). The total damage cost to health and quality of life (mortality, morbidity and quality of life) due to water pollution is estimated at about 0.9% of gross domestic product (World Bank, 2007). In addition, the cost of damage to natural resources (ecosystems from municipal and industrial wastewater) is about 0.1% of gross domestic product. This is why water quality management (WQM) is crucial to sustain the social and economic life of Egypt as well as to preserve the nation's environment.

Water Supply and Demand

The sources of conventional fresh water in Egypt are the Nile, the renewable groundwater in the Nile Valley and Delta (recharged from irrigation and canal seepage), and the non-renewable groundwater in the Western Desert and Sinai. Limited rainfall and flash floods are also available. The Nile is the main and almost the exclusive source of renewable fresh water and stands for almost 98%. Since the construction of the High Aswan Dam (HAD), the country relies on stable releases from the available water stored in Lake Nasser. Egypt's annual share of water from the Nile is fixed at 55.5 billion cubic meters (BCM) according to an agreement signed with Sudan in 1959. The annual per-capita share of Nile water dropped from about 1,900 m^3 in 1960 to only 670 m^3 at present. Demand for water continuously increases and demand has already exceeded the available fresh water resources for some time. The water demands for different water uses in 1997 and the expected water demands for the same uses in 2017 are shown in Table 1.

The imbalance between supply and demand will continue to rise as long as the population continues to grow and economic and social development increases. To close the gap between supply and demand, Egypt depends on either non-conventional water resources or implementing a set of water management policies which save water through more efficient uses. Table 2 shows water supplies from both conventional and non-conventional sources in 1977 and 2017. Reuse and recycling water became increasingly important practices, but they entail water quality implications which increase the need for more careful WQM.

Table 1. Water demands for the years 1997 and 2017 (National Water Resources Plan (NWRP)).

Sector	Water demand (BCM/year)	
	1997	2017
Agriculture	52.1	67.1
Evaporation losses from the Nile and canals	2.1	2.3
Municipal uses	4.5	6.6
Industrial uses	7.4	10.6
Navigation	0.2	0.2
Total	66.3	86.8

Table 2. Water supplies in 1997 and their predictions for 2017 (National Water Resources Plan (NWRP)).

Conventional and non-conventional water resource	Water supply (BCM/year)	
	1997	2017
Nile water	55.5	55.5
Jonglie Canal Project, Phase (1)	–	2
Groundwater in the Nile Valley and Delta	4.8	7.5
Reuse of drainage water in the Nile Delta	4.4	8.4
Reduced fresh water to the sea	0.2	–
Shifts in cropping patterns	–	3
Irrigation-improvement projects savings	0.2	4
Deep groundwater in desert aquifers	0.6	3.8
Treated wastewater	0.2	2
Rainfall harvesting	1	1.5
Total	66.9	87.7

Sources of Pollution

The severity of water quality problems in Egypt varies among various water bodies depending on the amount of flow, the pattern of use, population density, the extent of industrialization, the availability of sanitation systems, and social and economic conditions. The main sources of pollution are return flows from agriculture, domestic uses and industry, as well as solid waste (Agricultural Policy Reform Program (APRP), 2002; World Bank, 2005). Water quality shows more deterioration signs near big cities and industrial areas. However, rural areas that mostly lack proper sanitation are major sources of pollution of surface and ground water.

Agricultural Wastewater

Apart from being the largest water consumer (80% of all available water), agriculture is also a main contributor to water pollution. Drainage water, estimated at 15 BCM, seeps from agricultural fields. Although considered as a nonpoint source of pollution, this drainage water is collected and concentrated in agricultural drains to become a point source of pollution for the Nile and the Northern Lakes, as well as irrigation canals in the case of mixing water for reuse in irrigation. Drainage water could become also a source of pollution of groundwater. Major pollutants in agricultural drains are salts, nutrients (phosphorus and nitrogen), pesticide residues (from irrigated fields), and pathogens (from domestic wastewater). Drains also receive untreated domestic and industrial waste water which are sources of toxic organic and inorganic pollutants.

Municipal Wastewater

Based on population studies and the rates of water consumption, the total wastewater flows generated by all governorates is estimated to be 3.5 BCM/year. Approximately only 1.6 BCM/year receives treatment. By the year 2017, the target is an additional capacity of treatment plants equivalent to 1.7 BCM (APRP, 2002). Although the expected capacity increase is significant, it will not be sufficient to cope with the future increase in wastewater production from municipal sources due to population increase. Therefore, the untreated loads that will reach water bodies are not expected to decline in the coming years.

Industrial Wastewater

The industrial sector is an important user of natural resources and a contributor to the pollution of water and soil. There are some 24,000 industrial enterprises in Egypt, about 700 of which are major industrial facilities. The spatial distribution of industry in Egypt is influenced by the size of the employment pool, the availability of services, access to transportation networks, and proximity to principal markets. The manufacturing facilities are therefore often located within the boundaries of major cities, in areas with readily available utilities and supporting services. In general, the majority of heavy industry is concentrated in Greater Cairo and Alexandria.

Industrial demand for water in the year 1997 has been reported to be 7.4 BCM/year. By 2017 the industrial demand for water is expected to reach 10.6 BCM/year. Not all industrial facilities, especially the small ones, are provided with wastewater treatment facilities. Consequently, a corresponding increase in the volume of industrial wastewater is expected (Ministry of Water Resources & Irrigation (MWRI), 2005).

Municipal Solid Waste

Homes and businesses produce municipal solid waste as a consequence of everyday activities. Organic waste constitutes about 60% of municipal solid waste, paper averages 11%, and glass, plastic, textiles, bones and metal, etc. account for the rest. The average daily solid waste generation rate varies from 0.3 kg/person/day in rural areas to 1.0 kg/person/day in large cities. Specific establishments like hotels and tourist resorts may have generation rates as high as 1.5 kg/person/day (Ministry of State for Environmental Affairs (MSEA), 2001).

Of the 15 million tons of municipal solid waste generated annually, about 65% is in urban areas and the rest is from rural areas. Waste collection efficiency ranges from 90% in high-income areas to 10% in rural low-income areas where solid waste is, at best, dumped in government-designated sites or in adjacent land where it putrefies and self-ignites. Waste is also sometimes burned or disposed into agricultural drains and irrigation canals, thus significantly degrading water quality and aggravating health problems.

The Concept of Water Quality Management

WQM can be defined as an activity to realize water quality objectives and the implications of adopting them for a particular catchment, aquifer or coastal water. The process of WQM uses the concept of environmental values to set local water quality targets. Government establishes these targets, either directly or in partnership with the stakeholders. Thus, policies, planning and action should be linked to achieve an agreed vision or outcome. The product of a WQM process is sustainable water resources of a quality that meets the community's needs (Ibrekk, 2003).

Water quality has a direct effect on health and the environment. Ultimately, it could have a serious impact on the social and economic life of individuals, communities and nations. During recent years there has been increasing awareness of, and concern about, water pollution all over the world, and new approaches towards achieving sustainable exploitation of water resources have been developed internationally. Several countries have elaborated sustainable development strategies, which encompasses sustainable use of the nation's water resources by protecting and enhancing their quality while maintaining economic and social development. This broader perspective on WQM requires water managers to plan and manage the use of water holistically (Ibrekk, 2003).

The strategic objective of environmental policy in Egypt is to introduce and integrate environmental concerns relevant to protecting human health and managing natural resources including water into all national policies, plans, programmes and projects. The medium-term objective is to preserve natural resources, biological diversity and national heritage within a context of sustainable development. The short-term objective is to reduce current pollution levels and minimize health hazards to improve the quality of life in Egypt. Within this context, the National Water Resources Plan (NWRP) 2005–2017 sets three strategic objectives for WQM: (1) preventing pollutants from reaching the water bodies; (2) treating pollutants when it is not possible to avoid the disposal of waste water into water bodies; and (3) controlling the effect of pollution on health and the environment. These objectives were considered in the context of a holistic approach for integrated water resources management developed by the Ministry of Water Resources and Irrigation (MWRI, 2005).

The major national policies and action plans with bearing on WQM are:

- National Water Resources Plan 2017: Water for the Future.
- National Environmental Action Plan (NEAP) 2002.
- Agricultural Policy.
- National Policy for inhabitation of deserts.

The drivers for implementing a successful WQM plan include capable institutions, appropriate water quality standards and legislations, effective monitoring, adequate law enforcement capacity, and effective awareness programmes. WQM is best implemented by integrating national, state, and regional powers and responsibilities, and by using complementary water quality planning and policy tools.

Institutional Aspects of Water Quality Management

The NWRP recognized that implementation of a policy for WQM is dependent on the country's ability to set up appropriate institutional structures with clear mandates and the necessary infrastructure to carry out these mandates. Institutions need to have clear mission statements, an adequate personnel management system, a sound budgeting system, a continuous training programme for internal capacity building, as well as access to the environmental and other public information systems.

Several ministries in Egypt are involved in water quality activities for operational, research, monitoring and regulation purposes. Among these are:

- Ministry of Water Resources and Irrigation (MWRI).
- State Ministry for Environmental Affairs (MSEA) through the Egyptian Environmental Affairs Agency (EEAA).
- Ministry of Health (MOH).

- Ministry of Housing, Utilities and New Communities (MHUNC).
- Ministry of Agriculture and Land Reclamation (MALR).
- Ministry of Industry (MOI).
- Ministry of Interior (MOInt).

The roles of the different Egyptian ministries and institutions in the main control three functions of WQM: legislation, monitoring and pollution control, which are summarized in Figure 2 (APRP, 2002).

In order to ensure coordination among the several institutions involved in water management, a Ministerial Committee headed by the Prime Minister was established in 2005 as a high-level policy and decision-making body. The Ministerial Committee was established within the context of the National Water Resources plan, which advocates integrated water resources management. It took couple of years for the Ministerial Committee to meet, but it intensified the number of meetings during 2009 and 2010. It was not unusual that the first meeting of the Ministerial Committee focused on causes, effects, and solutions of the pollution problem of water streams and aquifers. The two institutions with main responsibilities in WQM are MWRI and EEAA. More highlights on their responsibilities are given below.

Ministry of Water Resources and Irrigation (MWRI)

The MWRI is formulating the national water policy for water resources development and management to face the problem of water scarcity and water quality deterioration. The overall policy's objective is to utilize the available conventional and non-conventional

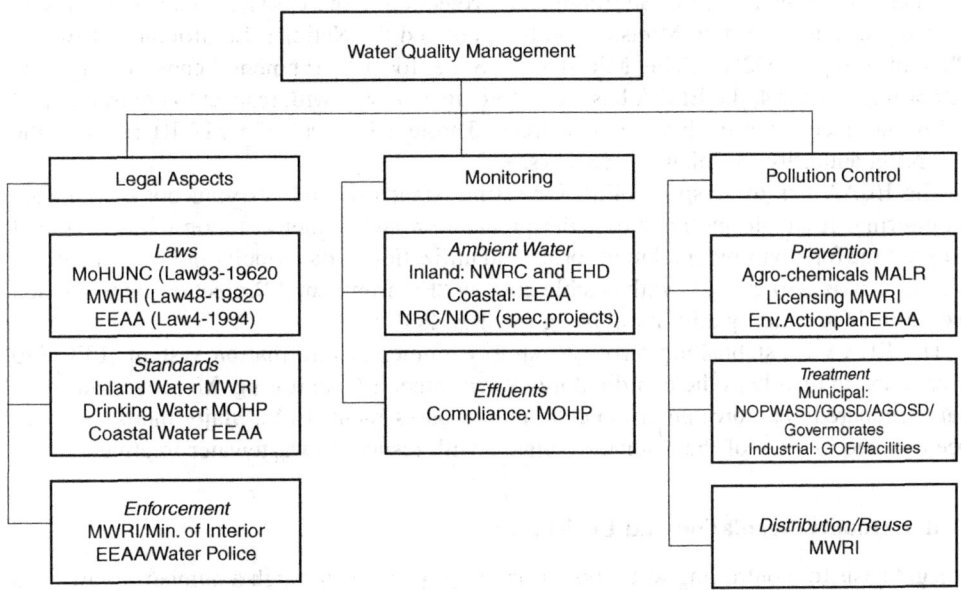

Figure 2. Institutional framework of water quality management in Egypt. *Source*: Agricultural Policy Reform Program (APRP) (2002).

water resources to meet the socio-economic and environmental needs of the country. Under Law No. 12 of 1984, MWRI retains the overall responsibility for the management of all water resources, including available surface water resources of the Nile system, irrigation water, drainage water and groundwater.

The MWRI is the central institution for WQM. The main instrument for WQM is Law 48. The MWRI is responsible for providing suitable water to all users. It has been given authority to issue licences for domestic and industrial discharges, and to monitor compliance to these licences through the analyses of discharges by the MOH.

The National Water Research Centre (NWRC) supports the MWRI's management efforts. Within the NWRC three institutes focus on the Nile, the irrigation and drainage canals, and groundwater: Nile Research Institute (NRI), Drainage Research Institute (DRI) and Research Institute for Groundwater (RIGW). The NWRC and its institutes established and maintain a national water quality monitoring network (see the section on *National Water Quality Monitoring Network*). The NWRC operates a modern, well-equipped water-quality laboratory and operates a database where all MWRI water quality data are consolidated.

The irrigation and drainage departments of MWRI monitor violations of Law 48 and its by-laws and enforce its regulations with the support from and specific tasks given to the ministries of Health and the Interior. To bridge the gap between the research institutes and the operational departments of the ministry and coordinate WQM with other government institutions outside the ministry, a Water Quality Management Unit was established at the central level and it reports to senior management of the ministry.

Egyptian Environmental Affairs Agency (EEAA)

The central organization for environmental protection is the EEAA. This agency has an advisory task to the Prime Minister and has prepared the National Environmental Action Plan of Egypt 2002/17. The Minister of State for Environment heads the agency. According to Law 4, the EEAA has the enforcing authority with respect to environmental pollution, except for fresh water resources. Through Law 48, the MWRI remains the enforcing authority for inland waterways.

The EEAA has the responsibility for setting standards and carrying out compliance monitoring. It should participate in the preparation and implementation of the national programme for environmental monitoring and utilization of data (including water quality). The agency is also charged with establishing an "Environmental Protection Fund" which would include water quality monitoring (MSEA, 2001).

The EEAA is establishing an Egyptian environmental information system (EEIS) to give shape to its role as the coordinator of environmental monitoring. Moreover, staff are being prepared to enforce environmental impact assessment (EIA). Major industries have been visited in view of their non-compliance with respect to wastewater treatment.

Water Quality Regulation and Legislation

A legal basis for controlling water pollution in Egypt exists through a number of laws and decrees. Law 48/1982 regarding the protection of the Nile and other inland waterways from pollution, and Law 4/1994 on environmental protection are the most important ones, as discussed below.

Law 48/1982 and Decree 8/1983

Law 48 of 1982 specifically deals with discharge into water bodies. This law prohibits discharges into the Nile, irrigation canals, drains, lakes and groundwater without a licence issued by the MWRI. Licences can be issued as long as the effluents meet the standards of the laws. The licence includes both the quantity and quality that are permitted to be discharged. Discharging without a licence can result in a fine. Licences may be withdrawn in the case of failure immediately to reduce discharge, in the case of pollution danger, or in the case of failure to install appropriate treatment facilities within a period of three months.

Under the law, the MOInt has police power while the MOH is the organization responsible for giving binding advice on water quality standards and monitoring effluents/discharges. Law 48 does not cover ambient quality monitoring of receiving water bodies, although some standards are given. Ambient quality standards are given for potable resources that are intended as raw water supplies for drinking water purposes. The implementation Decree 8 of 1983 specifies the water quality standards for different categories of water bodies and volumes of discharges.

Law 48 recognizes several categories of water body functions:

- Fresh water bodies as the Nile and irrigation canals.
- Non-fresh or brackish water bodies as drains.
- Lakes, ponds and groundwater aquifers.

Law 48 and Decree 8 prohibit discharge of treated sanitary effluents to the Nile and canals at all and any discharge of sanitary waste into other water bodies should first be chlorinated. The water quality standards are generally based on the drinking water standards and are not linked to all other functions a water body may have. The use of agrochemicals for weed control is also regulated in the law, although it has been banned due to public health concerns.

Law 4/1994

Law 4/1994 for environmental protection regulates the activities causing pollution and degradation of the Egyptian environment. The Egyptian Environmental Affairs Agency (EEAA) is the key responsible party for executing Law 4 and has the authority to monitor its implementation. Law 4 mainly covers coastal and seawater aspects. It states that for pollution control of the water environment all provisions of Law 48/1982 remain effective and are not affected. The MWRI remains the responsible authority for water quality and water pollution issues. The EEAA is responsible for coordinating the pollution-monitoring networks. A full chapter of this Law regulates the management of hazardous substances including chemicals. This has an important effect on properly regulating agrochemicals which are main cause of water pollution.

Law 4 states that all facilities discharging to surface water are required to obtain a licence and maintain a register indicating the impact of the establishment's activity on the environment. The register should include data on emissions, efficiency, and outflow from treatment units and periodic measurements. The EEAA inspects the facilities yearly and follows up any non-compliance. This provision is confusing or creating duplication because Law 48/1982 also includes certain standards for effluents with the MOH as the compliance monitoring organization and only MOH laboratory results are considered to be official. Both laws create funds where fines are collected and which are used to fund monitoring and other activities which creates another area of confusion.

Other Relevant Laws

Other laws that have bearings on WQM are:

- Law 2/1993, which is concerned with organizing tourism activities and their usage executed by the MOH.
- Law 3/1982 for the proper planning of industrial zones and the responsible party for its execution is the Ministry of Planning.
- Law 59/1960, which limits radioactive usage and is executed by the MOH.

The National Water Quality Monitoring Network

Programmes of water quality monitoring started early during the second half of the 20th century in the Nile, irrigation canals and agriculture drains. They were carried out by the National Water Research Center (NWRC) of the MWRI through three of its institutes.

Starting in 1991, the NRI has carried out a modified monitoring programme at a reduced number of sampling sites, one site on Lake Nasser, 30 sites along the Nile and two on each Delta branch. The number of parameters was extended to more than 45 depending on the type of water body to be investigated (major river sites, agricultural drain outlets, industrial outfalls). The network included 34 sampling points for the Nile and its downstream branches. A further 84 major point sources, agricultural drains and industrial outfalls were sampled during each campaign. Since 1991, ten campaigns have been conducted, but the programme is not being executed as frequently as designed. Most analysis was done in the NRI Laboratory.

The monitoring of drainage water started in 1984 when 100 monitoring sites were equipped with automatic recorders to measure water flow and salinity. A monitoring network consisting of 100 sites provided with automatic recorders for measuring the flow and salinity of drainage water in the main drains was established and they became fully operational by the Drainage Research Institute (DRI) in 1984. Water samples were collected monthly for chemical analysis at the laboratories of the DRI to determine the major ions of the soluble salts in the drainage water. In 1995–2000, the Monitoring and Analyses of Drainage Water Quality (MADWQ) Project was launched with the objective of setting up and implementing an integrated monitoring network to monitor drainage water quality that covers all critical sites which are of significance to drainage water quality. The main purpose was to determine where and when drainage water can be reused in irrigation. The project started with approximately 230 locations during the reconnaissance survey period (1995–1997) and ended with 120 locations representing the critical locations.

During the last two decades of the 20th century, the Research Institute for Groundwater (RIGW) set the foundation for an integrated groundwater management approach by using a number of diverse activities executed within the Environmental Management of Groundwater Resources (EMGR) project. This project was carried out jointly by the governments of Egypt and the Netherlands from 1994 to 1999. The main purpose of the establishment of a groundwater quality monitoring network in Egypt was to quantify quality change in the long-term, either caused by pollution or by salt water intrusion. The second purpose was to describe the overall current groundwater quality status of the main aquifers on a national scale. The monitoring network covers the entire country with wells distributed among the main aquifers in a low-density network (RIGW/IWACO, 1994). The monitoring effort is more concentrated in areas where aquifers are extensively used or have a high potential for drinking and irrigation purposes.

However, all these programmes were not fully coordinated to describe the overall water quality status of Egypt. Additionally, irrigation canals and groundwater were not properly included. Other institutions have their own water monitoring activities. In total, 25 agencies belonging to the seven ministries listed above are involved in water quality monitoring programmes (APRP, 2002). However, most of these monitoring activities are not conducted on regular bases and in a coordinated manner. Most are conducted for a single objective dictated by the mandate and mission of the involved institution. Also, there are many gaps in the geographical coverage as well as in the monitored parameters.

To remedy the situation, the Environmental Action Plan of Egypt (1992) stressed the need for an integration of the existing water quality activities and a coordination of institutional efforts in the understanding of water quality issues. Several initiatives in the late 1980s and early 1990s were jointly funded with the Dutch government provided a conceptual design of an integrated national water quality monitoring network in Egypt.

The National Water Quality and Availability Management Project (NAWQAM) was initiated by the National Water Research Center (NWRC) of the MWRI with technical assistance from the Canadian International Development Agency (CIDA) in 1998 and completed in 2007 (MSEA, 2001; National Water Research Center (NWRC) & Agriculture & Agri-Food Canada (PFRA), 2008). The project consisted of a number of components. The NAWQAM Objective is to assist the government of Egypt, through the MWRI, with the overall management of its water resources within the context of responsive and coordinated systems for sustainable water resources management.

The objective of the National Water Quality Monitoring component of the NAWQAM was to rationalize water quality monitoring activities into a sustainable national monitoring programme. The first step in designing a monitoring programme was the determination of an optimum number of sites (sampling points), the selection of the water quality parameters to be measured, and the sampling frequencies within the context of a unified national monitoring system.

The design of the national water quality monitoring programme was based on three different but interrelated approaches, which included (NWRC & PFRA, 2008):

- a pragmatic justification for the existing monitoring sites (number and distribution);
- end-user feedback on the current and future needs of the national programme (sites, parameters and frequency of sampling); and
- statistical analyses to obtain the optimum sampling frequency.

The design approach for the national water quality monitoring programme is shown in Figure 3.

Site justification considers the type and degree of catchment characteristics. Site justification was based on separate evaluation schemes for the irrigation canals, drains,

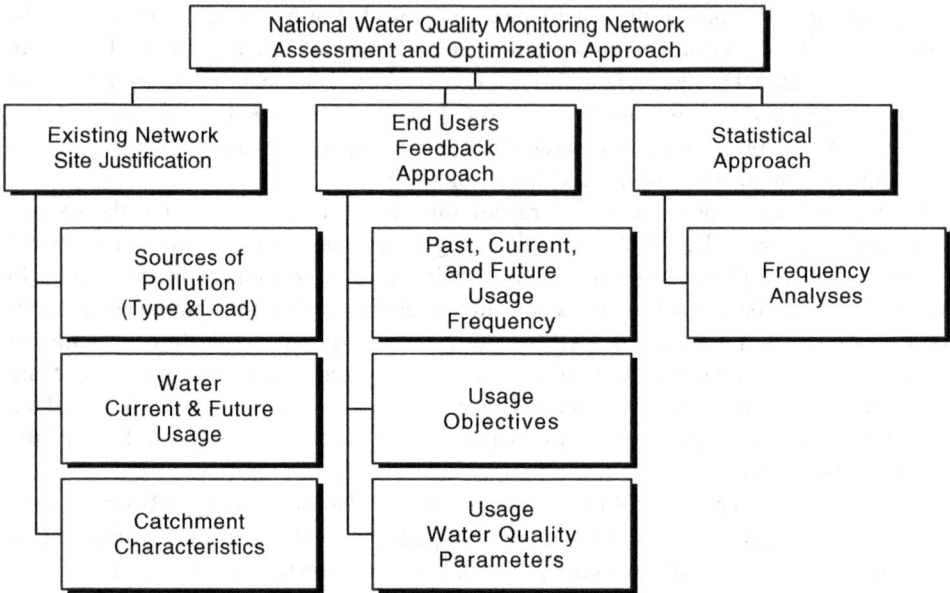

Figure 3. Design approach of the national monitoring network. *Source*: National Water Research Center (NWRC) & Agriculture & Agri-Food Canada (PFRA) (2008).

main stem of the Nile and its two branches, and Lake Nasser. Based on a score (points) assigned, a ranking for each site was determined.

A programme justification scoring matrix provided a primary assessment that was used as a guideline when defining the priority locations of the national monitoring programme. Generally, the sites linked to an active reuse project demonstrated potential for high domestic and industrial pollution, and those located at the outfalls or pump stations received a high priority ranking. A total of 50 irrigation sites and 110 drainage sites were recommended in the Delta (Figure 4a) and Fayoum (Figure 4b). Meanwhile, a total of 70 sites were recommended for the main Nile and irrigation canals and drains in Upper Egypt Figure (4c).

The water quality monitoring study is designed to obtain quantitative information on the physical, chemical and biological characteristics of water. Due to the stochastic nature of the collected information, the design, optimization and evaluation of the monitoring programme relies on a statistical approach to statistical analyses (Sanders & Anderian, 1987). The programme's design and evaluation were based on the selection of the optimum number and distribution of sampling sites for a given frequency of sampling and the number of important parameters.

Water quality parameters determined by the monitoring programme along Lake Nasser include water temperature, pH, dissolved oxygen (DO), turbidity, nitrate, phosphorus, biological oxygen demand (BOD), total dissolved salts (TDS), and faecal coliform (FC). In the main Nile and its two branches traces of heavy metals are also detected. Water quality data from the lake and river are generated from sampling campaigns carried out by the NRI during high- and low-flow periods.

The National Water Quality Monitoring network covers 54 sites on the irrigation canals of the Delta. Monthly data collection is carried out on a continuous basis by the DRI during each year. More than 34 water quality parameters used were regularly collected. They were classified into six main categories: physical (turbidity, total suspended solids, total volatile solids, pH, temperature); biological (BOD, DO, COD); bacteriological (FC, total coliform); nutrients (NH_4, NO_3, NO_2, P); heavy metals (Fe, Pb, Cu, Zn, Mn, Cr, Hg, etc.); and salinity (electrical conductivity (EC), TDS, soluble ions).

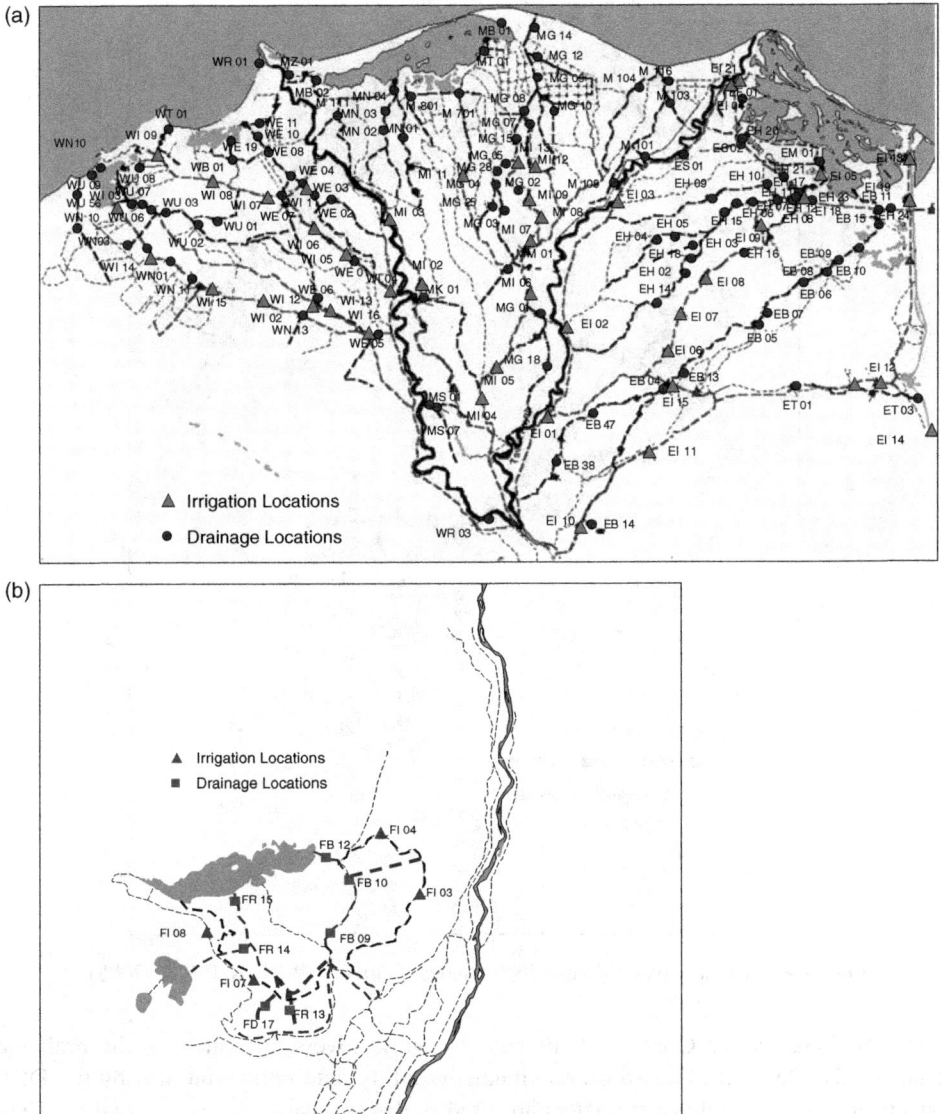

Figure 4. Surface water monitoring locations: (a) irrigation and drainage networks in the Delta; (b) irrigation and drainage canals in Fayoum. *Continued.*

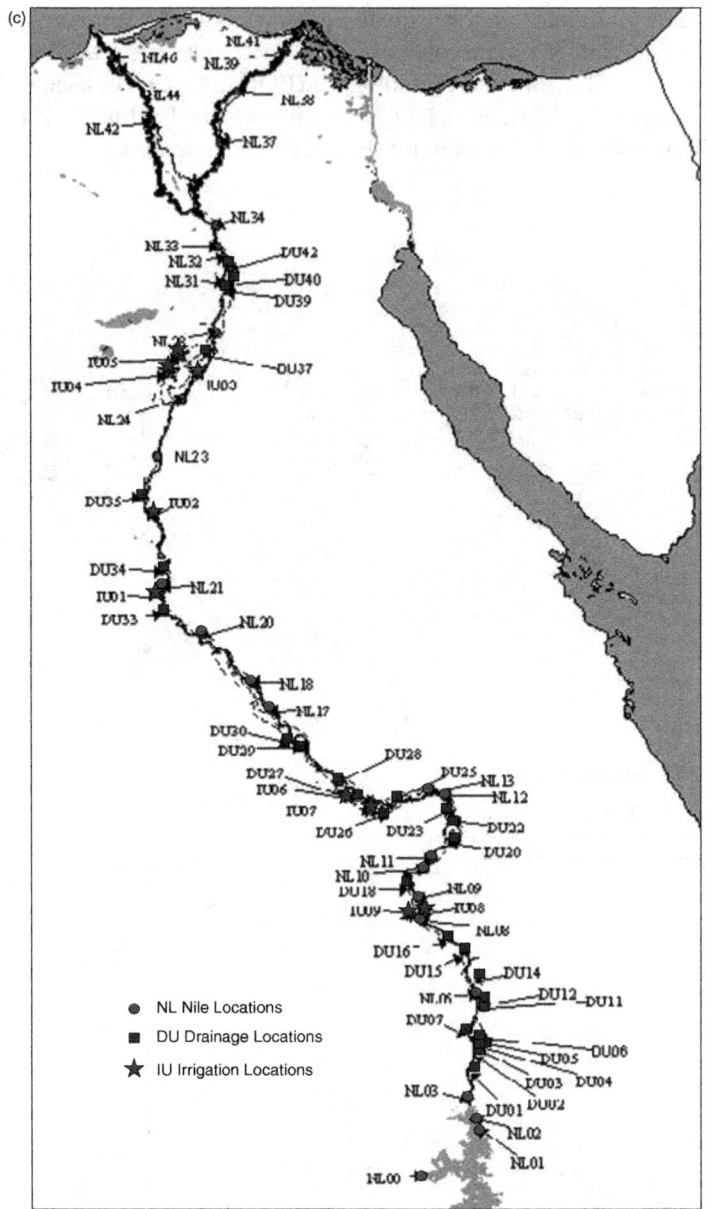

Figure 4. (c) the River Nile and its branches. *Source*: NWRC & PFRA (2008).

The National Water Quality Monitoring Network covers 106 sites on the drainage canals of the Delta. It is based on continuous monthly data collection also by the DRI. Twenty-four water quality parameters are used in the assessment of water quality. They are classified into the same main categories as for irrigation canals. Different modes of visualization and analysis are exercised to furnish enough information to judge the overall water quality status of the Delta drainage system.

Groundwater quality is monitored in the Delta, Valley, Eastern and Western Deswer, and Sinai Desert once a year. The major parameters for assessing groundwater quality are the major cations (K^+, Na^+, Mg^{2+}, Ca^{2+}) and the major anions (Cl^-, SO_4^{2-}, CO_3^{2-}, HCO_3^-, NO_3^-). TDS is a summation of all major solid constituents. pH and EC are important parameters providing indications for groundwater quality assessment.

Results of Water Quality Assessment

Assessment of the water quality data during the whole period of the NAWQAM project (1999–2007) indicated that Lake Nasser exhibits excellent water condition in both low- and high-flow conditions. However, nutrient samples indicate a trend towards atrophic conditions. The Nile from Aswan to the Delta Barrage remains healthy and suitable for present beneficial uses. Most water quality parameters were within the standards set by Law 48/1982, the exceptions being near Kom Ombo (Upper Egypt), around Greater Cairo and Giza. This is due to high total counts of coliform bacteria caused by the discharge of insufficiently treated sanitary wastes and high organic load discharged from drains. The self-cleaning capacity of the Nile, however, seems adequate to manage this problem. In addition, water quality analysis provides an indication of the high organic contamination and deficiency of dissolved oxygen in the waters of both the Damietta and Rosetta branches.

The water quality parameters that range in variation along Lake Nasser and the Nile for the period (2000–2006) during high- and low-flow conditions are presented in Table 3. The measured water quality parameters are averages over the study period. The concentrations of water quality parameters in the two branches of the Nile are always higher than in the main river. The spatial distribution of water quality parameters varies downstream of the HAD due to discharges of return flows from agricultural, domestic and industrial uses. Figure 5a,b gives the quality of drainage water at the outlet of agricultural drains into the Nile (NWRC & PFRA, 2008). The high pH at Km 9.9 (where Km 0 is the HAD) is related to industrial waste discharge by the CHEMA factory. The dissolved oxygen concentrations for most of the drains comply with Law 48/82 (above 5 mg/l), with few exceptions. Nitrates and total phosphates are also within the recommended limits (less than 45 mg/l), except for two drains. The BOD is generally within the allowed limits, except for a number of drains. TDS salts are generally low in the drainage water of Upper

Table 3 Variation of water quality parameters in Lake Naser, Nile River and the two branches.

Parameter	Lake Naser		Nile and branches		Law 48
	Low flow	High flow	Low flow	High flow	
Temperature (°C)	11.0–21.2	17.5–31.0	11.0–23.0	22.0–32.0	
Ph	7.1–8.9	7.1–8.7	7.4–8.7	7.16–8.74	7.0–8.5
Dissolved oxygen (DO) (mg/l)	4.1–9.8	2.4–7.7	3.0–13.78	2.1–9.3	>5
Turbidity (NTU)	0.6–13	0.4–41.9	0.75–21.50	0.9–22.3	
Nitrate (mg/l)	0.4–2.2	0.7–6.9	0.2–26.70	0.2–9.2	<45
Phosphorus (mg/l)	0.07–0.2	0.06–0.62	3.2[a]	1.89[a]	
BOD (mg/l)	0.4–2.2	0.77–2.47	0.8–22.0[a]	1.0–11.0	6
TDS (mg/l)	135–170	148–173	148–483	161–524	<500
Faecal coliform (cfu/100 ml)	0–830	5–1,500	10–3,834	10–3,834	2,000[b]

Notes: [a] High values are detected in the Nile branches.
[b] Recommended value in raw water intended for the potable water supply.

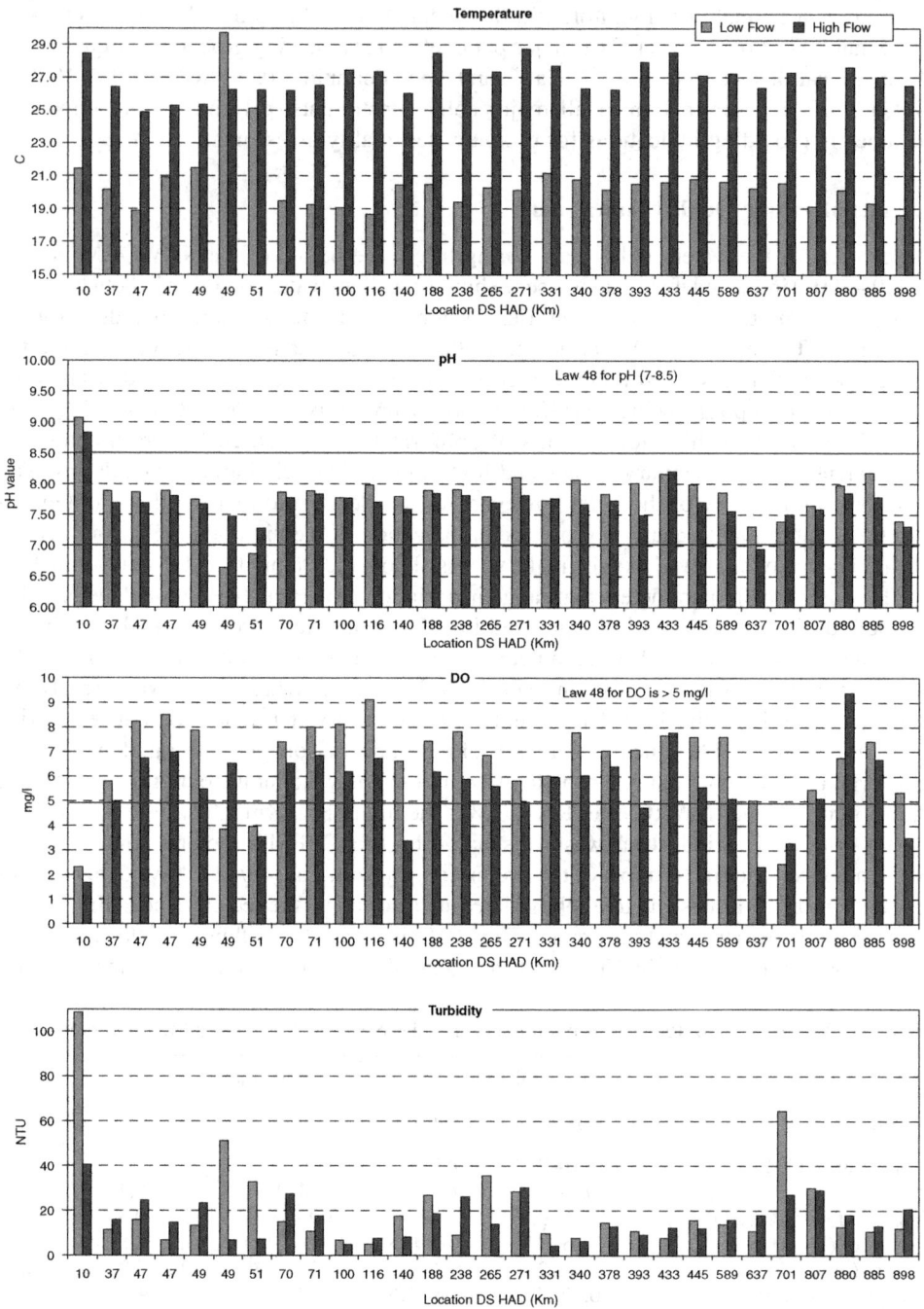

Figure 5a. Water quality parameters of drainage water (temperature, pH, dissolved oxygen (DO) and turbidity) along the Nile Downstream High Aswan Dam (HAD). *Source*: NWRC & PFRA (2008).

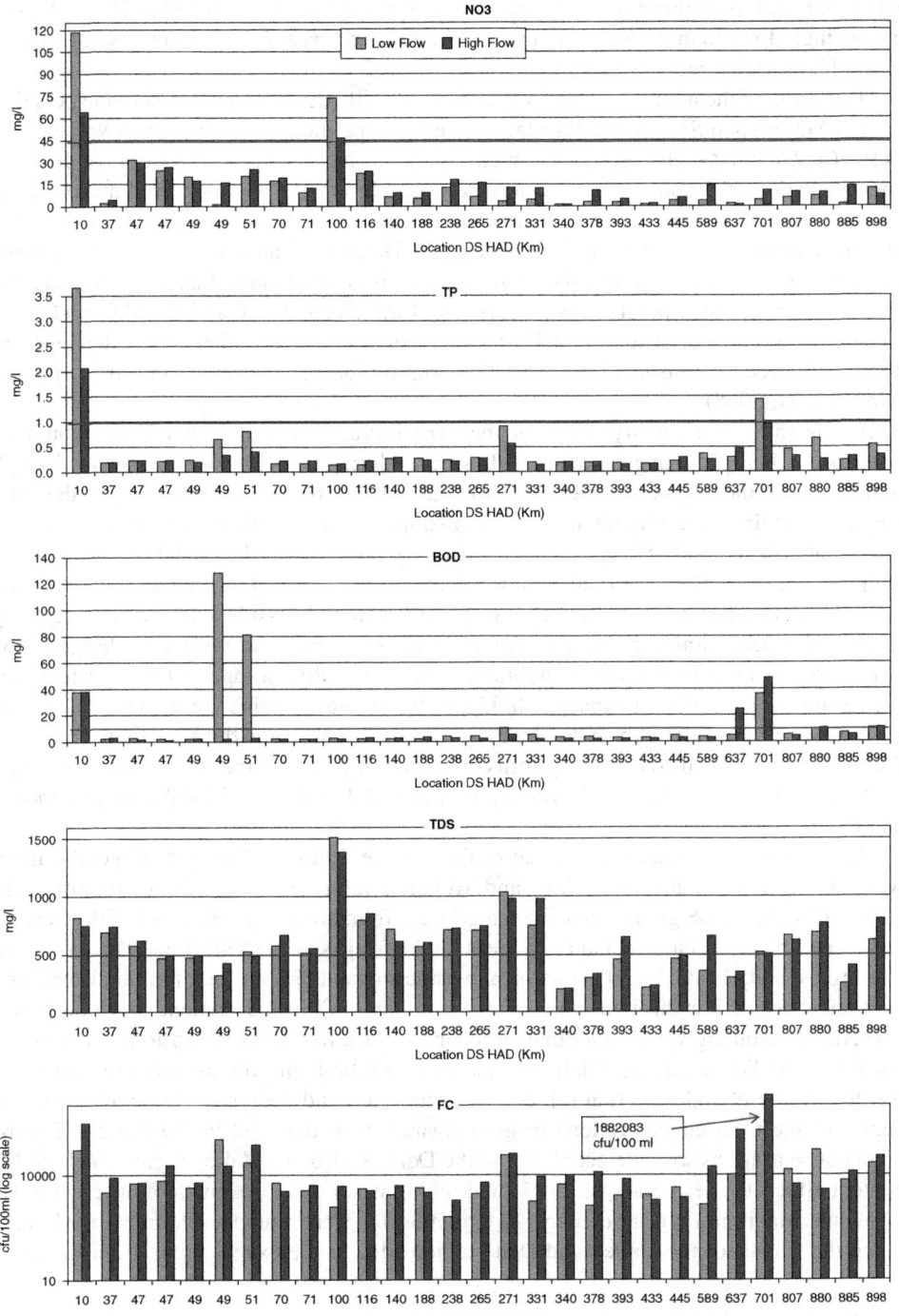

Figure 5b. Water quality parameters of drainage water (NO_3, total phosphorus (TP), biological oxygen demand (BOD), total dissolved salts (TDS) and faecal coliform (FC)) along the Nile Downstream High Aswan Dam (HAD). *Source*: NWRC & PFRA (2008).

Egypt but slightly higher for the drains serving new land reclamation areas. FC are higher during high flows than in low flows in most drains, perhaps because of the flushing effects of the high discharges.

The results of the analysis of heavy metals in the Nile demonstrated at certain spots that Al, Fe, Mn, Ba and Sr reached detectable values. Some metals have very low values such as Pb, Cu, Cr and Zn. However, other metals (Cd, Co, Ni, Sb, Se, Sn, V) were not detected because their concentrations were below detection limits. All the measured metals along the Nile were found below the recommended standards of Law 48/1982. However, only two sites located at Km 1058 and 1096 along the Damietta branch have Cr concentrations exceeding the standards. From the above results, it is recognized that the concentration levels of all monitored parameters comply with Law 48/1982, except DO, BOD and FC at some sites along the Rosetta and Damietta branches. On the other hand, the reported figures showed a seasonal and spatial variation along the Nile and its branches (NAWQAM, 2008).

The overall water quality status of the agricultural drains in Upper Egypt could be described as being low in general terms. It is better during low-flow conditions than in high flows. During the assessment period (1999–2007) low flows of about 48% of the total monitored drains were classified as being medium quality level, 45% were classified as poor quality level, and 7% were classified as very poor quality level. Meanwhile, during high flow, about 17% of the total sampled drains were classified as medium quality level, 80% were classified as poor quality level, and 3% were classified as very poor quality level. In the Delta, drains suffer from domestic and industrial waste water discharges. This is in addition to microbiological pollution along with diffuse agricultural waste water exceeding the assimilation capacity. In Upper Egypt, three drains are identified as major sources of pollution along the Nile: the El Sail drain at Aswan and the Kom Ombo and Etsa drains at El-Minya. In the Delta priority attention is given to the drains receiving high loads of pollution such as Bahr El Baqar, Bahr Hadus, El-Gharbia, El Rahawy, El Umoum and El Moheet drains.

Most of the ground water-monitored aquifers in the Delta and Valley have good-quality water that qualifies them for drinking and irrigation purposes with minimum treatment. In some wells, high TDS greater than 3000 mg/l were found in the groundwater of the Eastern Desert region and in land-reclamation areas on the fringes of the Valley and Delta regions. The high salinity is attributed to chemical weathering and dissolution of minerals within a geological formation through which the water flows. Chloride content in groundwater exceeds the drinking water guidelines frequently in Cairo and in the eastern and western fringes of the Delta regions. High chloride concentrations in groundwater are caused by the dissolution of soil salts (halite), evapotranspiration and deep percolation of irrigation water. Highest Na concentrations in groundwater were detected in the Eastern Desert. High concentrations were detected also in the Delta, Cairo and Valley regions. A high Na content affects the aesthetic quality of the drinking water and the growth of different types of plants. Maximum nitrate contents in groundwater were found in the reclaimed areas along the fringes of the Valley and Delta regions due to the excess input of fertilizers.

Data and Information Sharing

Obtaining data from various governmental agencies is not an easy task and is time consuming. Sharing and disseminating data and information is the weakest part of the

WQM process in Egypt. The Drainage Research Institute prepares an annual report which includes the collected data and information regarding drain discharges, their temporal and spatial variation, and the quantity and quality of reused drainage water. Distribution of this report is limited to a few clients mainly within the MWRI. The NAWQAM project through its national water quality monitoring component, since its inception, has considered a clear objective of providing relevant and timely water quality data and information on the state of Egypt's water resources. Several products have been provided during the project's life time, including the annual water quality status report and various reports of special water quality studies. The NWRC continued this practice after completion of the project.

Investments in Pollution Control

In addition to the effort to control pollution of fresh water bodies by monitoring and enforcing standards and regulations, the government of Egypt is investing to improve the infrastructure of collecting and treating municipal wastewater and capacities for solid waste management.

Improved Sanitation Infrastructure

New sewerage systems have been constructed or are under construction in Egyptian cities and villages. More financial resources are allocated to scale up the construction of new projects. The annual investments in water supply and sanitation increased from 5 billion Egyptian pounds (US$1 billion) in 2004–2005 to 15.7 billion Egyptian pounds (US$2.8 billion) in 2008–2009. Investment is expected to increase in the near future. Investments in the water supply and sanitation sector during the next ten years (2010–2020) are expected to be 200 billion Egyptian pounds (US$36 billion). Improved sanitation services will be extended by completing the on-going construction of new projects in 103 cities and 424 villages out of which 151 villages are counted among the poorest villages in Egypt. Figure 6 shows the progress in extending the waste water treatment services in the Egyptian cities and villages since 2004 until expected full coverage.

Integrated Rural Sanitation and Sewerage Infrastructure Project

The lack of adequate sanitation facilities and solid waste management tools in rural areas of Egypt is counted as a main source of pollution of canals and drains. In order to achieve

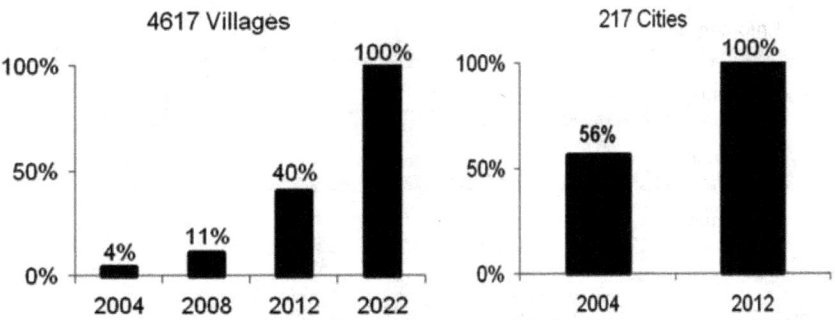

Figure 6. Waste water treatment infrastructures in villages (left) and cities (right).

the goals of integrated water resources management in rural areas, a new project is launched recently to provide adequate measures for sanitation in villages that were not enrolled in current national plans for improving sanitation. The project will be implanted in the Behaiyra, Kafr El-Sheikh and El-Gharbia governorates in the northern Delta where the Integrated Irrigation Improvement Irrigation Project (IIIMP) is being implemented within the principles of integrated water resources management.

Piloting Low-cost/Unconventional Technologies

Economy of scale makes conventional wastewater collection and treatment cost-prohibitive in small/sparse rural settlements. Several alternatives that vary in efficiency and cost have been experimented at the pilot level in Egypt. For instance, upgrading the self-purification capacity of open drains (by construction of in-stream wetlands or by aeration gabions) proved to be a viable non-conventional approach. EEAA's construction (with funds from Global Environmental Facility (GEF)/United Nations Development Programme (UNDP)) of a 20-hectare engineered wetland has been in operation since March 2004. The project had the objective of treating 25,000 m^3/day of drainage water as a demonstration for a low-cost technique to protect the ecology of Lake Manzala and the Mediterranean Sea.

Other low-cost and unconventional technologies are also being piloted in Egypt's rural areas including the application of double septic tanks, small-bore sewers, biogas technology, a gravel bed hydroponics (GBH) system and bioremediation. Scaling up these pilots requires improved financial and institutional arrangements.

The State of Water Quality Management in Egypt

WQM includes complex processes and actions that do not necessarily go side by side. It includes setting policies, developing institutions, issuing and enforcing laws and regulations, and many other aspects as explained above. Progress along one of these

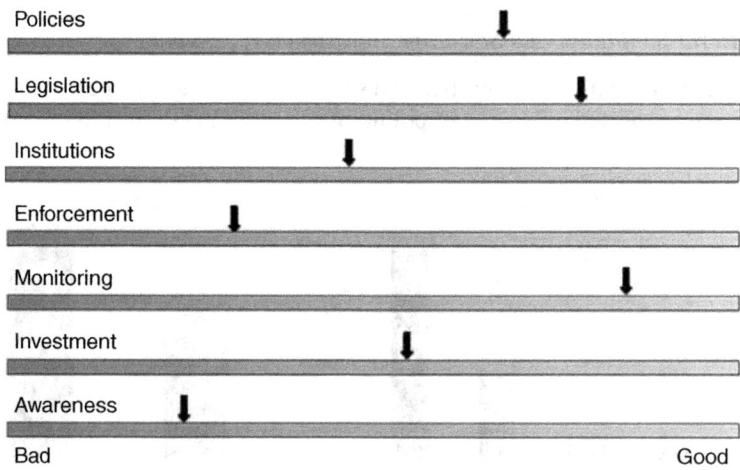

Figure 7. Progress in improving different aspects of water quality management.

aspects could be well ahead, but in others progress may lag behind (Figure 7). Therefore, it is extremely difficult to say that the overall WQM in a country or in a basin is good or bad. It is only possible to monitor the progress achieved with a certain aspect against a given set of indicators.

Improved WQM in Egypt is recognized as an indispensible way to sustainable development and a means to combat the challenge of water scarcity. It has been set as a primary development objective in the government short- and long-term plans. However, the increasing pressure on water resources due to the rising demand of a fast-growing population combined with a heavy financial burden to achieve this goal is a real challenge. Nevertheless, the efforts continue and the review presented in this paper shows that the country is on the right track. As a conclusion, the state of improving WQM can be summarized in terms of implemented, in-progress and to-be-initiated actions that can be summarized as follows:

- Implemented actions.
- Water quality standards are developed.
- Laws and by-laws exist.
- Monitoring networks are established.
- Quality control and quality assurance of monitoring programme are developed.
- Environmental law and regulation have been passed.
- Capacities in fields of WQM expertise are developed.
- Agrochemicals are properly regulated.
- Wastewater management is among the high priorities.

Actions in Progress

- A clear strategy and action plan for integrated management of resources.
- Industrial pollution is among the high priorities.
- Policy with respect to wastewater management.
- Monitoring objectives are clearly defined and programmes are established on the basis of sustainability.
- Up-to-date water law and related standards.
- Data and information on the status of water quality are shared with concerned institutions.
- Stakeholders are involved in water resources planning and management.
- Awareness of the cause-and-effect relationship in water quality status is increased.
- Awareness of facilities and possibilities is raised.
- Integrated approach for WQM.

Actions to be Implemented

- Options for involvement of the private sector.
- Responsibilities are coordinated among concerned authorities with minimum overlaps.
- Management instruments beyond those commonly used and controlled by bureaucracy.
- Decentralization of planning and decision-making.
- A complete set of standards and guidelines is available for the public.

- There is an appropriate legal framework to implement alternative ways and means for pollution abatement and control means.
- Standards are satisfactory and applicable to the specific situation in the country.
- Financial resources allocated for WQM are based on actual cost or on cost-recovery principles.
- Data and information on the status of water quality are freely accessed by the public.

Conclusion

Egypt has made noticeable progress in WQM over the past three decades. Policies, institutions, legislations and monitoring networks are in place. The water quality of Lake Naser and the main stem of the Nile are good and safe for all uses. The situation is not the same for some canals and most drains. The groundwater quality of most aquifers is good and safe for drinking, except in some agricultural areas. Overall, the trend of pollution in many water bodies is alarming, particularly for a water-scarce country like Egypt. Pressure created by a fast-growing population and economic development seems to exceed the existing capacities to reach an effective level of control on the several sources of pollution. More investment is needed to improve the wastewater treatment and disposal infrastructure. Compliance and enforcement of laws for pollution control and environmental protection are problems that need more awareness and better understanding. NGOs should take a bigger share of responsibility in improving this situation. Sharing and disseminating information about water quality is an effective tool to improve awareness and gain public support for pollution control programmes.

References

Agricultural Policy Reform Program (APRP) (2002) *Survey of Nile System Pollution Sources.* Report No. 64 (Cairo: Ministry of Water Resources and Irrigation)).

Arab Water Council (AWC) (2009) *Middle East and North Africa Region/Arab Countries Regional Report.* 5th World Water Forum, Istanbul (Cairo: AWC).

Ibrekk, H. O. (2003) *Water Quality Management.* Draft Issue Paper (Washington, DC: Environment Department, The World Bank).

Ministry of State for Environmental Affairs (MSEA) (2001) The National Environmental Action Plan of Egypt 2002/17: Environment at the Center of Modernizing Egypt, Unpublished Discussion Paper (Cairo: MSEA).

Ministry of Water Resources and Irrigation (MWRI) (2005) *National Water Resources Plan 2017: Water for the Future* (Cairo: MWRI).

National Water Research Center (NWRC) & Agriculture and Agri-Food Canada (PFRA) (2008) *National Water Quality and Availability Management: National Water Quality Monitoring Component 100. Final Report* (Cairo: NWRC & PFRA).

National Water Quality and Availability Management Project (NAWQAM) (2008) 2008 Component 1000 final technical report. Report no. WQ-TE-0811-023-FN, National Water Research Center, Egypt (Cairo: NAWQAM).

RIGW/IWACO B.V. (1994) *Inception Report of the Environmental Management of Groundwater Resources* (Cairo: Research Institute for Groundwater).

Sanders, T. G. & Anderian, D. D. (1987) Sampling frequency of river quality monitoring, *Water Resources Research*, 14, pp. 569–576.

World Bank (2005) *Country Environmental Analysis (1992–2002)* (Washington, DC: Water and Environment Department, the Middle East and North Africa Region, The World Bank).

World Bank (2007) *Making the Most of Scarcity: Accountability for Better Water Management Results in the Middle East and North Africa.* Middle East and North African (MENA) Development Report (Washington, DC: The World Bank).

A New Mindset for Integrated Water Quality Management for South Africa

L. BOYD* & R. TOMPKINS**

*Golder Associates Africa, Halfway House, South Africa; **Jeffares and Green, Cascades, Pietermaritzburg, South Africa

ABSTRACT *The aim is to develop a conceptual model for aligning water resource quality and drinking water quality management. The model is based on the premise that good water quality is in everyone's best interests. Current management approaches attach responsibility for good water quality at a high level. The integrated water quality management approach "breaks down" the management of water quality into smaller management units and establishes a horizontal and vertical reporting framework through the application of a generic business process. The business process requires the identification of critical risk factors and critical control points in each management unit. The results of the project were the development of the water use cycle, as a context for the model, and the development of a basic integrated water quality management. A case study is currently underway in the Breede River catchment of the Western Cape Province of South Africa in order to refine the proposed model.*

Introduction

The study presented in this paper was undertaken with sponsorship from the Water Research Commission. The paper advocates a mindset change and is therefore presented in simple language to make it accessible to a broad audience at varying levels of scientific understanding. The aim of the project was to develop a conceptual framework for aligning the management of the quality of water resources with that of drinking water quality in order to support the effective management of water use, in the interests of all water users. The project encompassed an international literature review as well as a review of the legislative instruments and management processes which relate to water resource and drinking water quality in South Africa, notably: the Water Services Act (1997) (Department of Water Affairs & Forestry (DWAF), 1998b); the National Water Act (1998) (DWAF, 1998a); the National Water Resources Strategy, Edition 1 (2004) (DWAF, 2004b); Regulations under Section 9(1) of the Water Services Act (1997) (DWAF, 2001); A 5-Year Water Resource Quality Monitoring Plan (2004) (DWAF, 2004a); and A Drinking Water Quality Framework for South Africa, Edition 2 (2007) (DWAF, 2007).

The literature review showed that some countries have moved towards a catchment-wide approach. However, many strategies, particularly in developing countries, remain focused on the quality of water at the waterworks and within the distribution system. Evaluated against the integrated catchment-to-consumer cycle used during the project as a context for the management model, the management framework established under current legislation is inadequate as it splits the legislative and institutional frameworks between raw water quality and drinking water quality. Furthermore, it is focused primarily on monitoring rather than managing water quality. The current management framework in South Africa is therefore reactive, rather than proactive. Furthermore, in developing countries in general, and particularly in South Africa, implementation of national legislation, and enforcement of its provisions, is an acknowledged area of weakness.

Currently, national level systems for the management of water quality are highly complex and positioned at a very high level: nationally (in the case of water resource quality) and municipally (in the case of drinking water quality). An added complication is that regulation of drinking water quality takes place at a national level. Accountability for good water quality is therefore also at a high level, at the end of a relatively long management chain.

The conceptual model developed is based on the premise that good water quality is in everyone's best interests. Current management approaches, however, attach responsibility for good water quality at a level that does not identify this premise. The management approach is therefore institutionally based at relatively high levels of government, and does not include potential community structures which should have responsibility for the water they use both consumptively and non-consumptively. It should be noted that the term "community" is used here in the sense of a group of people or organizations with common interests; in this case, regarding the quality and quantity of the water within their geographical area.

Problem Statement

There is a need for a management framework to integrate the management of raw water resources and drinking water at a level that goes beyond both the current high-level institutional approach, and the reactive approach to the protection of water quality and quantity.

Methodology

The approach to the development of the conceptual framework for integrated water quality management (IWQM) was to use an interactive forum with relevant stakeholders from both the water supply and sanitation (water services) and water resources sectors throughout South Africa. This was accomplished through a series of focused workshops held throughout South Africa over the period of two years. The inputs and comments from stakeholders were then assimilated by the team and categorized into the following themes:

- Guiding principles that underpin the model.
- Background conditions that support the implementation of the model.
- The management framework that determines the way management is structured (who manages).
- Management requirements that determine the way water quality and quantity is managed (how it is managed).
- Broader context issues that determine the issues management must address (what is managed; see "The Water Use Cycle").

The model was developed by linking the themes systematically to move toward the objective of IWQM. The various iterations of the model that were developed as a result of these assimilations were re-presented to the broader workshop group, and refined into a final conceptual model at the end of Phase 1 of the project. This process ensured that from project inception, the stakeholders and other interested and affected parties had a good understanding of the project and its progression.

Moreover, this active participation led to a group of stakeholders volunteering to be a test area for the second phase of the project currently underway, in which the model is being tested and refined in a small area (the Hex River Valley) of the Breede Catchment area of the Western Cape Province of South Africa. The testing process has resulted in significant refinement of the original conceptual model and a workable format of the model is now emerging from this process.

Integrated Water Quality Management Model

The model itself is based on the overarching philosophy that "Everyone is downstream"—everyone's use of water impacts someone else's use of water. When we talk about a mindset change, we mean this in the context that water management has traditionally been seen as an institutional responsibility: responsibility conferred on a particular institution by a particular Act or regulation. If the model is based on the philosophy that everyone is downstream, however, the possibility is set up that every water user is a water manager. This statement gave us the place to start: water uses and water users.

The Water Use Cycle

The Water Use Cycle developed from the catchment-to-consumer cycle; the basis of the approach used in the South African Department of Water Affairs Drinking Water Quality Framework (DWAF, 2007). However, the structure of the catchment-to-consumer cycle relates more specifically to the management of drinking water quality by water services delivery institutions than to IWQM. In this respect several iterations of the catchment-to-consumer cycle were developed and workshopped with stakeholders and The Water Use Cycle (Figure 1) became the context of the management model.

The Water Use Cycle includes the impact of land use as a context to the cycle; and also takes into account the activities of raw water consumers—both raw water abstraction and discharge without treatment, such as may occur from informal settlements. The Water Use Cycle continues to evolve and it has been recognized that precipitation as a consumptive use should also be taken into account in The Water Use Cycle. The Water Use Cycle therefore is the context for the IWQM framework in the sense that these are the elements of water use that need to be managed.

Having now clarified and contextualized the model, the elements of the model itself are described below. Following on from the overarching philosophy, the model is comprised of a number of components that will be discussed in the following sections.

Defining Principles

The following principles were prioritized based on the frequency with which they were raised in the consultation process. In this report principles are defined as being generalizations that are accepted as true and that can be used as a basis for reasoning or conduct. These following principles therefore underpin the conceptual model for IWQM

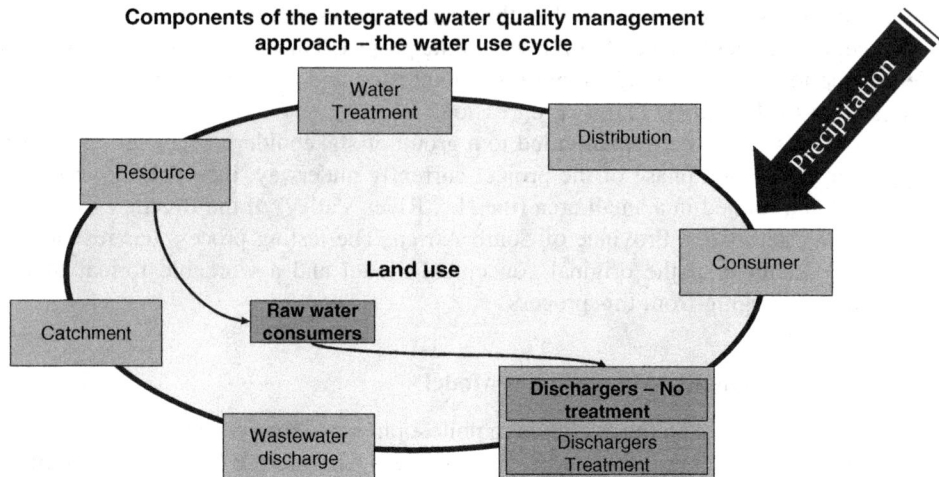

Figure 1. Water use cycle including precipitation as a consumptive use.

in the South African context:

- Water must be properly valued (there is not enough water).
- Institutions responsible for managing water must be accountable for water quality.
- Water quantity and water quality are inextricably linked.
- The polluter-pays principle must be applied to the true cost of water pollution.
- Short-term economic gain at the cost of increasingly deteriorating water quality is not acceptable.
- Everyone should have access to water quality information (not data).

Water must be properly valued: it is not only important to ascribe value to water based on water availability and increasing water scarcity. The concept of value in the context of water should include the downstream costs of pollution; the social and economic value of water; the value of wastewater; the significance of clean water in terms of public health; and even the price of not having water. Therefore, the principle of there not being enough water should encompass an understanding of the different values of water, and not be limited to the fact that there is not enough water. Appropriate valuing of water (and wastewater) also raises the issue of ring-fencing the revenue from water sales (both raw and treated water), so that the funds derived can be used for the management of water as opposed to elsewhere in the organization.

Institutions responsible for managing water must be accountable for water quality: accountability is the obligation to demonstrate and take responsibility for performance in light of commitments and expected outcomes. In the case of water quality, under our current framework accountability is not clear because of the complex institutional framework and the current understanding of cooperative governance.[1] Accountability implies that someone is accountable to someone else, for something. It is therefore important to ensure that responsibilities are clearly defined, and that those to whom institutions are accountable clearly understand the standards at which water must be managed in order that they can assess whether institutions are fulfilling their obligations with regard to water quality. Finally, a commitment to management practices that will ensure good-quality water must be evident at all levels both within and across the spectrum of water management institutions.

Water quantity and water quality are inextricably linked: it is important that the above statement is consistently recognized in all aspects of water management. "Poor"-quality water will reduce the quantity of water available for use, and less water will increase the impact of contaminants in water. While this seems to be stating the obvious, much of the documentation, legislation, regulation and research addresses either water quantity or water quality.

The polluter-pays principle must be applied to the true cost of water pollution: the principle is a well known and widely accepted environmental policy principle applied internationally through various mechanisms. It does, however, raise the question, "pays what?". In the case of water pollution, there are always "downstream costs" of a pollution incident. The term "downstream costs" must be understood in both its literal and figurative sense. There may be costs to water users physically downstream of a pollution incident, and there may be significant costs over time owing to environmental deterioration at the site and physically downstream of an incident. Furthermore, "downstream costs" could refer to indirect costs such as the cost of a community not being able to develop as a result of a lack of availability of clean water. It is important, therefore, that the polluter-pays principle encompasses the expanded definition of "pays what?".

Short-term economic gain at the cost of increasingly deteriorating water quality is not acceptable: this principle refers mainly to the fees levied on dischargers of wastewater to sewer, the discharge which then has an impact on the wastewater treatment works and its capacity to operate optimally. It is not acceptable that the discharger simply pays increasing fees when the "downstream" cost of discharging is creating a serious long-term impact on the water resource. The short-term economic gain received by those levying charges must be balanced against the total cost of wastewater entering the resource. This principle is closely related to the appropriate valuing of water.

Everyone should have access to water quality information (not data): everyone who uses water has some responsibility for water quality. Because water quality is a largely technical issue, most of the "information" disseminated about it is technical. While this is necessary at certain levels of responsibility, new and innovative ways to package information about water quality need to be found. It is important that there is some understanding about water quality at all levels, and this will require a "rolling-up" of water quality data into more broadly understood formats.

Background Conditions

In the context of the proposed model, the term "background conditions" is defined as those conditions external to water quality that support the implementation of this framework and therefore indirectly impact on water quality. They have been categorized under eight major themes, three of which are indirectly related to water quality and five more specific to water quality issues. The external issues that impact on IWQM include:

- Education across the board on water issues, using The Water Use Cycle as the basis for education and awareness.
- Effective strategic planning at various levels which is an acknowledged challenge in most developing countries.
- Funding that is seen as an important supporting condition for IWQM.

The following general categories under background conditions identify broad programmes or initiatives that need to be put in place to improve the implementation of an IWQM framework:

- The value of water (including wastewater) incorporating issues such as cost–benefit incentives and recycling initiatives.
- Management systems and tools (applicable to the various "levels") such as River Health Programmes and other existing water management systems or Water Safety Plans.
- Communication between management units as described below and also public access to information (which includes thinking about how to package water quality information for public consumption).
- Accountability including aspects such as the implementation of the polluter-pays principle, enforcement mechanisms and the implementation of a government watchdog.
- Improving institutional capacity.

Two additional conditions that would have an impact on IWQM that however do not fit into the eight main categories are:

- Understanding the final Catchment Management structure within the current 19 Water Management Areas in South Africa and how it relates to roles and responsibilities.
- Research that would include research into alternative and appropriate technologies as well as reassessment of certain established parameters such as Resource Water Quality Objectives (RWQOs) which may not apply to the whole catchment.

It is recognized that certain of the background conditions would need to be addressed by large, sometimes national, programmes and therefore may not be in place for some time. It is important to note though that the implementation of the model is not dependent on these background conditions being in place, but that the ongoing functioning of the model would be significantly enhanced if they were. It would also be possible to implement the model over a significantly larger area, at a much lower cost.

With the defining principles and background conditions identified, the focus must now be on who are the "managers". Part of the problem is that the management responsibility for water quality is generally at a high level and based in various institutions. To break down this issue, the concept of management units, and a framework into which those units fit, was developed.

Management Units and the Management Framework

A management unit in this context is a geographical area—not necessarily homogenous or continuous—that could be managed as a unit owing to common water-use characteristics at the "lower" levels and to institutional responsibilities with regard to the management of water quality at the "higher" levels. Many of the management units identified align with existing established institutions such as municipalities, catchment management agencies or water user associations. However, it is important to note that the establishment of a management unit at whatever scale is not dependent on whether a legislatively established institution exists at that level. There are four management "levels", which would correspond to management unit types, as follows:

- Community: remembering that "community" is used here in the sense of a group of people or organizations with common interests regarding the quality and quantity of the water within their geographical area, a community-type management could be anything from a single factory to a small settlement (informal or otherwise), to a large group of farmers who participate in an irrigation scheme.[2]
- Municipality: in South Africa municipalities with status as a Water Services Authority (WSA) have the responsibility of ensuring the delivery of water supply to people in their area of jurisdiction and many are also responsible for the treatment and discharge of wastewater. While this means that they have institutional accountability for water quality standards, there is often a lack of capacity to manage these processes effectively. Furthermore, there is no formal mechanism for reporting on water quality between adjacent municipalities, only the Department of Water Affairs (DWA). It must be noted that this is not the case in all municipalities and some do have a strong management capacity (and performance results) and have also implemented informal reporting mechanisms with other municipalities and with catchment management agencies, however this is not the norm.
- Catchment: South Africa is currently divided into 19 Water Management Areas (WMAs), which are comprised of one or more catchments (some have as high as eight catchments). Under the National Water Act (1998) (NWA), these areas are administered by Catchment Management Agencies (CMAs) or the DWA if a CMA is not yet established. The CMA has institutional responsibility for managing water quality in the catchment(s) under their jurisdiction through the implementation of Resource Water Quality Objectives (RWQOs) which should be identified for all the water resources in their area. However, the process of establishing all RWQOs is a long one, and implementation has been slow. A CMA or even a group of water users at catchment level, which becomes a management unit in this model, can therefore begin the process of managing water quality even if the legislative process is incomplete. The management unit can apply to one catchment or to a group of catchments as delineated by the WMA boundary.
- Regional/national: this level refers to the regional (or provincial) boundary (which does not always conform to the catchment boundaries) and the national boundary. Here there is definite institutional responsibility under both the NWA (1998) and the Water Services Act (1997) and at this level of the model the background conditions become increasingly important. Given the relative infancy of the concept, the model has not been tested at this management level.

The management framework indicated in Figure 2 shows how the various management units (made up of water users or water user groups) relate to each other. This structure also addresses those instances where management units may occur across municipal or catchment boundaries. Figure 2 indicates how the water user groups (management units) are represented in the integrated management context and indicates the overlapping management "chains" from the smallest management unit to the largest at a national level. A single, full IWQM chain is illustrated in Figure 3.

The basic premise of the management framework is to break down the challenge of IWQM into manageable areas in order to reduce the reporting between management units

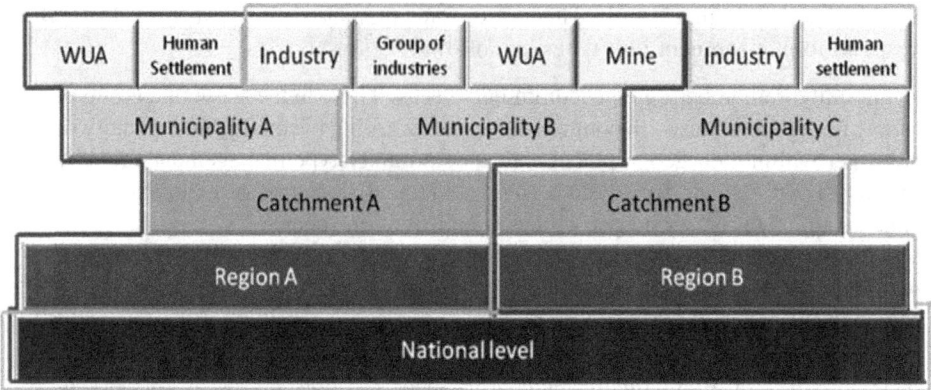

Figure 2. IWQM framework.

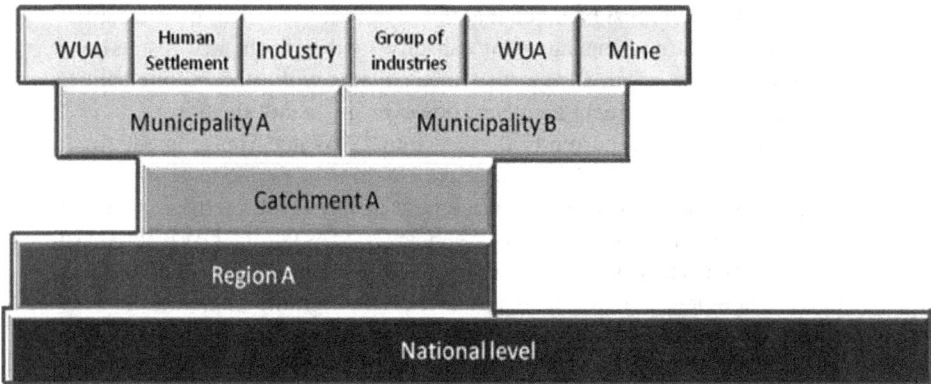

Figure 3. One management "chain" in the IWQM framework.

to a simple "Yes" (quality and quantity parameters are being met) or "No" (they are not). This approach demands effective auditing, but is structured in such a way that adjacent management units audit each other. That is, the management unit is responsible for auditing the quality and quantity, as required, of water entering its geographical area and then reporting on that to the next level of management; as well as to the adjacent upstream management unit, from where the water originated.

It is at this point that the "how" becomes the focus of the model, through a simple generic business process which can be applied at every level of the model.

The IWQM Business Process

A business process can be defined as a process for carrying out a particular activity. In this case the activity is IWQM. The IWQM business process is generic in the sense that its various elements apply at every "level" of management, or rather, to every management unit, and therefore each aspect of the business process must be in place in every management unit. However, the detail of each element will vary according to the type of management unit. The generic IWQM business process is indicated in Figure 4.

The ultimate goal of IWQM is to achieve specific objectives at a particular management unit, taking into consideration the defining principles and background conditions relevant to that management unit. How this is done may be through various tools that may include, for example, a Water Safety Plan (World Health Organization (WHO), 2005) for a municipality or an integrated water and waste management plan (IWWMP) for an industry (Department of Water Affairs (DWA), 2008).

Firstly, it is important to establish a management mechanism. With regard to IWQM, the management mechanism must contain specific elements. These are derived from the WHO Water Safety Plan. However, as mentioned previously, the Water Safety Plan is not applicable to every case in the management context.

There are, however, specific elements that must be included in each tool:

- Hazard assessment/risk analysis.
- Risk management.
- Contingency planning.

The hazard and risk assessment will determine the critical risk factors and critical controls points which are central to the implementation of the business process in every management unit. The business process is based around performance targets (the specific objectives for each management unit) which must be set for Critical Risk Factors (CRFs)

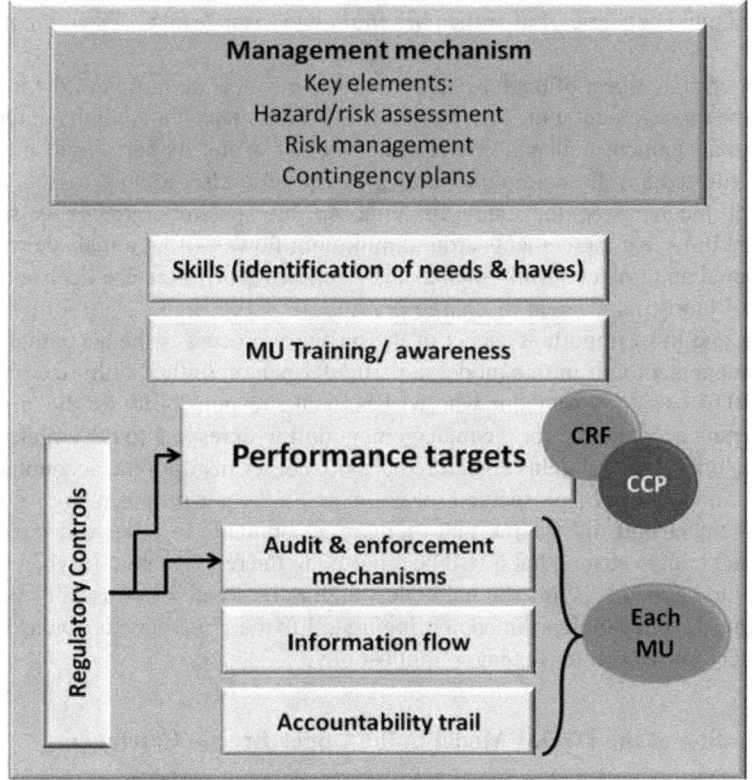

Figure 4. Generic IWQM business process.

and Critical Control Points (CCPs) which must be identified for each management unit. A CRF is defined as a point or process at which, if a failure occurs, the CCP performance targets will not be met. A CCP is defined as a hazard and requires technical target measures/parameter ranges. It is usually defined by the regulatory controls and will usually be a monitoring point (though not necessarily a water sample monitoring point).

Examples of areas within management units that may give rise to the establishment of CRFs or CCPs are a sewage works (or various points within the sewage works), reservoirs or pump stations at the municipal level, the discharge point or drain in a washing bay at a factory or even the management of the use of fertilizers across a number of farms.

To perform the tasks essential to the management mechanism, certain skills are required. If these skills do not exist within the management unit, it is important that the next-level management unit provides support if it is not viable to provide training. In all cases, when a management unit is established, it will be necessary to raise awareness around both IWQM generally, and around the processes being implemented in the particular management unit. A skills assessment will determine where training and support are required, and the form which the awareness campaign will take.

The specifics of the management unit will also determine the nature of the management tools which will be required to manage water quality to the performance targets that must be established for each CCP in the physical area of the management unit. In the larger management units, the CCPs will occur at the boundaries of the smaller management units within it. For example, a municipal management unit will set CCPs at community or WUA management unit boundaries, rather than be required to manage all CCPs within the smaller units.

A further critical aspect of the business process is the clear definition of the information flow from the management unit. This identifies the other management units in the chain to which this management unit will report as to whether or not its performance targets are met. The information flow into the management unit also relates to the audit and enforcement mechanisms that may be imposed by specific agreements with other management units, e.g. agreements around minimum flows exiting a management unit or by institutional audit mechanisms such as the monitoring of water-use license conditions or municipal industrial effluent discharge permits.

The final and most important aspect of the business process is the accountability trail. To be a management unit in this model, a particular person (either authorized by a group such as a WUA or simply someone who wishes to take responsibility for the targets) must sign as a person accountable for the management unit with respect to the targets. It is very important to understand that this signature does not confer institutional accountability—in other words, no one can be prosecuted for not meeting the performance targets. However, each management unit must agree that they are accountable to other water users in the management chain to ensure that it will be reported to the relevant people when the targets are unlikely to be met. It is this mutual understanding between water users of the impacts of their own uses and which is aimed at bringing to life the "Everyone is downstream" and "Every water user is a water manager" philosophy.

Implementation of the IWQM Model in the Upper Breede Catchment

The model is currently being tested, and will subsequently be refined, in the Breede River catchment in the Western Cape Province of South Africa. This catchment falls within the

Breede Water Management Area (WMA) and the Breede River Local Municipality. Figure 5 shows the skewed overlap of the Upper Breede River catchment, which has the mandate to deal with raw water issues, and the Breede River Local Municipality, which is responsible for water service provision, highlighting the concerns described in the Introduction.

The main water users in the catchment are table grape farms, vineyards, urban use including informal settlements and sewage works, various other industries such as pet food producers and distilleries, and chicken farms and piggeries. The main issues related to these users are therefore diffuse pollution from agriculture and industries; diffuse pollution from urban formal and informal areas; and point discharges from industries and sewage works.

The Business Process Implemented

In implementing the business process in the study area, the forms depicted in Tables 1 and 2 are used. The following management units have been chosen as test cases for implementation:

- Chicken farms.
- The Hex River Valley Water User Association (irrigation scheme).
- The Breede River Municipality.
- The Breede/Overberg Catchment Management Agency.

Each management unit is in the process of undertaking a risk assessment to identify the CRFs and CCPs as well as answering the questions relating to the form indicated in Table 1.

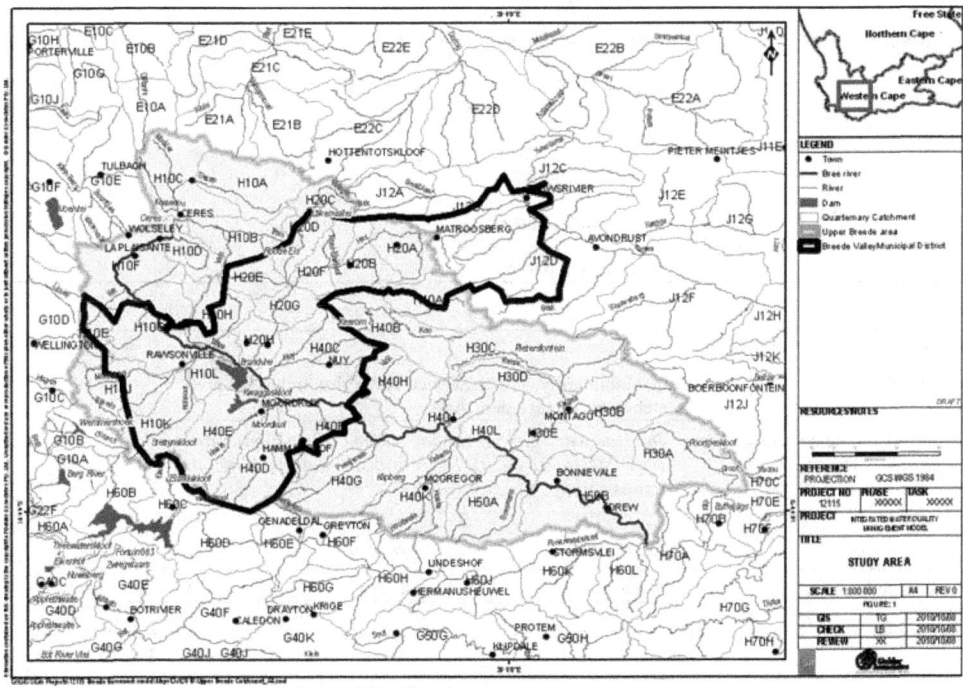

Figure 5. Study area showing the institutional boundary overlap.

WATER QUALITY MANAGEMENT

Table 1. Business Process form

MU Name	
MU Type	E.g. Community, Water Service Authority, Industrial area, Catchment Management Agency
Management mechanism	*If there is an existing management mechanism (e.g. WaSP, EMS or CMS), please indicate:*
	Management Mechanism/(s)
CCPs/CRFs	Refer to Table 6
Existing management tools	*Indicate whether any tools are currently being used to manage water quality, and if yes, what tools:*
	Management Tool/(s)
Regulatory Controls	*Indicate whether any tools are currently being used to manage water quality, and if yes, what tools:*
	Regulatory Control/(s)
Available skills:	*Give details of the people involved currently in management of water quality (of factors which may impact water quality)*
	Personnel Name Position Skill Responsible for
Training requirements:	*Identify where training is necessary to improve achievement of management objectives*
	Training Type Objective of training
Reporting framework	*Give details of other organizations you report to and on what basis:*
	Report name
	Date of last report
	Period (e.g. monthly)
	Who is the report for
	What is reported on

MU Name			
MU Type	E.g. Community, Water Service Authority, Industrial area, Catchment Management Agency		
Audit or Enforcement mechanisms	*Indicate whether your area is audited through any process (e.g. ISO 14001 audits, DWA drinking water quality audit, Blue Drop assessment):*		
	Description *Frequency*		
Accountable person for the MU	*Name*	Organisation	Contact Details
			Email
			Tel
	Signature:		

Table 2. Questions to be asked to define Critical Risk Factors (CRFs) and Critical Control Points (CRPs).

	Defining Critical Risk Factors and Critical Control Points	Critical risk factor/(s)	Critical control point/(s)
1	Description: What is it?		
2	Why is it a CRF/CCP? (e.g. a discharge point, or where fertilizer is being used)		
3	Where is it? (Coordinate or location description—CCPs MUST have coordinates)		
4	What are its targets?		
5	Is mitigation in place? If yes, what? If no, what is being done to improve the situation?		

Table 3 shows some examples of the CRFs and CCPs at each management unit and illustrates the relationship between the CRFs and CCPs as defined above. For example, should the targets at CRF1 not be met, then potential groundwater contamination could occur creating a potential human health risk for those users abstracting groundwater for domestic purposes. Similarly, should the targets at CRF2 not be met, then the potential for pollution of the Hex River will be enhanced (CCP2 targets will not be met) and the potential for eutrophication will increase the impact on downstream users.

Table 3 also illustrates that a CCP at a lower-level management unit (such as CCP2) may become a CRF at a higher level management unit (CRF4). This shows how the reporting structure would be reduced because only CCPs are reported on externally, i.e. between management units. Note that the CRFs must be reported on internally within the management unit. This becomes particularly relevant when management units occur across a larger geographical area than, say, one chicken farm.

In addition, Table 3 shows how one management unit would "audit" another: as in the case above, should a CCP at a lower-level management unit become a CRF at a higher-level management unit, then the lower management unit would report on the results of the CCP to the next level. The next level would not necessarily need to audit these results each time, but only as required. In this way sampling is also reduced.

Conclusions

The IWQM approach "breaks down" water management into smaller management units while establishing both a horizontal and a vertical reporting framework. A further benefit of the model is that responsibility for water quality is based on significantly smaller geographical areas, and accountability to the adjoining areas (horizontal accountability) and to the next level of management (vertical accountability) is established with the establishment of the management unit. This allows accountability for water quality to be focused on smaller management units, rather than diffused up ever higher levels of management. In other words, by making all water users aware of their own responsibility for the protection of South Africa's water resources and accountable for the impacts that they have on the resource.

The identification of CRFs and CCPs in the various management unit types in the Upper Breede catchment shows that the business process is scalable at the various levels of the

Table 3. Example Critical Risk Factors (CRFs) and Critical Control Points (CRPs) for the four management unit levels identified

	CRFs				CCPs			
	CRF1	CRF2	CRF3	CRF4	CCP1	CCP2	CCP3	CCP4
Management Unit	Chicken farms in the area	The Hex River Valley WUA (irrigation scheme)	The Breede River Municipality: Sewage works	The Breede/Overberg CMA	Chicken farms in the area	The Hex River Valley WUA (irrigation scheme)	The Breede River Municipality: Sewage works	The Breede/Overberg CMA
Description: What is it?	Chicken houses	Addition of fertilizer	Maturation ponds	Hex River downstream of the irrigation farms	Groundwater (borehole)	Hex River downstream of the irrigation farms	Up- and downstream of the sewage works	Surface water point in the Breede River
Why is it a CRF/CCP?	Potential groundwater pollution from wash-down water	Diffuse pollution from run-off from areas where excess fertilizer is added	Overflow from the maturation ponds will contribute to pollution load in the Hex River	Surface water from the Hex River is abstracted for other downstream uses	Groundwater is used by other domestic water users in the area	Surface water from the Hex River is abstracted for other downstream uses	Surface water from the Hex River is abstracted for other downstream uses	Surface water from the Breede River is abstracted for other downstream uses
Where is it?	Chicken houses: 33°37′36.30″S, 19°29′34.12″E	All farms in the Water User Association area	Maturation ponds: 33°28′39.21″S, 19°39′03.12″E	*Upstream 1:* 33°24′45.89″S, 19°45′33.01″E *Downstream 1:* 33°32′48.19″S, 19°31′42.36″E	Borehole at: 33°37′34.08″S, 19°29′34.92″E	*Upstream 1:* 33°24′45.89″S, 19°45′33.01″E *Downstream:* 33°32′48.19″S, 19°31′42.36″E	*Upstream:* 33°28′31.14″S, 19°39′09.93″E *Downstream:* 33°28′38.64″S, 19°38′55.45″E	Downstream in the Breede River at the border of the Upper Breede and the two adjacent catchments
What are its targets?	Dry sweeping for removal of solids before wash down	Optimal volume of fertilizer per hectare added (to be researched)	No overflow of ponds	Resource Water Quality Objectives for the Hex River	SANS 241 standards for drinking water	Resource Water Quality Objectives for the Hex River	Resource Water Quality Objectives for the Hex River	Resource Water Quality Objectives for the Upper Breede River
What mitigation is in place?	None; plan to install collection sump to collect contaminated water for disposal/treatment	None; fertilizer use will be measured and added at optimal concentrations	Pond design, evaporation and irrigation	Monthly monitoring	Quarterly monitoring	Monthly monitoring	Monthly monitoring	Monthly monitoring

management framework and across different types of management units. The complexity of the process rests in the detail rather than in the process itself, and therefore larger management units would necessarily have more CRFs and (possibly) CCPs and different types of management units would have varying degrees of technical expertise required for measurement of their CCPs.

What has emerged from the initial implementation phase is that implementation will be greatly enhanced if converted into a web-based system into which the various management units can report. It is understood that not all management units will have access to the technology required to enter data into a system and receive reports from it, but the upper levels of the model certainly will. A system for collecting data from those without access to technology (e.g. management-oriented monitoring system, or MOMS) should be integrated into the implementation process of the model and adequate feedback loops created to ensure that reports reach all management units even if they are not operating a web-based system.

Finally, the IWQM approach allows water quality information to be packaged for a broader audience, as reporting is simplified to provide information on whether or not a management unit is within specifications of its CCPs, rather than extensive technical reports to a national level through the management chain. This addresses the issue of the raising of awareness in the broader community, of the basic premise that good-water quality is in everyone's best interests, while providing for "everyone's" involvement in its management through the allocation of responsibility at more localized levels.

Notes

1. A formal definition for cooperative governance could not be found, although principles exist that define cooperative government (rather than governance). These can be accessed at: http://www.info.gov.za/documents/constitution/1996/96cons3.htm/. The Intergovernmental Relations Act of 2005 does not contain an interpretation of the phrase "cooperative governance". The Act can be downloaded at: http://www.info.gov.za/aboutgovt/coopgov/structures.htm/. Because it is important that this phrase is defined, since the management framework will only be effectively implemented under a cooperative governance framework, cooperative governance is defined in the context of this project, as follows: "South African government institutions or organs of state acting or operating jointly to make decisions that define expectations, grant power or verify performance." This would imply that there must be accountability between institutions or organizations responsible for carrying out management activities.
2. A group such as this is often referred to as a Water User Association (WUA) in South Africa. This is a statutory body established under the South African National Water Act (NWA) No. 36 of 1998 and must be established by a proposal to the Minister of Water Affairs. This means that there are specific provisions regarding what a WUA must undertake to put in place when they are established and also what must be reported on. However, the model presented does not require a group of farmers to be organized in an institution such as a WUA.

References

Department of Water Affairs (DWA) (2008) *Guidelines for the Development of Integrated Water and Waste Management Plans* (Pretoria: DWA).

Department of Water Affairs and Forestry (DWAF) (1998a) *National Water Act (Act 36 of 1998)* (Pretoria: DWAF).

Department of Water Affairs and Forestry (DWAF) (1998b) *Water Services Act (Act 108 of 1997)* (Pretoria: DWAF).

Department of Water Affairs and Forestry (DWAF) (2001) *Regulations under Section 9(1) of the Water Services Act, 1997* (Pretoria: DWAF).

Department of Water Affairs and Forestry (DWAF) (2004a) *A 5-Year Water Resource Quality Monitoring Plan* (Pretoria: DWAF).

Department of Water Affairs and Forestry (DWAF) (2004b) *National Water Resources Strategy, Edition 1* (Pretoria: DWAF).

Department of Water Affairs and Forestry (DWAF) (2007) *A Drinking Water Quality Framework for South Africa, Edition 2* (Pretoria: DWAF).

World Health Organization (WHO) (2005) *Water Safety Plans, Managing Drinking-Water Quality from Catchment-to-Consumer.* WHO/SDE/WSH/05.06 (Geneva: WHO).

Water Quality and Health in Poor Urban Areas of Latin America

MARIA ONESTINI

Centro de Estudios Ambientales (CEDEA), Buenos Aires, Argentina

ABSTRACT *Water quality is largely absent in the water policy debate and analysis in the Latin American region. Although there is no disagreement as to the negative impact of unsafe and poor-quality water on human health, there is very little scrutiny and policy discussion on the matter. Considering data from different case studies on health and environment in poor urban areas in the region, this paper reviews some of the knowledge on water quality and human health in Latin American cities. Furthermore, conclusions as well as recommendations are drawn highlighting policy-oriented approaches to this problem.*

Introduction

The health impact of environmental degradation associated with poor-quality water in poor urban areas of Latin America is a critical issue from many points of view. For equity reasons, on environmental grounds or as a policy rationale (as well as for any combination of these factors), the matter is a vital one.[1]

This article draws comparative, integral and wide-ranging conclusions related to water issues from a series of case studies carried out in several Latin American countries and examines (from an integrated perspective) the interconnected issues of health, environment and poverty in these urban areas.

The Latin American region, regrettably, has a long history of unsafe and unhealthy conditions associated with urban deficits, in particular for the poor (Onestini, 2000). While it seems to be widely recognized that the problems of health and the environment have their own specificities in the urban environment (as opposed to the rural environment) in the so-called irregular settlements, poverty usually appears at the beginning of the chain of multiple determinations (Cuenya & Rodríguez, 2006). Lack of access to safe water (safe being the key quality indicator) is and has been one of the keystone issues for the urban poor. Newer research and analysis that incorporates recent trends have not only confirmed these determinants, but also added new concerns. Drawing on studies carried out in urban areas of Argentina, Bolivia, Brazil and Peru, this article considers these new findings. Furthermore, a series of policy and general recommendations from a systems' perspective also emerge from the comparison of the cases and progresses have been identified in several specific situations.

Urban Development, Water Issues and the Poor in Latin America in Recent Decades

In Latin America, as in other parts of the developing world, accelerated urbanization trends and processes have occurred without corresponding adequate provision of sanitation infrastructure and of drinking water. It has also been argued that efforts by national and local governments in order to provide sustainable solutions to these problems have been deficient. This situation has been aggravated by policies implemented in the 1990s which involved, *inter alia*, state-reduction strategies as well as a lack of capacity and policies by local governments in order to deal with these issues. Together with these matters, deficient health systems have created a combination of factors that are the structural causes for health problems for the urban poor. Major pathologies associated with the living conditions of low-income settlements, and many associated with deficient access to water of adequate quality, are:

- gastrointestinal, and infectious diseases often referred to as "poverty diseases" (which have disappeared in general in developed countries but are a major cause of morbidity and mortality in the developing world);
- chronic and degenerative diseases associated with poor life and work conditions; and
- pathogenic disorders related to stress, social isolation and cultural conflicts.

Case studies show that health problems associated with health impairments may be up to three times higher in urban poor areas that their rural equivalent (World Health Organization (WHO), 1989).

Based on this context, newer patterns emerge and newer risks develop hazardously linking the poor's living conditions in urban areas due to a lack of adequate environmental conditions, including unsafe water. The latest issues to emerge (and the trends expected) include climate change and its associated impacts (especially on water quality) as well as large infrastructure development such as, for example, dam construction and associated vector-borne diseases (Rapport & Singh, 2005). For the Latin American region other trends identified have been: increased urbanization in vulnerable areas; a lack of adequate water and sanitation infrastructure; and a lack of investment as well as disinvestment in infrastructure, especially in poor neighbourhoods.

Case Studies in Argentina, Bolivia, Brazil and Peru

Drawing comparative and wide-ranging conclusions related to water problems from a series of case studies carried out in the urban areas in Argentina, Bolivia, Brazil and Peru, and which examined (from an integrated and systems perspective) the interconnected issues of water, health, environment and poverty in these cities, these tendencies are confirmed and highlighted. Nevertheless, these studies (as well as others) can only provide brushstrokes as to the whole image of what are the water-related problems in the region. It is well accepted that there are no comprehensive data on water quality and access in the region, let alone differentiated by more fine indicators.

Case studies in Argentina have indicated, for example, that up to 10% of the population lives in areas where the water contains high level of arsenic (Ministerio de Salud, 2006). This occurs where most people do not receive water that has been treated in order to make drinking water potable (Onestini *et al.*, 2001) and are, therefore, chronically exposed to disease associated with arsenic such as skin lesions and cancer. In many cases in Argentina it has been found that even water provided by municipal systems is non-potable due to a high content of arsenic and nitrates, among other contaminants, as well as bacteria in unsafe water.

Studies in the city of Moreno (Province of Buenos Aires, Argentina), an urban area within the capital's metropolitan ring, indicates that not even one-third of the population has access to piped water in their homes. Nevertheless, this is done in precarious manner. Parting from the uncertain premise that the water provided is of adequate quality, it is habitually piped inadequately, often with hoses, contaminating the water in the process. About 70% of the population does not even have this provision, using groundwater which is polluted with biological as well as other contaminants such as nitrates. This translates in specific health issues associated with unsafe water quality given that water-related gastrointestinal pathologies represent nearly 40% of all pathologies reported within the municipal health system (Rofman *et al.*, 2006).

The Cochabamba urban area in Bolivia was analysed. The metropolitan area presents an overwhelming spatial concentration of infrastructure, equipment, services, and commerce in the Northeast and historic centre. Yet, the provision of services continues to be highly deficient in poor neighbourhoods. These neighbourhoods are of an irregular nature and tend to be located in areas with an extremely high environmental vulnerability. Regarding infrastructure and services, one of the most important issues is water provision for human consumption. In poor neighbourhoods this is largely done by the *"aguatero"* method of provision where informal supply by trucks, carts and such sell water to urban dwellers who do not have access to piped water systems. This water is of dubious quality and has not sanitary control. Furthermore, this issue is compounded given not only these doubtful systems of water distribution, but also the increased discarding of waste and wastewater from household and industries into water courses. The inequity in this situation is highlighted by two dynamics. First, in poor districts, over 85% of households have no connection to public water services, neither at the household level nor at the immediate areas. The availability of this service is closely related to the socio-economic level of the population. In more affluent districts, only 11% of households face a lack of piped water. Second, those areas without piped water purchase it from trucks and carts, paying up to 400% more than households with in-house provisions. In Cochabamba, the case study also identified the perception of illness by the urban poor related to the lack of safe water (Table 1). These surveys indicate that 62.4% of poor males and 66.6% of females in deprived districts perceive that the lack of safe water is associated with illness (Ledo, 2006).

In Brazil, São Paulo was the urban area where a study was undertaken. Water access is high in this urban area, even in poor neighbourhoods, with better provision in the city proper and less so in the periphery. Access to sanitation, however, is still deficient, causing water quality issues and water stress problems, particularly in poor neighbourhoods. Mapping areas where consumption of the water supply due to its poor quality is not advisable gives a graphic rendering of this problem (Figure 1). In the city's periphery water is often obtained through wells. These wells located on the outskirts largely coincide with slum areas. Even fractional studies indicate that 20% of wells were not within recommended standards and they have a high degree of pollution with direct impacts on the health of the population. It was found that these sources were mainly polluted with ammonia (about 71%), nitrates (about 72%) and coliforms (about 60%). Again in the São

Table 1. Southern periphery: households' perceptions of problems caused by lack of potable water, by gender of head of household, 2004

What problems are caused by not having potable water?	Male	Female	Total
1 Illnesses	62.4	66.6	63.5
2 Getting up at dawn	13.4	14.7	13.7
3 Environmental pollution	3.8	3.9	3.8
4 Buying water	17.0	12.3	15.8
5 Increased time spent to fetch water	3.4	2.6	3.2
Cochabamba	100.0	100.0	100.0

Source: Own generation with data from CEPLAG-UNIFEM (2004).

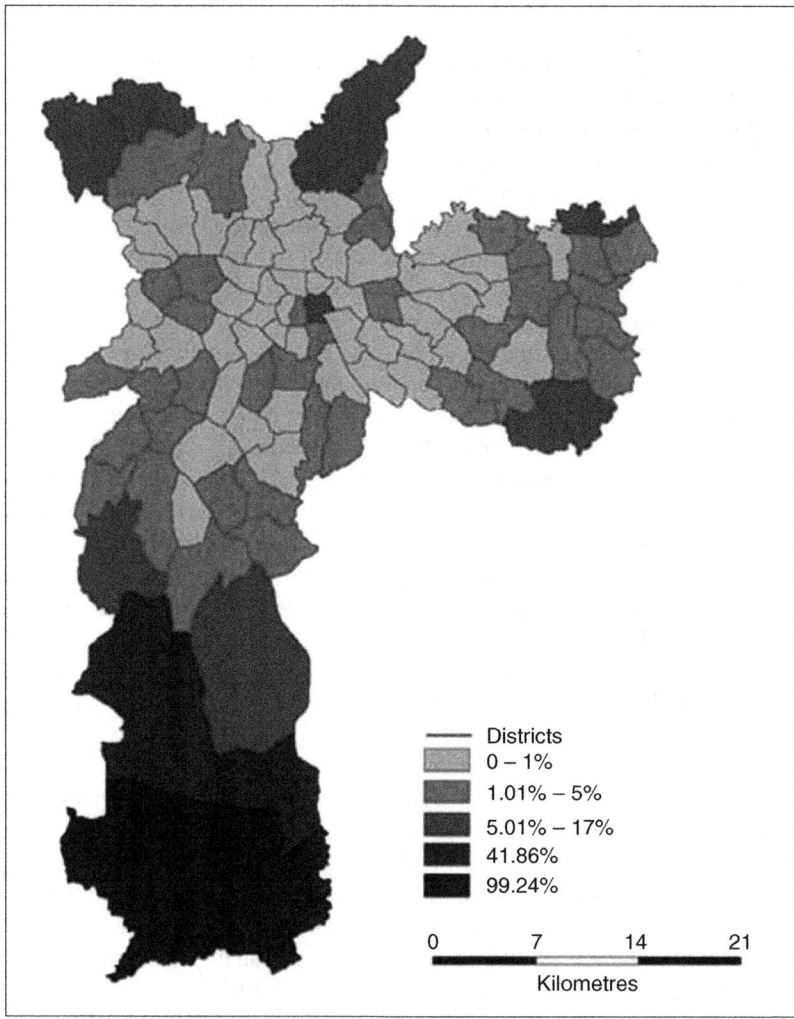

Figure 1. São Paulo: areas where consumption of water supply is not advisable due to its poor quality. *Source*: 2000 Census.

Paulo case as in other smaller cities in the region there are little data on water quality and its impact on human health. Data on illnesses associated with water are not precise enough to establish holistic patterns. Nevertheless, albeit sketchy, some data offer information on the impact of unsafe poor water quality on human health. In some peripheral areas of the city it was found that between 2002 and 2003 there was an increase of nearly 15% in admissions by intestinal infectious diseases in local health centres. Furthermore, the water quality problems persist due to the tendency of poor urban settlements to establish and grow along the banks and close to highly contaminated courses of water, exposing these dwellers to harmful health threats. For example, the incidence of leptospirosis, a zoonosis transmitted by contact with infected waters, has increased from about an 1.8% incidence rate in 1995 to 4.5% in 2004. In addition to this level of morbidity, mortality from leptospirosis has also leaped from 7.6% in 1995 to 18% in 2002 (Jacobi & Besen, 2006).

In Peru the case study concentrated in the metropolitan region of Lima, specifically its southern section. These so-called "newer neighbourhoods" are within an area which is classified as either poor or very poor. Only about 15% of neighbourhoods have piped water services. In the absence of water services, dwellers are supplied via tanker trucks. There are few or no assurances regarding the origin and quality of distributed waters, nor whether treatment or disinfection (if any) is carried out to the tanks. Often, water arrives to consumers with strange particles and even with insect larvae. Furthermore, this water costs up to 800% more than water supplied by municipal systems. As an aggregated problem to this supply of water of dubious quality, the issue of intra-household water storage is problematic. Water storage tanks are often unsuitable and affect water quality, due to their oxidation, deterioration or being exposed to polluting agents. Many were previously used for storing toxic materials. Although families in surveys manifest that they process water, this is doubtful due to their income. Household income analysis indicates that the great majority of households cannot afford fuel and decontamination costs. Consumption of contaminated waters is, according to health records, the cause for most gastrointestinal diseases. These diseases occupy second place among conditions suffered by dwellers in these neighbourhoods increasing to reach an estimated 22% of all consults in the last few years (García & Riofrío, 2006).

Knowledge Gaps and Policy Implications

As indicated, a major knowledge gap identified by all the case studies is the absolute lack of broad-spectrum knowledge on water quality accessed in the region's poor urban areas. All the case studies indicated vague and erratic data for this issue. Not only for water itself provided by urban dwellers, but also for water provided by vendors and, furthermore, for water provided by municipal systems. When information was available, it indicated that the poor consume water that is unsafe or of dubious quality. This lack of data points to a lack of acknowledgement of the problem by the science and policy-oriented communities.

Furthermore, in all cases analysed, unsafe water was identified as a major cause for illnesses, corroborating the global analysis that the environmental burden of disease for the poor is very high and in many instances associated with unsafe water (Prüss-Üstün & Corvalán, 2006). Notwithstanding this issue, the real incidence of disease related to the consumption of unsafe water and water of hazardous quality by the poor is, by all estimates, much higher than what is reported within the health system(s). Some water-related diseases, such as gastrointestinal ailments, are not reported obligatorily in some countries (Jacobi & Besen, 2006). Yet, most importantly, the lack of access by the poor to adequate health systems is indicated to be the main problem for this issue. Therefore, it is estimated that underreporting of water-related sicknesses is widespread for the urban poor in Latin American cities.

These two knowledge gaps (from a strictly analytical point of view as well as from a health–unsafe water standpoint) hinder an awareness of the problem and consequently obstruct at some level the application of equitable and proactive policies in this issue.

Proactive Urban Patterns in Poor Neighbourhoods in Latin America: Safe Water Access

Notwithstanding the conditions and lack of access to safe water and the lack of integrated policies and analysis regarding water quality, dwellers of poor neighbourhoods in Latin America face and promote proactive urban arrangements to improve living conditions in general and access safe environmental services. Some patterns emerge from the case studies which will be detailed below.

Some positive factors in the provision of water of adequate quality are the new processes facing the region, including some sluggish yet constructive patterns such as devolving urban development policies to local communities. Although this is an encouraging pattern, these processes are slow and discontinuous since they face resource scarcity and a lack of policy coordination between and among stakeholders (Cuenya & Rodríguez, 2006).

At the different governmental levels, many programmes and projects have been identified that promote access to water. Many of these are municipal and, in a sense, these local stakeholders are new actors in urban infrastructure and habitat production in the region. Other activities identified, such as in the São Paulo case study as indicated by Jacobi & Besen (2006), are also environmental health-vigilance programmes that, among their pursuits, include surveying water-related diseases as associated with water quality issues. Moreover, in the case studies carried out in Bolivia, Brazil and Peru, a high degree of actions by civil society organizations have been identified with a positive impact on the living conditions and health of the community, including obtaining safe drinking water. Although many of these experiences are random and not intersectorial, they do provide adequate learning, replication and up-scaling opportunities. In some case studies a positive direct link has been found between these two dynamics, i.e. between municipal water provision policies and community participation programmes. As indicated by García & Riofrío (2006) a positive trend is present in water infrastructure when poor urban dwellers demand it. As indicated in this case study, the provision of water services has not been carried out in a comprehensive policy context, but as public responses to the demands of stakeholders themselves. That is, in neighbourhoods where social actors have demanded from public authorities the provision of water, this has been carried out to a better extent than in new neighbourhoods with similar characteristics where this utility was not demanded.

An emerging question in the region, which found a parallel in most of the case studies here analysed, is the issue of tenure. As in much of the literature on urban infrastructure in the region, the case studies have determined that safe water access is substantially higher where land and/or housing tenure is secured. In the case study carried out in the Lima Region, for example, it was found that secure water provision is up to three times higher (300%) for households with legal entitlement to their home, when the comparison is made for the same type of socio-economic background. That is, this proportion is even this high when the same sort of poor neighbourhoods are compared. The same sorts of patterns were found in the case study carried out in Bolivia. The dynamic appraised indicates that municipal water policies and programmes tend to be implemented where tenure is secure. Furthermore, even households will tend to demand infrastructure, mainly water provision, when their tenures is securely held.

Conclusions

A series of policy and general recommendations also emerge from the comparison of the case studies and progress has been identified in several specific situations. The case studies point to new and crucial problems related to water and health in Latin America's poor urban areas, with the evident divergences between and among different urban configurations within the metropolitan areas that have been analysed. The general conclusion, however, remains that environmental burden of disease for the poor is very high and growing in Latin America. Of this, a great deal is related to the access (or lack of access) to safe water and of water-borne illnesses and the lack of integrated systems perspective in water management issues.

Some general conclusions vis-à-vis the knowledge that links issues such as water, health and poverty in urban Latin America can be schematized as follows (following the findings in the aforementioned case studies). Under-reporting of water-related illnesses for the poor is a weighty issue, not only from a research point of view, but (importantly) from an equity and policy standpoint. It is recommended that progress in this matter be pressed by analysis to be carried out that confronts under-reporting as well as unreliable official data in many circumstances. A lack of holistic and integrated urban planning continues to plague the poorer urban regions in Latin American cities, increasing informality, meagre infrastructure and their associated health problems. It is recommended that proactive policy alternatives that link health issues for the poor as related to water problems, and particularly water quality, be furthered. A lack of safe water provision has been associated in many cities with the informal property rights on land, and that progress is evident when tenure is secured. It has been recommended that a move forward to more formal situations of land tenure can lead to better water-provision infrastructures and (as a result) to situations where access to safe water is also improved and progress evidenced.

Positive experiences and progress have also been identified in the case studies, and it is suggested that several of these experiences (programmes of infrastructure in shanty towns, community projects, health vigilance programmes, systematic management and urbanization plans, etc.) should also be analysed in the future to recognize the opportunities that these experiences bring the better to provide for water-related issues in relation with health as well as their possible replication in other urban areas.

Note

1. The case studies on environment and health in poor urban areas of Latin America as well as a comparative analysis on the conceptualization of health and environment dynamics in poor urban areas in the region were commissioned by the International Development Research Centre (IDRC) of Canada and coordinated by Centro de Estudios Ambientales (CEDEA).

References

CEPLAG-UNIFEM (2004) Uses of safe drinking water for households, with a gender dimension–women's right to water. Survey by CEPLAG-UNIFEM, Cochabamba, Bolivia.

Cuenya, B. & Rodríguez, G. (2006) Asentamientos irregulares y salud humana en América Latina y el Caribe. Aspectos conceptuales y evidencias empíricas, in: Proceedings of Workshop Health and Habitat in Poor Urban Areas of Latin America, IDRC–CEDEA, November 2006, Buenos Aires, Argentina.

García, L. & Riofrío, G. (2006) Estudio exploratorio sobre asentamientos irregulares y salud humana en América Latina y El Caribe desde un enfoque de eco-salud. El caso de Lima, in: Proceedings of Workshop Health and Habitat in Poor Urban Areas of Latin America, IDRC–CEDEA, November 2006, Buenos Aires, Argentina.

Jacobi, P. & Besen, R. (2006) Ambiente e saude em assentamentos irregulares na cidade de São Paulo, in: Proceedings of Workshop Health and Habitat in Poor Urban Areas of Latin America, IDRC–CEDEA, November 2006, Buenos Aires, Argentina.

Ledo, C. (2006) Estudio exploratorio sobre asentamientos irregulares y salud humana en América Latina y El Caribe desde un enfoque de eco-salud. Estudio de caso: Ciudad de Cochabamba, algunas evidencias empíricas, in: Proceedings of Workshop Health and Habitat in Poor Urban Areas of Latin America, IDRC–CEDEA, November 2006, Buenos Aires, Argentina.

Ministerio de Salud (2006) *Epidemiología del Hidroarsenicismo Crónico Regional Endémico en la República Argentina* (Buenos Aires: Ministerio de Salud).

Onestini, M. (2000) *Desarrollo y Ambiente en la Práctica Urbana de América Latina* (Buenos Aires: Corbal).

Onestini, M., Palos, C. & Casal, J. (2001) *El Agua y su Problemática Integrada: El Caso del Municipio de Junín, Pcia. De Buenos Aires* (Buenos Aires: Centro de Estudios Ambientales (CEDEA)).

Prüss-Üstün, A. & Corvalán, C. (2006) *Preventing Disease through Healthy Environments: Towards an Estimate of the Environmental Burden of Disease* (Geneva: World Health Organization (WHO)).

Rapport, D. & Singh, A. (2005) *An EcoHealth-based Framework for State of Environment Reporting*. Available from: http://www.elsevier.com/locate/ecolind/.

Rofman, A., Rofman, A., Alsina, G. & Suárez, A. (2006) Estudio exploratorio sobre asentamientos irregulares y salud humana en América Latina y El Caribe desde un enfoque de eco-salud. El caso de Moreno, Buenos Aires, in: Proceedings of Workshop Health and Habitat in Poor Urban Areas of Latin America, IDRC–CEDEA, November 2006, Buenos Aires, Argentina.

World Health Organization (WHO) (1989) *La urbanización y sus repercusiones en la salud infantil. Posibilidades de acción* (Geneva: WHO).

Conceptual Framework for Protecting Groundwater Quality

C. MARTÍNEZ-NAVARRETE*, A. JIMÉNEZ-MADRID*,
I. SÁNCHEZ-NAVARRO**, F. CARRASCO-CANTOS† &
L. MORENO-MERINO*

*Geological Survey of Spain, Madrid, Spain; **Ministry of the Environment and Rural and Marine Affairs, General Directorate of Water Resources, Madrid, Spain; †University of Malaga Hydrogeology Group, Málaga, Spain

ABSTRACT *A conceptual framework is defined to establish safeguard zones in groundwater bodies intended for drinking water according to the requirements of the Water Framework Directive. For this, the foundations of a three-phase methodology within a dynamic process are proposed. The results of the first two phases are presented, which contemplate the distribution of groundwater body abstraction points as well as hydrogeological criteria, evaluation of pressures and aquifer vulnerability, in addition to the defined wellhead protection areas of abstraction points. As a final proposal, it will be necessary for competent authorities to create a recommendations and restrictions guideline, which should be integrated into the rest of the policies related to land-use planning in order to protect groundwater adequately.*

Introduction

Groundwater constitutes a basic resource for the urban water supply in Europe. In countries such as Austria, Germany, Italy or Denmark, more than 70% of the population's water supply comes from groundwater (Martínez-Navarrete *et al.*, 2008). In Spain, the percentage of groundwater used as the main source of water supply varies between 19% for populations greater than 20,000 inhabitants and 70% for those with fewer than 20,000 inhabitants (MIMAM, 2000).

The protection of groundwater has become one of the high-priority environmental objectives in European policies entering in force in the year 2000 through the Water Framework Directive (WFD), Directive 2000/60/CE of the European Parliament and the Council (European Union, 2000), and more specifically in 2006 through Directive 2006/118/CE of the European Parliament and the Council (European Union, 2006), dealing with the protection of groundwaters against pollution and deterioration. For this reason, the most convenient measures should be undertaken by Member States of the European Union to protect water quality so that, at abstraction points previous to treatment, a significant deterioration in water quality does not occur, which would require an increase in purification treatments (Carceller-Layel *et al.*, 2007).

According to the WFD, a "pressure" is defined as an activity caused by man that could have a negative environmental impact on water. In relation to groundwater, pressures are classified into five types: point contamination sources, diffuse contamination sources, water abstractions, artificial recharge, and seawater intrusion. As a consequence of these activities, pollutants may be introduced into the water (e.g. from agricultural activities, livestock, urban settlements, industrial activities, hazardous waste storage and landfills), the physico-chemical properties of water may be modified (referring to artificial recharge or landfills) or there could be a reduction of water resources (referring to water abstractions).

Vulnerability to contamination is an aquifer's susceptibility to groundwater contamination due to the impact of human activity. It is possible to distinguish between intrinsic and specific vulnerability (Foster, 1987; Margat & Suasis-Parascandola, 1987; Vrba & Civita, 1994). Intrinsic vulnerability is the susceptibility of groundwater to contamination generated by human activity taking into account the geological and hydrogeological characteristics of a given area, although it is independent of the nature of the contaminating agents. On the other hand, specific vulnerability is the susceptibility of groundwater to a contaminating agent or group of specific contaminating agents taking into account the characteristics of them and their relation to the intrinsic vulnerability components (Zwahlen, 2004).

The necessity to make socio-economic activity compatible with the protection of the quality and quantity of groundwaters has historically been addressed though land-use planning, carried out based on the characterization of the physical geography (Harper *et al.*, 1992). Several works exist on the necessary decisions needed to protect groundwater (Wang, 1997; Berg *et al.*, 1999; Eliasson *et al.*, 2003; Mödinge & Kobus, 2005). To date, the most frequently used measures for groundwater protection have been wellhead protection areas and the evaluation of the intrinsic vulnerability to contamination, which is a preventive action with respect to the use of resources (Martínez-Navarrete *et al.*, 2008). Wellhead protection areas are focused on protecting the source or water abstractions intended for drinking water, whereas the objective of vulnerability cartography is the protection of the resource, which is the groundwater body (COST 65, 1995; Hötzl, 1996, 2002).

The wellhead protection area of water abstraction points for drinking water is a measure broadly reflected in the water protection legislation of different countries in the European Union, where the definition of its boundary has been addressed differently (García García & Martínez-Navarrete, 2005). However, the establishment of the protection measures varies greatly between European Union countries, e.g. in Germany wellhead protection areas cover 20% of the territory's extension (Vorreyer, 1998), whilst in Spain its establishment is practically nil (Martínez-Navarrete & García García, 2003). In Spain the most frequently used zoning system distinguishes three areas around an abstraction point (MIMAM, 2002; Martínez-Navarrete & García García, 2003; Diputación Provincial de Alicante, 2008).

To delineate wellhead protection areas in aquifers with intergranular porosity, hydrogeological and analytical methods as well as mathematical models can be used whose characteristics have been analysed in diverse works (Environment Agency, 1998; Lallemand-Barrès & Roux, 1999; Martínez-Navarrete & García García, 2003). There is a restricted number of methods available for application in karst and fissured environments due to their great heterogeneity (Daly *et al.*, 2002). For these types of aquifers the evaluation of the intrinsic vulnerability is the most advanced protection tool (Döerfliger, 1996; Goldscheider et al., 2000; Zwahlen, 2004; Vías et al., 2006a).

Wellhead protection areas and the cartography of intrinsic vulnerability are useful tools for the protection of resources; however, they need to be complemented with the assessment of other criteria to establish an effective global protection measure for groundwater bodies (Martínez et al., 2008). In particular, it is highly advisable to consider contamination as a combination of pressures and vulnerability to contamination (Jiménez-Madrid et al., 2008, 2010).

Article 6 of the WFD enforces that all water bodies used for the abstraction of drinking water that provide an average of more than $10\,m^3/day$ or serve more than 50 persons, and those water bodies destined to such use in the future, are included in a protected zones registry (Drinking Water Protected Areas (DWPAs)). Due to these rigorous requirements, in numerous Member States the majority of groundwater bodies must be considered under this protection and therefore cover a great extent of their territory.

Article 7.3 of the WFD enforces that Member States ensure the necessary protection for the DWPAs "with the aim of avoiding deterioration in their quality in order to reduce the level of purification treatment required in the production of drinking water".

Safeguard zones are areas (that can be established optionally as contemplated in the WFD) focused on measures to protect groundwaters with the aim of avoiding the deterioration of water quality and reducing the level of purification treatments required for drinking water. This is a highly recommendable option as it has been extensively used for the delineation of numerous water bodies in diverse Member States. They are therefore equivalent to the "wellhead protection areas" of groundwater bodies intended for drinking water according to Article 7.3 of the WFD. At present these have not yet been incorporated into Spanish legislation. In this context, the purpose of this work is to define a conceptual framework that allows the delineation of safeguard zones as a global protection measure in all water bodies intended for drinking water.

Methodology

The methodology that has been proposed in Spain to delineate groundwater body safeguard zones in inter-community basins is shown in this work. The methodology is divided into three sequential working phases. For this, a first step made was to analyse the different layers of information to be considered, and following this the working phases were defined.

Definition of the Criteria To Be Considered

To delineate safeguard zones, the following information layers were taken into account.

Location and characteristics of groundwater abstractions intended for drinking water. According to the requirements of the WFD (providing an average of more than $10\,m^3/day$ and/or serving more than 50 persons), and of previously defined wellhead protection areas in the water body to be protected with regards to abstractions for water supply and mineral waters. It is necessary to emphasize that it is a dynamic process (Figure 1).

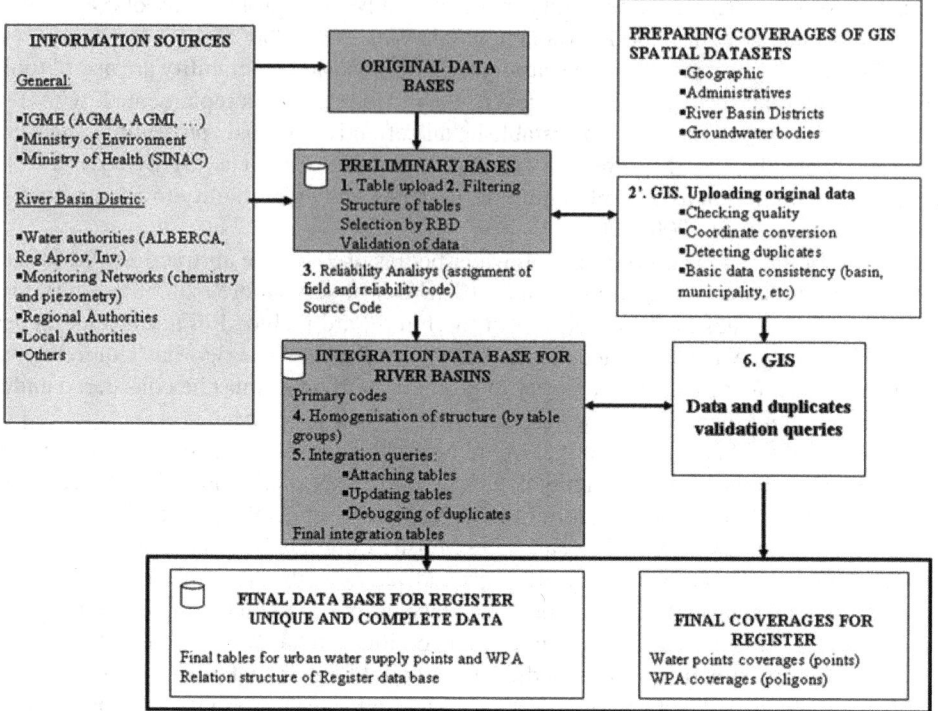

Figure 1. Making of the water abstractions and wellhead protection area registry.

Evaluation of the intrinsic vulnerability to the contamination of groundwater bodies. The evaluation of intrinsic vulnerability is a highly useful tool for taking preventive or corrective measures with respect to the land use and exploitation of water resources. Its objective is to subdivide the area into categories according to the capacity that the subsurface material has to protect the groundwater. Over the last few decades different methods have been developed to create contamination vulnerability maps following various approaches (Aller *et al.*, 1987; Foster, 1987; Döerfliger, 1996; Civita & de Maio, 1997; Goldscheider *et al.*, 2000; Zwahlen, 2004; Vías *et al.*, 2006a).

The COP method (Vías *et al.*, 2006a) and Reduced DRASTIC method (DGOHCA, IGME, 2002; DGOHCA, CEDEX, 2002) were selected here to evaluate the vulnerability of carbonate aquifers and detritic/mixed aquifers, respectively.

The COP method was been successfully applied in different carbonate aquifers in southern Spain (Andreo *et al.*, 2006, Vías *et al.*, 2005, 2006a). It has also been used in other countries (Vías *et al.*, 2006b; Ravbar, 2007; Baldi *et al.*, 2009), which tested its suitability for this type of medium in comparison with other methods that deal with carbonate aquifers. This method employs three factors: C (flow concentration), O (overlying layers) and P (precipitation). The C factor considers the existence of flow concentrations along with rapid infiltration across karstic conduits that reduce the natural protection capacity of the aquifer; the O factor takes into account the capacity that an aquifer has to protect itself according to the layers that form the vadose zone, and the P factor depends on the pluviometric characteristics that determine the

Table 1. Vulnerability classes for the COP index (carbonate aquifers) and Reduced DRASTIC index (detritic aquifers)

Vulnerability classes / Method	Reduced DRASTIC	COP
	Range of the index of vulnerability	
Very low vulnerability	16–30	4–15
	30–44	
Low vulnerability	44–58	2–4
	58–72	
Moderate vulnerability	72–86	1–2
	86–100	
High vulnerability	100–114	0.5–1
	114–128	
Very high vulnerability	128–142	0–0.5
	143–156	

transport of pollutants through rainwater from the ground surface to the water table. The COP index is calculated by the product of these three factors. Values range from 0 to 15, which are grouped into five classes of vulnerability (Table 1) where lower values indicate higher vulnerability (Vías et al., 2006a).

The DRASTIC method (Aller et al., 1987) was modified in studies carried out in two Spanish basins by governmental bodies (DGOHCA, IGME, 2002; DGOHCA, CEDEX, 2002). This modification proposes a reduction in the seven original factors of the DRASTIC method to four factors (Reduced DRASTIC) to avoid repetition of common information in the preparation of various thematic maps. The four factors in the Reduced DRASTIC method correspond to basic factors that one must consider when assessing intrinsic vulnerability: topsoil, lithology of the vadose zone, vadose zone thickness and net recharge (Jiménez-Madrid et al., 2009). The parameters are valued from 1 to 10 from the least to the most vulnerable and are weighted differently so that the index values range from 16 to 156, where ten vulnerability classes are established and are numbered from 1 to 10 representing minimum to maximum vulnerability, respectively (Table 1).

Pressures on groundwater bodies and risk evaluation. An inventory of pressures is required including, at a minimum, the systematic identification of each activity, its geographical location and its characterization, which will identify the areas where there are activities that may result in the deterioration of water quality or determine the cause of a known water quality degradation. Each pressure should be evaluated to establish its capacity to alter the status of groundwater.

Hydrogeological criteria. This is a detailed study of the behaviour of groundwater bodies by analysing piezometric maps and flow directions. Piezometric maps provide an insight into the direction of groundwater flow and its relationship to groundwater abstractions intended for drinking water. As a result of this phase, areas affected by abstractions are delineated and zoned appropriately in order to improve water body characterization and centralize the protection measures to be established.

Other factors to consider in a later working phase.

- Delineation of aquifer recharge areas. The delineation of recharge areas will establish which areas are most vulnerable to contaminant transport to groundwater bodies and therefore need to be protected.
- Establishment of control networks in protected areas (DWPAs) that assess trends in both the piezometric fluctuations (quantity) and the change in water chemistry over time (quality). They are a necessary management tool to address the protection of groundwater resources used to supply a population. The establishment of appropriate monitoring networks should be based on a detailed knowledge of the area's hydrogeology (Sargaonkar *et al.*, 2008; Mehan, 2004).
- Existing purification treatments, as a reference point to prevent the possible contamination of groundwater resulting from different pressures.
- Effects of seawater intrusion. According to the WFD, seawater intrusion is one type of groundwater pressure; it is necessary to take this into account in coastal aquifers where possibly groundwater is not used due to salinity.

Working Methodology Phases

Analysis of the previously described parameters and their joint assessment using Geographical Information System (GIS) tools (Figure 2) allows for the establishment of a methodology to delimit safeguard zone boundaries as a measure of a global protection

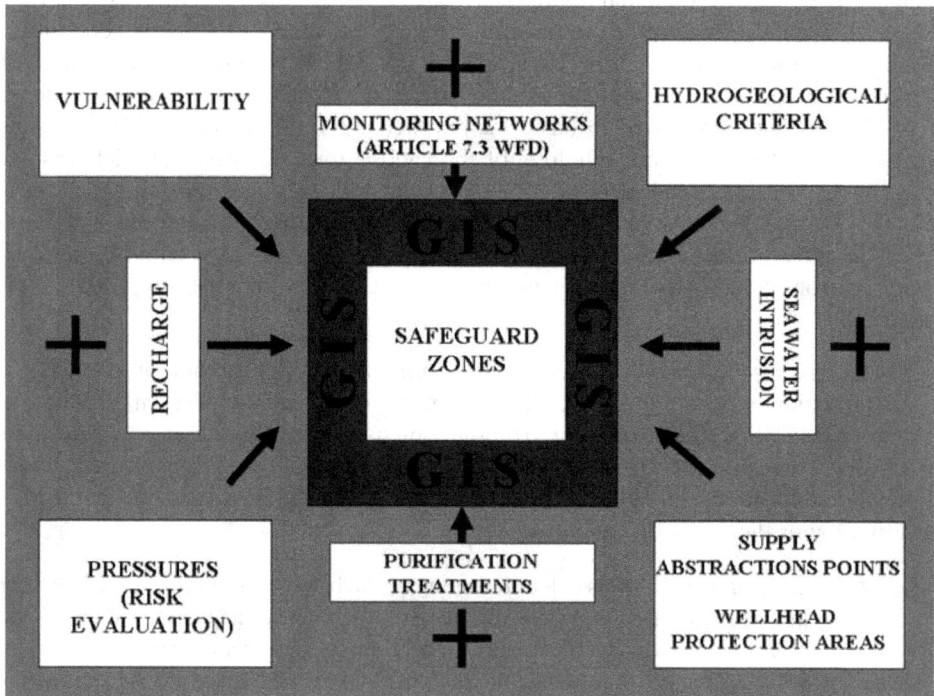

Figure 2. Conceptual framework for the delineation of safeguard zones.

of groundwater bodies intended for drinking water. This paper presents the results obtained after implementation of the first and second working phases and lays the groundwork to implement the third phase of work.

The proposed methodology consists of three successive working phases (Figure 3). To delimit safeguard zones in the first phase, vulnerability classes were set into two groups according to the methods applied, these groups being low and high vulnerability which were assigned values of 1 and 2, respectively. This group is combined with the presence or absence of significant pressures (Table 2).

Table 3 describes the different sources of information used to characterize the vulnerability assessment methods applied.

The size of the safeguard zones can be highly variable; in many cases, they will be smaller than the groundwater body and include areas of high vulnerability as well as pressures that can affect water abstractions or only those found within the same hydrogeological basin. Several safeguard zones can coexist in the same water body as well

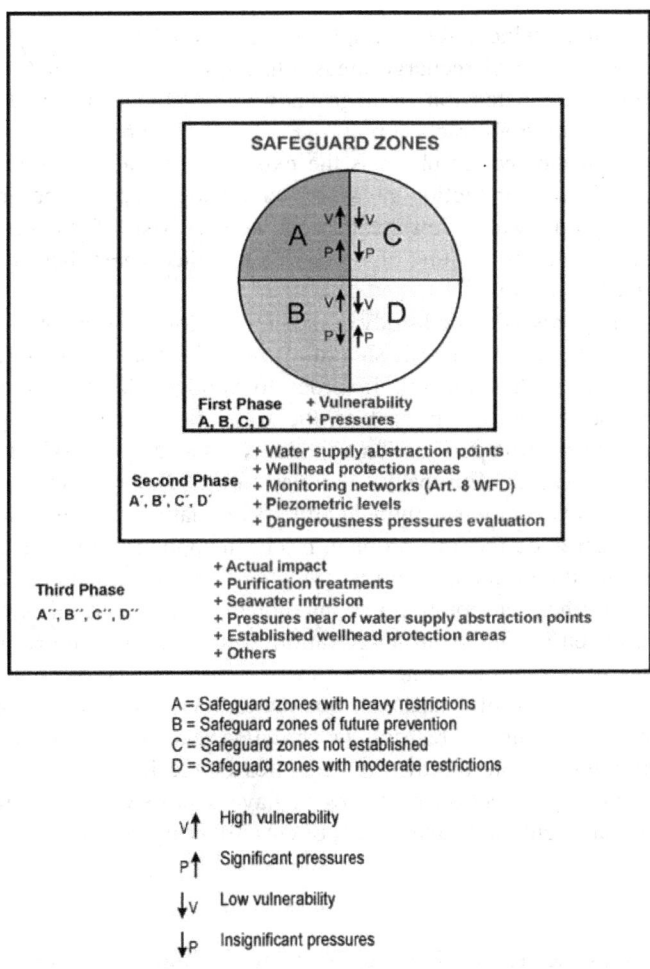

Figure 3. Working phases for the delineation of safeguard zones.

Table 2. Early progress in the delineation of safeguard zones in Spain (first phase)

Vulnerability (V)		Pressures (P)	
Group	Score	Group	Score
COP: between 0 and 1 = HIGH	2	Existence of significant pressures in the scope of the groundwater body	4
COP: between 1 and 15 = LOW	1		
Reduced DRASTIC: between 72 and 156 = HIGH	2	Non-existence of significant pressures in the scope of the groundwater body	2
Reduced DRASTIC: between 16 and 72 = LOW	1		

Notes: Zone A (V + P = 6): High vulnerability and significant pressures. Heavy restrictions.
Zone B (V + P = 4): High vulnerability and insignificant pressures. Future restrictions.
Zone C (V + P = 3): Reduced vulnerability and insignificant pressures. Without restrictions.
Zone D (V + P = 5): Reduced vulnerability and significant pressures. Moderate restrictions.

as extend outside of it, which, for example, occurs with karst materials due to their characteristics and location of recharge areas. On the other hand, safeguard zones can correspond to the entire extension of a groundwater body or surround the wellhead protection areas of existing abstraction points (Figure 4) (Jiménez-Madrid et al., 2008).

Also considered in the second phase is the existence of water abstractions used for drinking water, wellhead protection areas of both water supply and mineral water, piezometric maps, groundwater flow evaluations and the risk of pressures. Therefore, safeguard areas delineated in the first phase can be modified when these new factors are taken into account (Figure 4).

The third phase requires a detailed study of the piezometric map and pressures identified in the vicinity of abstraction points. It should also consider seawater intrusion in water bodies affected by this phenomenon. In order to validate the results obtained, it is necessary to analyse the current impact on the water body in general, particularly at abstraction points and existing purification treatments. As an example of the importance of considering these criteria, the European Commission (2007) shows an example where a problem of water quality that may originate either from natural causes or human activity exists, and a treatment to correct this problem has been applied so that the drinking water standards concerning the pollutant are met in the case of contaminant 1. This treatment may also correct a further deterioration in contaminant 2. However, this system hides the fact that there has been a significant deterioration in the quality of untreated water. The goal of preventing deterioration has not been met.

Once the final proposal of safeguard zone boundaries has been defined, it is necessary to elaborate a recommendations and restrictions guideline related to the introduction of new activities, conditioning factors to existing pressures or the location of new abstraction points. In order for the protection measures to have a positive effect, they should be included in the urban regulations and other policies affecting land use.

Results

This section presents the results obtained following the implementation of the first two working phases applied in intercommunity river basins of Spain.

Table 3. Information sources used to evaluate the intrinsic vulnerability to contamination

Reduced DRASTIC	Lithoestratigraphy and permeability map. Scale 1:200,000 (http://www.igme.es)
	Soil of Spain map. Scale 1:1,000,000 (http://www.ign.es)
	Soil map. SINAMBA Project. Scale 1:400,000 (http://www.juntadeandalucia.es)
	Piezometry databases of Geological Survey of Spain (IGME) (http://www.igme.es) and Ministry of the Environment and Rural and Marine Affairs (http://www.marm.es)
	Digital elevation model (25 × 25 m) of Center of Studies and Experimentation of Public Works (CEDEX) (http://www.cedex.es).
	SIMPA model (simulation rainfall—contribution) (http://www.cedex.es)
COP	Lithoestratigraphy and permeability map. Scale 1:200,000 (http://www.igme.es)
	Soil of Spain map. Scale 1:1,000,000 (http://www.ign.es)
	Soil map. SINAMBA Project. Scale 1:400,000 (http://www.juntadeandalucia.es)
	Piezometry databases of Geological Survey of Spain (IGME) (http://www.igme.es) and Ministry of the Environment and Rural and Marine Affairs (http://www.marm.es)
	Digital elevation model (25 × 25 m) of Center of Studies and Experimentation of Public Works (CEDEX) (http://www.cedex.es)
	Geomorphology map. Scale 1:1,000,000 (http://www.igme.es)
	SIMPA model (simulation rainfall—contribution) (http://www.cedex.es)
	CORINE Land Cover
	Database of the Spanish Company of Speleology (http://www.sedeck.org)
	Topography map. Scale 1:25,000 (http://www.ign.es)

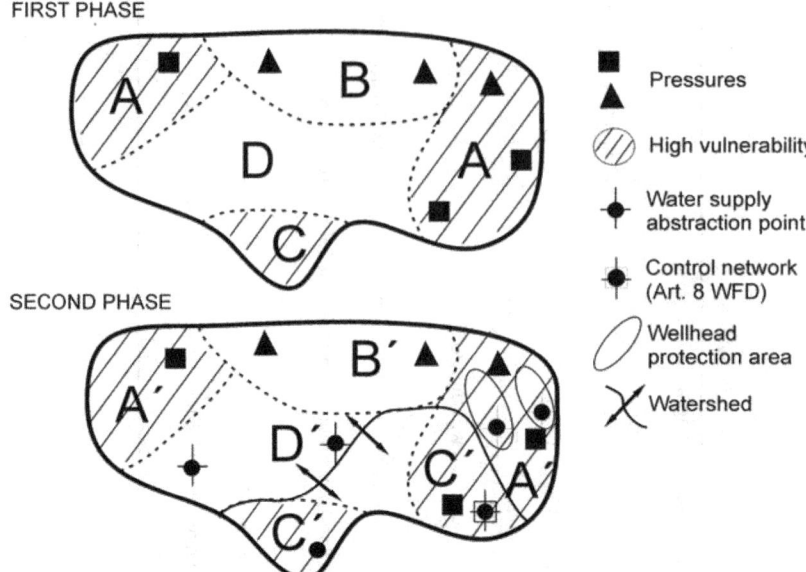

Figure 4. Example of delineating safeguard zones (first and second phases).

Registry of Groundwater Abstractions Used for Drinking Water and Wellhead Protection Areas

Table 4 summarizes the results obtained from the registry of abstraction points and wellhead protection areas following the procedure outlined in Figure 1. For this, databases containing general information were used from the Geological Survey of Spain (IGME), the Ministry of the Environment and Rural and Marine Affairs (water quality network data), and the Ministry of Health and Consumption (National Drinking Water Information System), as well as specific databases from each river basin.

Table 4. Registry of abstraction points and wellhead protection areas.

Basin	Water abstraction point		Wellhead protection areas	
	Water supply	Mineral water	Water supply	Mineral water
Cantabrico	2,640	24	17	9
Miño-sil	905	38	15	14
Duero	4,261	46	16	14
Ebro	4,329	103	54	33
Guadalquivir	2,834	22	192	13
Guadiana	2,481	5	13	4
Júcar	2,204	37	97	19
Segura	445	5	4	3
Tajo	3,144	28	103	13

The results show the large amount of water abstraction points meeting the requirements set out by the WFD. Due to this large number, it is virtually impossible to define wellhead protection areas individually, therefore water body safeguard zones are the most adequate measure of protection in Spain. Also, it is noted that the implementation of wellhead protection areas in Spanish territory is almost non-existent.

Intrinsic Vulnerability to Contamination

The results illustrated in Figure 5 show a predominantly low to moderate vulnerability to contamination in detritic and mixed-type water bodies. In marshes and alluvial areas the vulnerability is high to very high due to the shallow depth of the vadose zone. Note that this factor combined with the lithology factor exert a considerable influence on the final result.

Figure 6 shows the spatial distribution of the vulnerability classes for carbonate groundwater bodies. Areas of high and very high vulnerability, in most cases, coincide with areas where the O factor has values of greater vulnerability and in areas of preferential flow. Even though these areas occupy a small percentage of Spanish territory, they are the most important areas when considering vulnerability to contamination. Preventive measures should be taken in these areas to avoid the deterioration of natural water quality.

Pressures

This work exclusively considers the presence or absence of significant pressures. For this, the Integrated System for Water database, CORINE Land Cover database, and the IGME

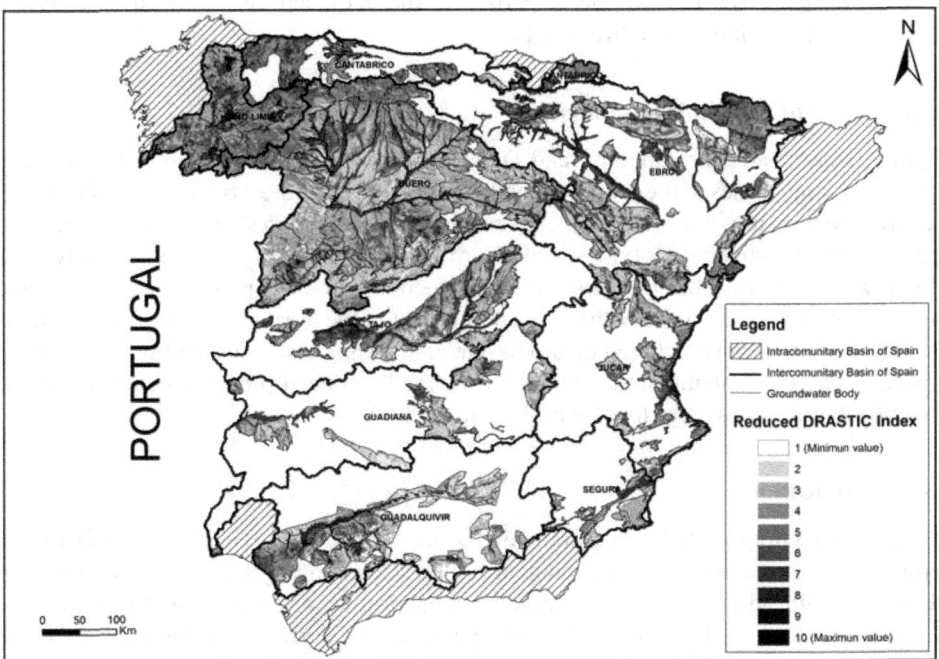

Figure 5. Vulnerability of detritic and mixed-type water bodies: Reduced DRASTIC index.

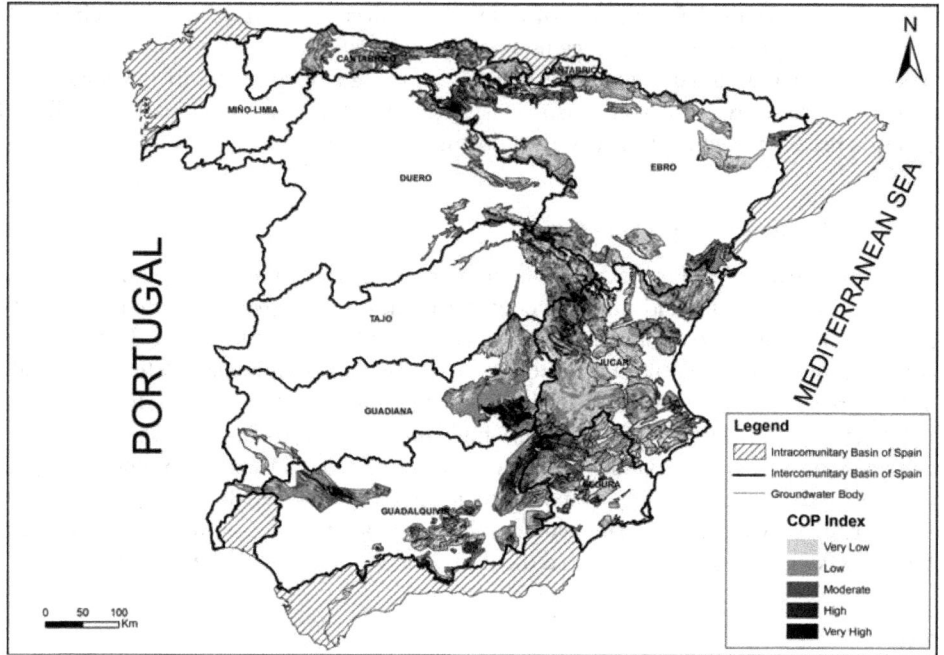

Figure 6. Vulnerability of carbonate water bodies: COP index.

mining and tailings ponds database were considered, as well as pressures inventoried in the report under Article 5, to meet the objectives of the WFD and results obtained after the initial characterization of the different basins.

Initial Delineation of Safeguard Zones

Figure 7 shows safeguard areas A′ and D′ following the implementation of the first and second working phases of the proposed methodology (an accurate assessment of the risk on the different pressures considered is yet to be completed). Table 5 shows the percentages of Spanish territory covered by type A′ safeguard zones with heavy constraints and type D′ which were obtained by implementing the second working phase. The Duero and Guadalquivir basins are those with the greatest extent of the territory of zone type A′. In the Duero basin this is due to alluvial materials that give the highest vulnerability, whilst in the Guadalquivir basin the numerous amount of defined wellhead protection areas make the percentage it occupies close to 10% of the total area.

Policy Implications

In order to comply with the requirements of the WFD, it provides the possibility of defining "safeguard zones". The technical projects for defining the safeguard zone (for which the wellhead protection areas could be defined) must be included in the programme of measures (Article 11.3 d, WFD) and be operational at the latest in December 2012, as they have to be included in the corresponding river basin management plan (Annex VII. A.7, WFD).

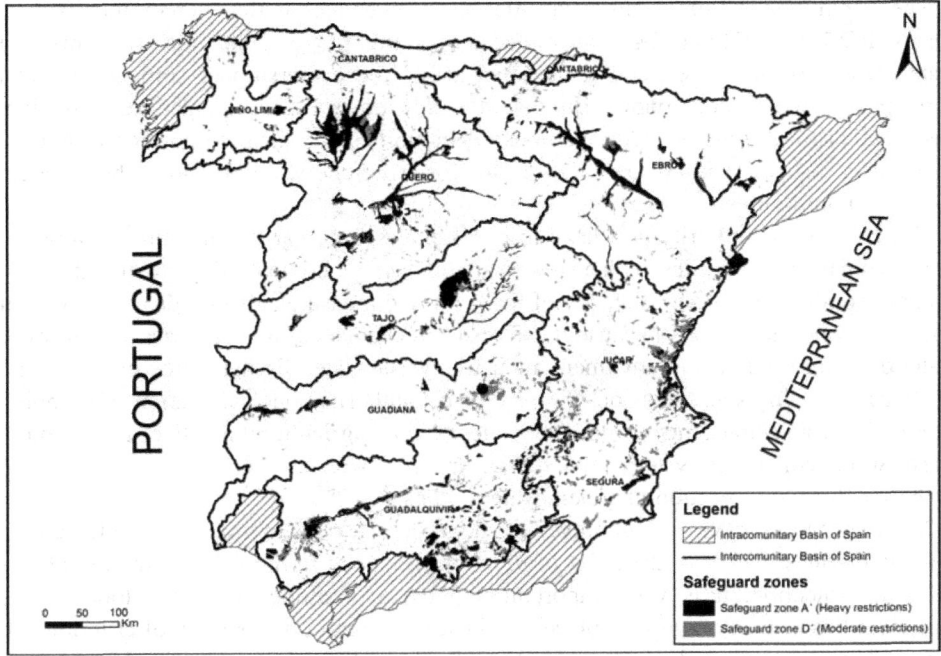

Figure 7. Safeguard zones defined after the second working phase (safeguard zones A′ in black and type D′ in grey).

The river basin management plan (include the register of protected areas) will coordinate and integrate the programme of basic and supplementary measures elaborated by the different authorities with competencies in water protection. These measures are necessary to guarantee the achievement of the objectives established in European Union water protection legislation that specifically includes regulation on the protection of waters used for the abstraction of drinking water. The programme of measures will be the result of a participatory process (Martínez-Navarrete et al., 2010).

Table 5. Per cent coverage of the territory by type A′ and type D′ safeguard zones.

		Percentage occupation of the territory	
River basin	Surface (km^2)	Safeguard zone type A′	Safeguard zone type D′
Cantabrico	20,885	1	0.9
Miño-sil	17,592	3	0.2
Duero	78,859	9	3
Ebro	85,566	6	1
Guadalquivir	57,228	8	3
Guadiana	55,388	3	3
Júcar	42,957	6	5
Segura	18,897	7	4
Tajo	55,764	5	1

The Committee of Competent Authorities has been created in the intercommunity river basins (RD 126/2007) in order to guarantee cooperation in the application of the different water protection measures, and it does not affect the ownership of competences that correspond to the different public administrations. It consists of representatives of the river basin authorities, ministries, Autonomous Communities and local authorities within the river basin district. It is also responsible for fostering cooperation among the different competent authorities.

The river basin authority will have to check the effects that the measures as a whole produce on the water bodies, with a view to guarantee their compatibility and find the most suitable combination. The efficiency of the proposed measures also needs an analysis of possible economic impacts because it can result in changes of the land uses, urban waste water discharge and manure treatment, industrial waste water discharge treatment, closure of illegal dumping sites, revision of waste water authorizations, adaptation of dumping sites and petrol stations, the setting up of codes of good agricultural practice, adaptation of waste water collecting systems and septic tanks, as well as the buying of some land properties for the protection of groundwater bodies.

Member States have a high level of flexibility for designing and implementing safeguard zones, but from the point of view of efficiency in the implementation of this zone of protection, it is very important to have a common, adaptable tool for the particularities of every complete place, but inside a common framework of criterion and action. The proposed methodology tries to be the basis for this common implementation. Furthermore, in future reviews of the river basin plans, the definition and implementation of safeguard zones, it can be considered (between other criteria) to delineate groundwater bodies according the requirements of the WFD.

In the studies carried out in Spain mentioned above, four types of safeguard zones with different protection levels are considered: A (Heavy restrictions. Specific measures to reduce the source of contamination as chemical substances, microbiological and radiological contamination); B (Future restrictions. Preventive measures); C (Without safeguard zones—restrictions. Codes of good agricultural practice applicable); and D (Moderate restrictions. Suitable preventive measures focused on the risks identified). In this respect, an ongoing doctoral thesis tries to perfect the proposed methodology. For it, between other aspects, a map of restricted activities is generated in relation to protecting groundwater destined for human consumption. In order for it to be effective, this cartography must be binding to land management.

The active cooperation among all actors involved is essential in defining the measures and restrictions that have to be taken, which will at a later stage be discussed and modified within the public participation processes. The implementation of the necessary measures involves the participation of Autonomous Communities and local authorities, but the general criteria and the objectives are set at a national level. In this process the role of the Committee of Competent Authorities mentioned above is therefore very important.

To finish the process, administrations with competences in land use and urban planning will have to take into account the definition and measures established for safeguarding zones in the drawing up of the planning instruments as well as in the granting of licences for possible water abstractions.

It is advisable to include these activity restrictions in the safeguard zones. It requires adjustments in the WFD in order to guarantee its common implementation in Europe.

Conclusions

The protection of groundwater intended for human consumption is one of the high-priority environmental objectives of the Water Framework Directive (WFD). For this reason, it is necessary to establish suitable protection measures for the different environments. With this purpose, wellhead protection areas have been delineated in many countries of the European Union to protect waters used for the abstraction of drinking water and the intrinsic vulnerability to contamination has been evaluated to protect groundwater bodies.

A conceptual framework for the establishment of safeguard zones in all groundwater bodies destined for human consumption has been defined in this work. Safeguard zones are areas within the Drinking Water Protected Areas (DWPAs) where the necessary measures are focused on conserving groundwater quality according to the requirements of the WFD, and so are equivalent to groundwater body wellhead protection areas. Therefore, the formulation of a methodology to establish safeguard zones as a global protection measure is intended, considering, among others aspects, the distribution of water abstractions used for drinking water, hydrogeological criteria, the evaluation of pressures (that have to include dangerousness pressures evaluation), the evaluation of vulnerability, the evaluation of recharge, and a study of the wellhead protection areas of existing abstraction points. The proposed methodology tries to be the base for a European common implementation in this subject. Finally, the paper has considered policy implications, economic impacts, their consequences for the legal framework and land management.

The proposed methodology is divided into three phases, of which the first two have been developed in this work. In the results obtained, the dynamic nature of the record of abstraction points and wellhead protection areas should be noted, as what is presented here is the current situation, and due to its impact on the delineation of safeguard zones it is advisable to keep these records updated. With regards to vulnerability, the results are limited to the working scales used for the different data layers. It would be desirable to develop a method to identify vulnerability that may be suitable for all types of aquifers, to unify criteria for the development of thematic layers and working scales. In relation to the pressures, the results presented evaluate their existence or absence qualitatively; it is necessary to quantify the risk of pressures through detailed fieldwork. The use of spatial analysis tools within a Geographical Information System (GIS) allowed one to deal with various factors at the same time which were to be considered in the proposed methodology.

The protection strategies presented in this work are to be integrated in a necessary programme of measures arising from a negotiation process between the different competent authorities and public participation. Only in this way will effective measures that are implemented in land-use planning and planning instruments be achieved. It is a pressing need that is currently ongoing in Spain.

References

Aller, L., Bennett, T., Leer, J., Petty, J. & Hacket, G. (1987) *DRASTIC: A Standardised System for Evaluating Groundwater Pollution Potential Using Hydrogeologic Settings* (Oklahoma: US Environmental Protection Agency).

Andreo, B., Goldscheider, N., Vadillo, I., Vías, J. M., Neukum, C., Brechenmacher, J., Carrasco, F., Hötzl, H., Jiménez, P., Perles, M. J. & Sinreich, M. (2006) Karst groundwater protection. Application of a Pan-European approach in the pilot site of Sierra de Líbar (South Spain), *Science of the Total Environment*, 357(1–3), pp. 54–73.

Baldi, E., Guastaldi, E. & Rossetto, R. (2009) Evaluation of intrinsic groundwater vulnerability to pollution: COP method for pilot area of Carrara hydrogeological system, (Northern Tuscany, Italy). Paper presented at the General Assembly 2009, European Geosciences Union, Vienna, Austria, Vol. 11, EGU2009-10405-2 (Vienna: Europa Geosciences Union).

Berg, R. C., Curry, B. B. & Olshansky, R. (1999) Tools for groundwater protection planning: an example from McHenry County, Illinois, USA, *Environmental Management*, 23(3), pp. 321–331.

Carceller-Layel, T., Costa-Alandí, C., Coloma-López, P., García-Vera, M. A. & San Román-Saldaña, J. (2007) Groundwater in the central sector of the Ebro Basin, *International Journal of Water Resources Development*, 23(1), pp. 165–187.

Civita, M. & de Maio, M. (1997) *SINTACS Un sistema paramétrico per la valutazione e la cartografia della vulnerabilità degli all'inquinamento. Metodologia e automazione* (Bologna: Pitagora).

COST 65 (1995) *Hydrogeological Aspects of Groundwater Protection in Karstic Areas. Final Report (COST Action 65)*. Report No. EUR 16547 (Brussels: European Commission, Directorate-General XII Science, Research and Development).

Daly, D., Dassargues, A., Drew, D., Dunne, S., Godscheider, N., Neale, S., Popescu, I. C. & Zwahlen, F. (2002) Main concepts of the European approach for (karst) groundwater vulnerability assessment and mapping, *Hydrogeology Journal*, 10(2), pp. 340–345.

DGOHCA, CEDEX (2002) *Vulnerability Mapping in Duero Basin* (Madrid: CEDEX).

DGOHCA, IGME (2002) *Vulnerability Mapping in Guadalquivir Basin* (Madrid: IGME).

Diputación Provincial de Alicante (2008) *Wellhead Protection Areas in Water Abstraction Points* (Alicante: Diputación provincial de Alicante).

Döerfliger, N. (1996) Advances in karst groundwater protection strategy using artificial tracer test analysis on a multiattribute vulnerability mapping (EPIK method), Doctoral thesis, Neuchâtel University, Neuchâtel.

Eliasson, A., Rinaldi, F. M. & Linde, N. (2003) Multicriteria decision aid in supporting decisions related to groundwater protection, *Environmental Management*, 32(5), pp. 589–601.

Environment Agency (1998) *Policy and Practice for the Protection of Groundwater* (Norwich: The Stationery Office).

European Commission (2007) *Common Implementation Strategy for the Water Framework Directive (2000/60/EC) Guidance Document No. 16. Guidance on Groundwater in Drinking Water Protected Areas* (Luxembourg: European Community).

European Union (2000) Water Framework Directive, Directive 2000/60/CE of the European Parliament and of the Council of 23 October 2000, establishing a framework for Community action in the field of water policy. DO L 327 of 22 December 2000.

European Union (2006) Directive 2006/118/CE of the European Parliament and of the Council of 12 December 2006 on the protection of groundwater against pollution and deterioration, DO L 372 of 27 December 2006.

Foster, S. (1987) Fundamental concepts in aquifer vulnerability, pollution risk and protection strategy, in: W. van Duijvenbooden & H. van Waegeningh (Eds) *Vulnerability of Soil and Groundwater to Pollution*, pp. 69–86 (Den Haag: TNO Committee on Hydrological Research).

García García, A. & Martínez-Navarrete, C. (2005) Protection of groundwater intended for human consumption in the water framework directive: strategies and regulations applied in some European countries, *Polish Geological Institute Special Papers*, 18, pp. 28–32.

Goldscheider, N., Klute, M., Sturm, S. & Höltz, H. (2000) The PI method—a GIS-based approach to mapping groundwater vulnerability with special consideration of karst aquifers, *Zeitschrift für Angewandte Geologie*, 46(3), pp. 157–166.

Harper, C. R., Goetz, W. J. & Willis, C. E. (1992) Groundwater protection in mixed land-use aquifers, *Environmental Management*, 16(6), pp. 777–783.

Hötzl, H. (1996) Scientific basis for karst groundwater protection: guidelines and regulations, Paper presented at Recursos Hídricos en Regiones Kársticas, Proceedings of the Conference in Vitoria-Gasteiz, October 1996, 147–158 (Vitoria-Gasteiz: Universidad del País Vasco).

Hötzl, H. (2002) Karst groundwater protection, in: F. Carrasco, J. J. Durán & B. Andreo (Eds) *Karst and Environment*, pp. 33–39 (Málaga: Fundación Cueva de Nerja).

Jiménez-Madrid, A., Carrasco-Cantos, F. & Martínez-Navarrete, C. (2009) Comparative analysis of the evaluation of the intrinsic vulnerability in carbonate aquifers (Cañete Mountain Range, province of Malaga), *Boletín Geológico y Minero*, 120(1), pp. 81–94.

Jiménez-Madrid, A., Martínez-Navarrete, C. & Carrasco-Cantos, F. (2008) Comparative analysis of analytical methods for wellhead protection areas implementation. Application to different types of aquifers in the south of Spain, in: B. Villunsen (Ed.) *GroPro—Groundwater Protection, ATV Jord og Grundvand,* Denmark, pp. 67–74 (Korsor: ATV Jord og Grundvand).

Jiménez-Madrid, A., Martínez-Navarrete, C. & Carrasco-Cantos, F. (2010) Groundwater risk intensity assessment. Application to carbonate aquifers of the western Mediterranean (Southern Spain), *Geodinamica Acta*, 23(1–3), pp. 101–111.

Lallemand-Barrès, A. & Roux, J. C. (1999) *Périmètres de protection des captages d'eau souterraine destinée a la consommation humaine* (Orleans: Éditions du BRGM).

Margat, J. & Suasis-Parascandola, M. F. (1987) Mapping the vulnerability of groundwater to pollution, some lessons from experience in France, in: W. van Duijvenbooden & H. G. van Waegeningh (Eds) *Vulnerability of Soil and Groundwater to Pollutants*, pp. 433–436 (The Hague: BRGM).

Martínez, C., Carrasco, F. & Jiménez, A. (2008) Methodologies for groundwater protection used for human consumption, in: I. Vadillo (Ed.) *Conceptos y Técnicas en Hidrogeología*, pp. 177–216 (Málaga: ICOGA).

Martínez-Navarrete, C. & García García, A. (2003) *Wellhead Protection Areas for Water Abstractions Point. Methodology and Application to the Territory* (Madrid: IGME).

Martínez-Navarrete, C., Grima Olmedo, J., Durán Valsero, J. J., Gómez Gómez, J. D., Luque Espinar, J. A. & De la Orden Gómez, J. A. (2008) Groundwater protection in Mediterranean countries after the European water Framework directive, *Environmental Geology*, 54, pp. 537–549.

Martínez-Navarrete, C., Sánchez-Navarro, I., Jiménez-Madrid, A., Carrasco-Cantos, C. & Moreno-Merino, L. (2010) Coordination–interaction between safeguard zones and land-use planning policies. Paper presented at the European Groundwater Conference, Madrid, Spain, pp. 47–56 (Madrid: IGME).

Mehan, G. T. (2004) Better monitoring for better water management, *Water Environment Research*, 76(1), pp. 3–4.

MIMAM (2000) *White Book of the Water in Spain* (Madrid: Ministerio de Medio Ambiente).

MIMAM (2002) *Guides for the Delimiting and Implantation of Wellhead Protection Areas of Groundwater for Public Supply* (Madrid: Ministerio de Medio Ambiente).

Mödinge, J. & Kobus, H. (2005) Approach and methods for the assessment of sustainable groundwater management in the Rhine-Neckar region, Germany, *International Journal of Water Resources Development*, 21(3), pp. 437–451.

Ravbar, N. (2007) *The Protection of Karst Waters: A Comprehensive Slovene Approach to Vulnerability and Contamination Risk Mapping* (Ljubljana: ZRC).

Sargaonkar, A. P., Gupta, A. & Devotta, S. (2008) Evaluation of monitoring sites for protection of groundwater in an urban area, *Water Environment Research*, 80(11), pp. 2157–2164.

Vías, J. M., Andreo, B. & Perles, M. J. (2005) A comparative study of tour schemes for groundwater vulnerability zapping in a diffuse flow carbonate aquifer under Mediterranean climatic conditions, *Environmental Geology*, 47(4), pp. 586–595.

Vías, J. M., Andreo, B., Perles, M. J., Carrasco, F., Vadillo, I. & Jiménez, P. (2006a) Proposed method for groundwater vulnerability zapping in carbonate (karstic) aquifers: the COP method. Application in two pilot sites in Southern Spain, *Hydrogeology Journal*, 14, pp. 912–925.

Vías, J. M., Neukum, Ch., Hötzl, H. & Andreo, B. (2006b) Statistical comparison and control of different vulnerability mapping methods in Bauschlotter Platte aquifer (Germany), *Proceedings of the 8th Conference on Limestone Hydrogeology*, Neuchâtel, Switzerland, September 2006 pp. 263–266.

Vorreyer, C. (1998) Delineating surface source water protection areas in Germany, in: Proceedings of Source Water Assessment and Protection 98, Dallas, TX, USA pp. 61–64.

Vrba, J. & Civita, M. (1994) Assessment of groundwater vulnerability, in: J. Vrba & A. Zaporozec (Eds) *Guidebook on Mapping Groundwater Vulnerability*, pp. 31–48 (Hannover: IAH, Heinz Heise).

Wang, X. (1997) Conceptual design of a system for selecting appropriate groundwater models in groundwater protection programs, *Environmental Management*, 21(4), pp. 607–615.

Zwahlen, F. (Ed.) (2004) *COST Action 620. Vulnerability and Risk Mapping for the Protection of Carbonate (Karstic) Aquifers* (Brussels: European Commission).

Evolution of Water Management in Mexico

FELIPE I. ARREGUÍN CORTÉS & ENRIQUE MEJÍA MARAVILLA
National Water Commission of Mexico, Mexico

ABSTRACT *The recognition of the need for an integrated and humane management of water resources has been gradually developed as a result of several major international conferences and forums. These, together with the World Water Vision, have reinforced the need for a comprehensive assessment of global freshwater resources as a basis for implementing a more integrated management of water. The recommendations suggested in the meetings and international forums have undoubtedly helped the development of water resources management in Mexico. In general, however, to implement them effectively and efficiently, it is necessary to develop financial mechanisms available to the payment capacity of each country and establish reasonable deadlines for meeting the goals. This paper analyses the impact of the recommendations arising from international meetings on water management in Mexico as well as their compliance in terms of water and wastewater management.*

Introduction

The territorial extension of Mexico is 1.964 million km^2, of which 1.959 million km^2 belong to the continental surface and the rest to insular areas. For its geographical location, the southern portion is located between the tropics and the north portion is in the temperate zone at the same latitude as the Sahara and Arabian deserts (latitude N 32°43'06" and S 14°32'27" for Mexico). Moreover, because of significant geographic features that characterize the national landscape, there is a variety of climates (Figure 1) (National Water Commission of Mexico, 2010).

Its geographical location and topography affect directly the availability of water resources. Two-thirds of the land is considered arid or semiarid, with annual rainfall of less than 500 mm/year, while the south-east is humid with average rainfall exceeding 2,000 mm/year. In most areas the rainfall is more intense in summer, mainly torrential. On the other hand, there are regions where rainfall or droughts occur cyclically causing extreme damage to the hydraulic urban infrastructure, the agricultural sector and human health.

Mexico comprises 31 states and one Federal District, consisting of 2,440 municipalities and 16 boroughs respectively. From 1950 to 2005 the country's population quadrupled, and went from being predominantly rural (57.3%) to predominantly urban (76.5%). Currently it has an estimated population of 108.81 million inhabitants (Figure 2).

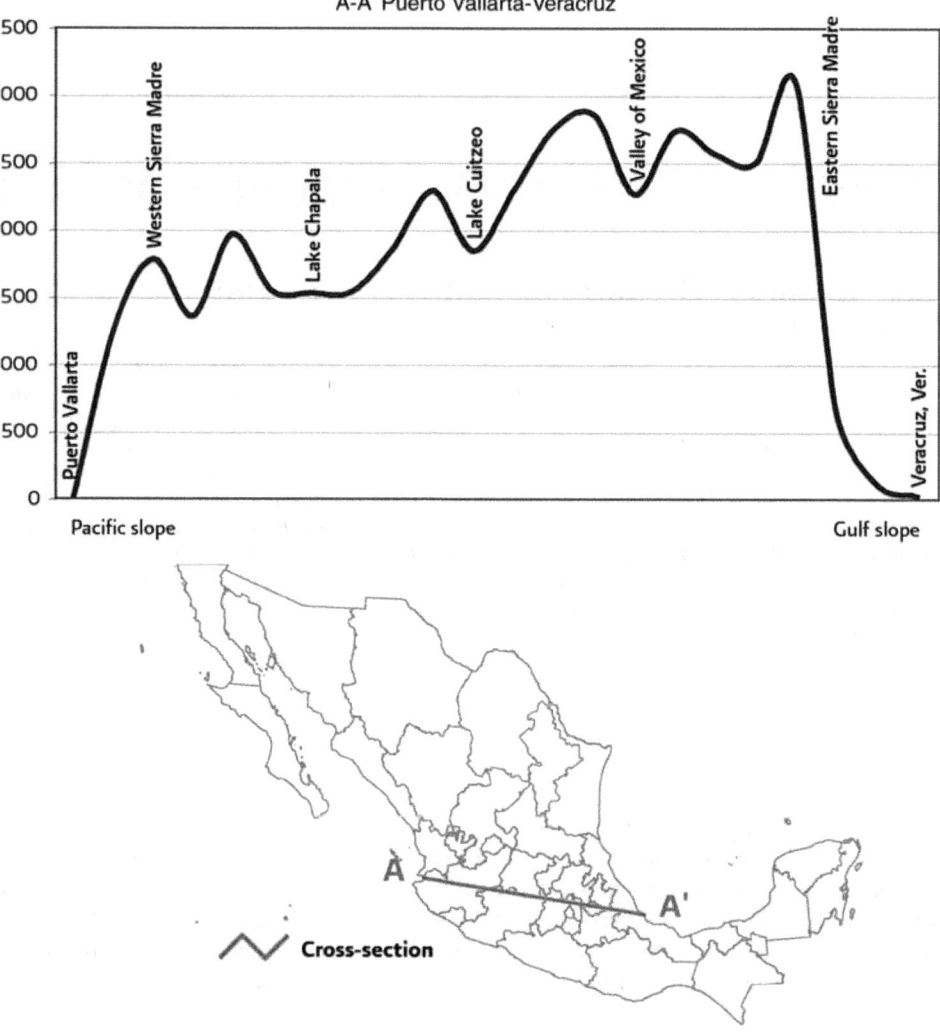

Figure 1. Profile of the elevation from Puerto Vallarta to Veracruz (metres above sea level).

For the management and conservation of the nation's waters since 1997, the country has been divided into 13 Hydrological-Administrative Regions, which consist of several basins and respective municipal boundaries. The Valley of Mexico Region XIII has a large population, a low amount of renewable water and provides a large proportion of gross domestic product. In contrast, other regions have different characteristics in the contribution to gross domestic product, population and renewable water, such as the Southern Border Region XI which has the greatest amount of renewable water but a relatively low population and contribution to gross domestic product (Figure 3).

In recent years the problem of water availability has increased because of imbalances created by continuing growing demand, the inefficient use of water and increased pollution levels.

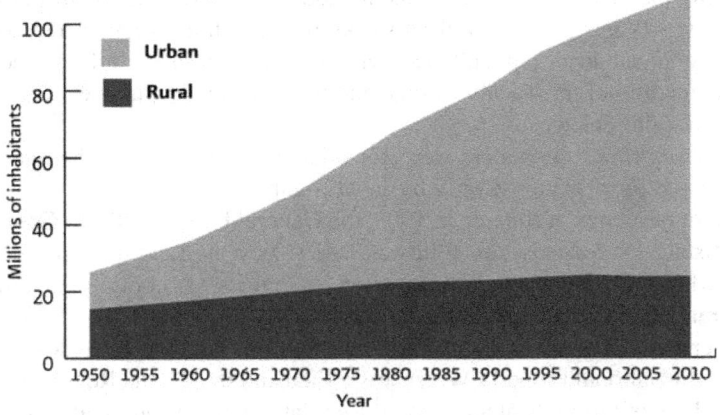

Figure 2. Evolution of Mexico's urban and rural population from 1950 to 2005, and projection to 2010 (millions of inhabitants).

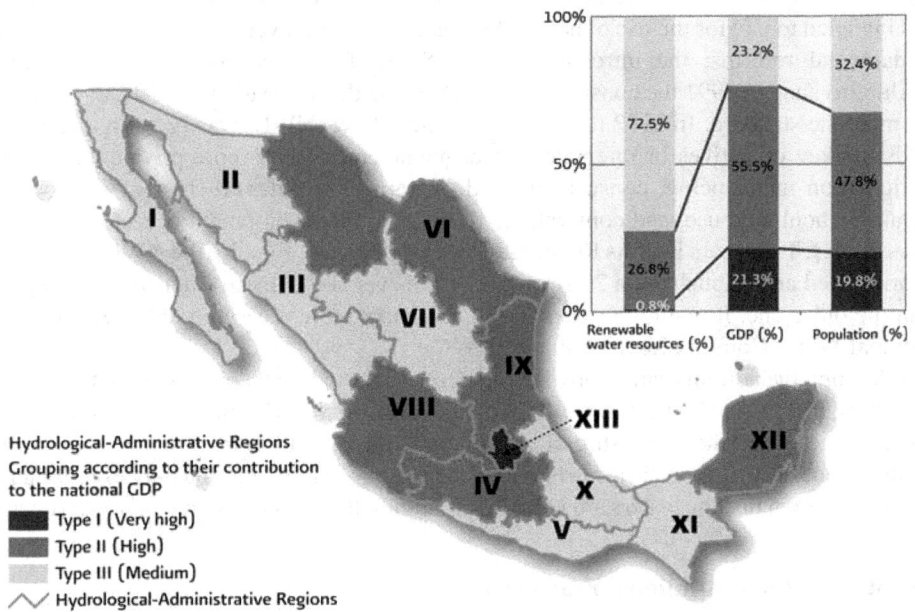

Figure 3. Regional comparison between infrastructure and availability of water.

Evolution of the Legal Framework

The hydrological management of water, dating back to pre-Hispanic times, includes the construction of waterworks for flood control, navigation, agriculture and fisheries in the Valley of Anahuac, which in Nahuatl means "between water". Towards the end of this period dams, roads and barricades divided the great lake of Mexico into compartments whose function was to control the flow of water from lakes and rivers to prevent flooding and to prevent gradual salinization of the system; two barricades separated Mexico from Texcoco Lake (National Water Commission of Mexico, 2009a).

Later, during the viceroyalty, highlights included the conveyance, distribution, storage and drainage works in cities, as well as works in agriculture, mining and seaports. An example is the aqueduct of Chapultepec, which springs from the hill and continues by arches to the junction of the Tacubaya road and then by the Chapultepec road, to finish at the fountain of Salto del Agua.

During colonial times the waters were considered the property of the Spanish Crown and were allowed for private use only by royal grants.

With the independence of Mexico in 1908, the General Law of Paths of Communication began to legislate for water issues, followed in 1910 by the Law of Federal Jurisdiction Water Use, which considered the classification of sources of supply, regulation of water use and formalization of the concessions system. In 1926 the Law of Irrigation with Federal Waters was enacted, which led to the construction of huge hydraulic works. In 1934 the Law of National Property Water was issued and in 1936 its Regulations. By 1956 and 1958 the Law of Groundwaters Use and Regulations was promulgated to regulate the extraction and use of water. In 1972 the Federal Water Law was issued; and in 1975 the National Water Plan was published, which continued being a model for planning. Another relevant legislation moment came in 1982 when in the Federal Law of Fees the obligation was included to pay for the use of national surface and groundwaters. This law has had two fundamental reforms: the introduction in 1986 of fees according to regional water availability; and in 1991 the consideration of charging those wastewater discharges that do not meet the standard. In 1992 the National Waters Law (NWL) was promulgated which holds the key objectives of integrated water management with more private and citizen participation in financing, constructing and operating the infrastructure of water, legal certainty about land use, and comprehensive sustainable development of water, its goods and services. Two years later its Regulations were published. Finally, the NWL was deeply reformulated and published on 29 April 2004. This version is more extensive and explicit than the older one. It mainly justifies the changes that were recently implemented by the National Water Commission of Mexico (NWC).

In Mexico the administration of water resources is based on the principle that "Water is owned by the Nation", since Article 27 of the Constitution (Main Act) of the Mexican United States gives the ownership of waters to the nation within the limits of its national territory and states that waters are originally the property of the state, and only exceptionally when it is shown that the waters are not national should they be considered private property.

Evolution of the Institutional Framework

Regarding the institutional framework, both the importance of irrigation and rainfall variability made the post-revolutionary governments give emphasis to the type and quantity of the country water resources. In 1917 with the re-organization of the Ministry of Agriculture and Development (MAD), the Water, Lands and Settlement Management Department was created. In 1926 under the Law of Irrigation with Federal Waters the National Irrigation Commission (NIC) was established. It was in charge of employing American engineers in order to direct the first huge irrigation projects and train Mexican engineers. In 1933 this Mexican–American group constituted the Technical Advisory Department, which was in charge of waterworks, conducting inspections, reviewing the technical conditions of works, conducting elaborate specialized studies and monitoring projects, all with the goal of obtaining high-quality hydraulic works. During that time several dams were built such as Santa Gertrudis, Tamaulipas, Don Martin between Nuevo Leon and Coahuila, El Mante, Tamaulipas, Guatimapé, Durango, Tepuxtepec in Michoacan and Plutarco Elias Calles in Aguascalientes. In 1946 due to the disappearance of the NIC and the creation of the Ministry of Water Resources (MWR), the Technical Advisory Department was assigned to the particular secretary of the Ministry. Between 1940 and 1950 many plans were developed and water committees at the regional and sectorial levels were created in order to establish goals and strategies for managing the use of water resources (Ministry of Agriculture and Water Resources, 1976). In 1989 the National Water Commission in Mexico (NWC) was created as the single authority on the subject. It was established like a decentralized entity under the Ministry of Agriculture and Water Resources. Since December 1994 it became part of the new Ministry of Environment, Natural Resources and Fisheries, which since 2007 became the Ministry of Environment and Natural Resources.

Water Management in Mexico after Mar del Plata

Thirty-three years ago in March 1977 the United Nations Conference on Water was held in Mar del Plata, Argentina (United Nations, 1977). One of its conclusions was that most countries of Latin America and the Caribbean had given relatively little importance to the management of water resources, which is why it was agreed that these countries should consider the following subjects:

- Improve the assessment of water resources.
- Promote efficient water use.
- Increase efficiency in the regulation and distribution of water resources.
- Develop public systems of water supply and waste disposal.
- Regulate water use in industry.
- Guarantee environmental protection and good water quality for health purposes.
- Take measures to control pollution.
- Establish national water policies.
- Strengthen institutional capacities.
- Adapt national laws.
- Promote public participation.
- Develop appropriate technology.
- Reduce losses from floods and droughts.
- Strengthen education, training and research.
- Promote regional cooperation between countries sharing international rivers.
- Improve technical cooperation among developing countries.

Following that conference, 17 conferences, several forums and/or summits on water subjects have been held, producing twelve pronouncements on water issues, six action plans or programmes, one International Decade for Water Supply and Sanitation, one Decade entitled "Water for All", one International Year of Freshwater and five World Water Forums (Table 1) (United Nations, 2009).

Legal Framework

Currently the legal framework that regulates the nation's waters is based on the Constitution of the Mexican United States (Articles 27, 28 and 115), the NWL which

Table 1. Main meetings in the last 33 years.

Year	Event
1977	United Nations Conference on Water in Mar del Plata
1990	Global Consultation on Drinking Water and Sanitation for the Nineties (New Delhi)
1990	World Summit for Children (New York)
1990	International Decade for Natural Disaster Reduction
1992	International Conference on Water and Environment (Dublin)
1992	United Nations Conference on Environment and Development "Earth Summit" (Rio de Janeiro)
1994	Ministerial Conference on Drinking Water Supply and Environmental Sanitation (Noordwijk)
1995	World Summit for Social Development (Copenhagen)
1995	Fourth World Conference of the United Nations on Women (Beijing)
1996	Second United Nations Conference on Human Settlements, Habitat II (Istanbul)
1997	First World Water Forum (Marrakech)
1998	International Conference on Water and Development
2000	Second World Water Forum (The Hague)
2000	Declaration of the Millennium by the United Nations
2001	International Conference on Freshwater, Dublin +10 (Bonn)
2002	World Summit on Sustainable Development, Rio +10 (Johannesburg)
2003	Third World Water Forum (Kyoto)
2006	Fourth World Water Forum (Mexico City)
2009	Fifth World Water Forum (Istanbul)

regulates Article 27 of the Constitution, the Federal Law of Fees, the Law of Contribution of Federal Public Works and Improvement of Water Infrastructure, and state and local laws regarding water.

General or domestic water use can only be conducted by users, private and public, through concessions and allowances granted by the NWC in accordance with the rules and conditions stipulated in the legislation.

Historically the agencies responsible for handling the water supply and sanitation facilities at national level have been: the MAD in 1917, the NIC in 1926, the MHR in 1948, the Ministry of Human Settlements and Public Works (MHSPW) in 1976, the Ministry of Urban Development and Ecology (MUDE) in 1982, the Ministry of Agriculture and Water Resources (MAWR) in 1982, and the NWC from 1989 to present. In 1948 the process of centralizing federal policy on water supply, sewerage and sanitation was consolidated when the MHR, a federal government agency, took over the management of these services and was responsible for planning, scheduling, designing and building the systems through Federal Drinking Water Boards. Those Federal Boards created some space for the participation of members from local governments and entrepreneurs, but basically were administered by staff of the MHR, which provided the bulk of funds for the building infrastructure. Later during the 1970s, before the great

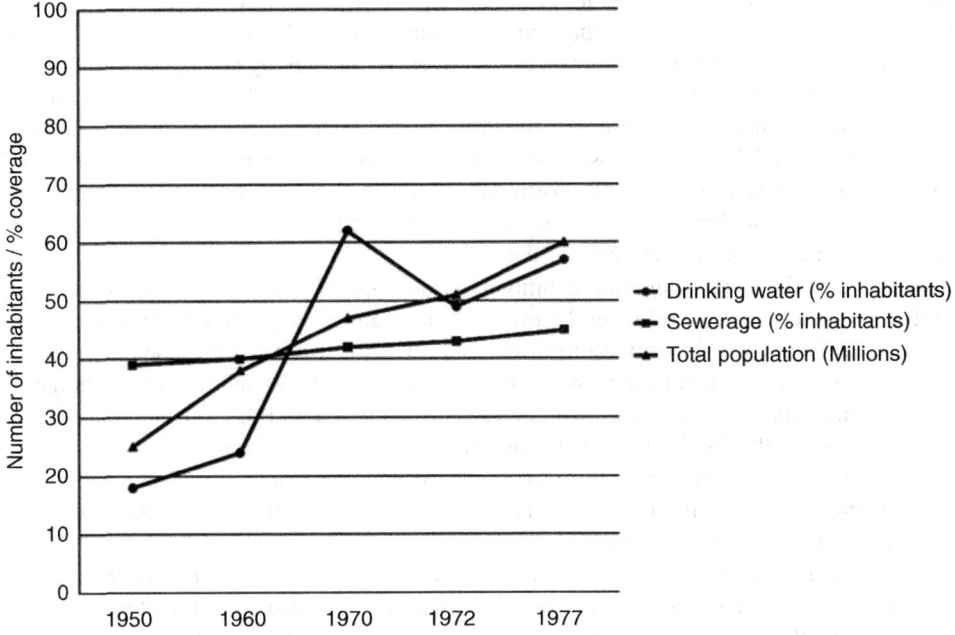

Figure 4. Coverage of water and wastewater before Mar del Plata.

increase in urban growth, centralized administration could not meet growing demand and maintain acceptable levels of service quality. Paradoxically, the low participation of states and municipalities up to that time had created the myth that the service was an obligation of the federal government.

In late 1976 the human resources and functions of the MHR were divided into two new departments: the MHSPW and the Ministry of Agriculture. The first was responsible for drinking water and sewage and sanitation systems, the second was in charge of works of larger magnitude and technical complexity. This created an artificial division between what are called "Works of bulk water supply" and those for drinking water, sewerage and sanitation.

Figure 4 shows the evolution of drinking water and wastewater coverage before the meeting at Mar del Plata. Note that an increasing population trend reached 60 million in 1977 when the MHR become the MHSPW, and also the increased coverage of the population served with drinking water from 1950 to 1970, rising from 18% to 60% with a 50% decline in 1972, the year of the United Nations Conference on Environment in Stockholm, and a small recovery by 1977, when the meeting was held at Mar del Plata. As for sanitation, it remained nearly constant with coverage close to 40%.

Water Management in Mexico from Mar del Plata to the End of the International Decade for Water Supply and Sanitation

The United Nations Conference on Water approved the initiative of the International Decade of Water Supply and Sanitation 1981–1990 (IDWSSD), which arose from a recommendation that governments around the world support the United Nations

Conference on Human Settlements (Habitat) in Vancouver, Canada, in 1976; it was proclaimed in November 1980 by the General Assembly of the United Nations. This global movement aimed to improve and promote the coverage of drinking water and sanitation to as many people as possible, especially the marginalized sectors of the cities and rural areas.

The World Health Organization and the Pan American Health Organization, were fully committed to IDWSSD's purposes and provided constant motivation, guidance and support to countries through policy formulation and the coordination of comprehensive health activities, technical assistance, the procurement of financial resources, and monitoring and evaluating progress of the Decade.

At the same time, following the recommendation of the Conference of Mar del Plata to establish a system that would meet the information needs of water and sanitation, the Pan American Information Network on Environmental Health was created at the beginning of the decade not only to meet the demand for information, but also mainly to encourage its use, exchange and dissemination throughout Latin America and the Caribbean.

In Mexico at the beginning of 1980s the MHSPW transferred the operation and maintenance of systems to state governments and, in some cases, to municipal governments. However, the lack of human and budgetary resources in the municipalities often led to the use of funds derived from the water service on other needs. The result of decreasing capital plus weak management to establish appropriate rates meant that the systems needed external funding continuously even for operation and maintenance.

Since 1982 the responsibility for federal intervention has passed to the MUDE, which also decentralized construction works and provided technical advice in applications for international loans.

The Ministry of Agriculture is responsible for large works undertaken for bulk water supply in those cases where, according to state governments, projects and works can be implemented by the Federal government.

Most of the period 1948–82 was characterized by the centralization of policy on water supply and sanitation by the Federal government, as it was the only institution that had the ability to meet the demands of investment. This period was also characterized by the low participation of local governments and low social participation. The effectiveness of this federal policy has declined as much as the demand for water has increased.

Until 1982 the policies and institutional structures adopted by Mexico for water management and sanitation showed great heterogeneity, and the administrative system was composed of various institutions with limited central coordination and limited or no delegation of responsibilities at the local level. This limited water resource management and knowledge about quantity and quality and it did not allow the country to meet the recommendations of the Plan of Action of Mar del Plata.

On 3 February 1983 the Legislature passed a bill that included additions and amendments to Article 115 of the Constitution, which regulates the activities of municipalities. Such modification explicitly defines the fact that drinking water and sanitation were the primary responsibility of municipalities, with the concurrence of the states, when necessary, and determined by local laws. To comply with this provision, state governments were instructed to undertake the necessary legal modifications for constitutional reform and then replicate them at state level, so the municipalities would have a comprehensive legal framework to provide these services. While the reform has delegated responsibility for services to municipalities, it has also left open the possibility that the state governments are responsible for them.

In order to comply with legal provisions on drinking water, an agreement between several federal agencies to transfer the construction and management of systems to state governments was published in the *Official Journal of the Federation* on 26 September 1983. This was the first step in the decentralization process that transferred services outside the federal sphere and located them at state level. The next step depended on the states, and it was decided whether to transfer the services to all or only some of the municipalities, or to keep them under the state administration or provide services jointly with the municipalities.

In 1988, water services and sanitation were administered at the state level in 21 of the 32 states (31 states and the Federal District), while for the remaining 11 states services were transferred to municipal governments.

In 1989 the NWC was established with the task of designing a consistent water policy, which was coherent and appropriate to the downsizing of state guidelines, and it promoted its own regulatory mechanisms of free markets. The NWC has emerged as a decentralized agency responsible for large waterworks, as well as for the regulation of irrigation districts managed by user committees and sanitation services managed by states and municipalities (National Water Commission of Mexico, 1989).

As a national regulator of water services, sewerage and sanitation, since its inception the NWC had made a definite and active role in the formulation of a new water policy, becoming the engine of change and realignment for operators of the resource. It published new guidelines for the water supply sector, sewerage and sanitation, with the premise of increasing the technical capacity and strengthening autonomy in the operating organizations, of promoting the establishment of service fees and income levels of appropriate stock, as well as of promoting their autonomy and private participation in services.

With the creation of the NWC, several of the recommendations of the Plan of Action of Mar del Plata were met, including the creation of a national department with broad responsibilities to collect water resources' data; the allocation of functions in a coordinated manner; the establishment of databases for the collection, processing, storage and systematic dissemination of information; the expansion of monitoring networks over the long-term, if possible according to the recommendations of the specialized agencies of the United Nations on the standardization of equipment, techniques and comparability of data; and the use of existing meteorological and hydrological series for the study of seasonal and annual variations in climate and water resources. In order to do this, the policy does not remain exclusively at the federal level. The guidelines of the NWC encourage state governments to update their laws for drinking water supply and sanitation, and to establish forms of organization and adequate financial systems for the new policy.

In response, state governments began to enact new laws on water and sanitation or, in some cases, to amend previous legislation to incorporate the guidelines and criteria of the NWC.

By 1996, 17 of the 31 Mexican states had enacted new laws or had modified the existing laws. The other 14 states had retained their old laws. By 1996 the policy adopted by the NWC had progressed substantially, despite not being implemented in all states.

Figure 5 shows the evolution of water coverage and wastewater after the meeting at Mar del Plata. The population reached 80 million in 1990, when the NWC was created as a single window to manage national waters, and it also increased coverage of the population served with drinking water between 1977 and 1990, rising from 60% to 79%. Regarding sewerage, coverage increased from 40% to 60%.

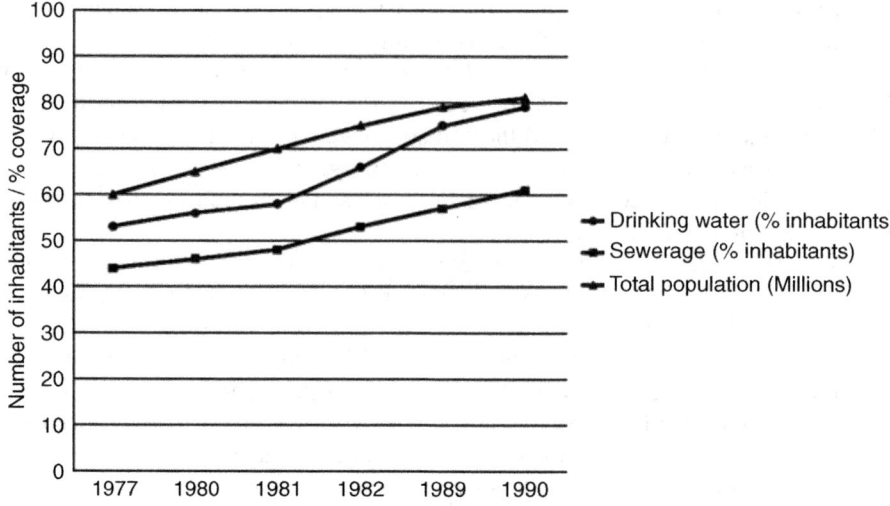

Figure 5. Evolution of the coverage of water supply and sewerage since the Mar del Plata meeting to the end of International Decade.

Moreover, with respect to the United Nations Conference on Water (Mar del Plata), the International Conference on Water and Environment (Dublin), the International Conference on Water and Sustainable Development (Paris), the United Nations Conference on Environment and Development Earth Summit (Rio de Janeiro), and the World Summit on Sustainable Development Rio +10 (Johannesburg), from the creation of the NWC, the following recommendations, among others, have been addressed:

- Verifying assessments of water resources and groundwater, including precipitation, evaporation and runoff, the volume of lakes and ponds in the national watershed, and publishing the availability of water basins and aquifers.
- Standardizing techniques and measurement tools and automated hydrometeorological stations, as needed and, if necessary, using standards and recommendations adopted by governments in various international organizations, and establishing training programmes and facilities for meteorologists, hydrologists, and hydrogeologists in the levels of specialization, master's and doctoral degrees.
- In coordination with the Ministry of Health, the NWC introduced the Clean Water and Clean Beaches Program for prevention of diseases related to water consumption and recreation activities as part of the assessment of water resources.
- As for sanitation, since 1994 private investments have been made under the Built, Operate and Transfer scheme for more than US$1 billion, mainly for the construction of plants for wastewater treatment with a combined treatment capacity of 24,122 L/s. However, the payment of fees has been a problem since the increase of fees to factories is based on the consumer price index; and the increase in fees for services to municipalities requires permission from the municipal council.
- Drinking water has been supplied to more than 5.5 million people and sewerage coverage has increased by nearly 9 percentage points, meaning that 13 million

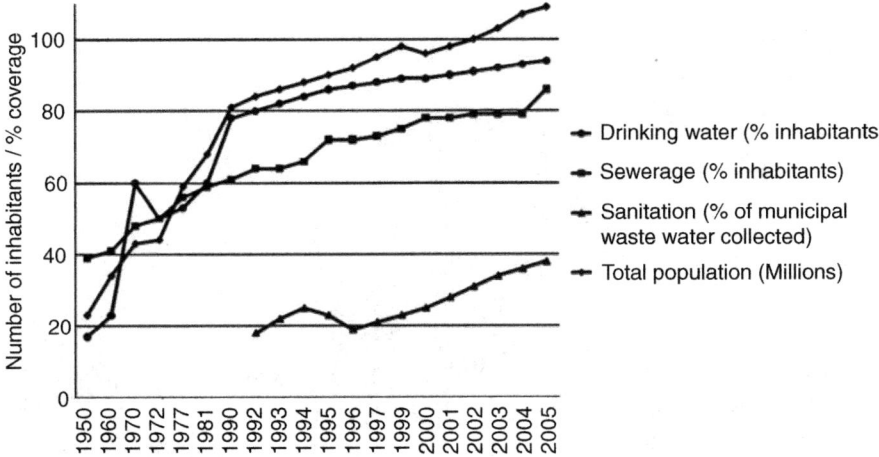

Figure 6. National coverage of water and wastewater treatment services.

people have access to this service for the first time. Figure 6 shows the development of water supply services, sewerage and wastewater treatment from 1950 to 2005, as well as the headquarters of the international meetings around the world.

Current Situation

The NWC, the administrative, regulatory, technical and advisory body responsible for water management in Mexico, performs its functions through 13 Basin Organizations, whose field of competence is the Hydrological-Administrative Regions.

The percentage of renewable water used for consumptive purposes is an indicator of the degree of pressure exerted on the water resources. If it is more than 40% it indicates a strong pressure on the resource. Overall, the whole country is experiencing a degree of pressure of 17.4%, which is considered to be a moderate level, but the central, north and northwest areas of the country have experienced a stronger degree of pressure (Figure 7).

The importance of groundwater is clear due to the magnitude of the volume employed by the main users; almost 37% (29.5 billion m^3/year in 2008) of the total volume allocated for off-stream uses is from groundwater sources. For the purpose of groundwater management, the country has been divided into 653 aquifers, the official names of which were published in the Official Government Gazette on 5 December 2001. From the 1970s onwards, the number of overexploited aquifers grew steadily, from 32 aquifers in 1975, to 80 in 1985, to 97 in 2001, and to 101 over-drafted aquifers as of 31 December 2008. From these over-drafted aquifers, 58% of groundwater is extracted for all uses. The results of recent studies have defined whether aquifers remain over-drafted or cease to be so, based on the ratio of withdrawal/recharge. For 2008, saltwater intrusion was found in 16 aquifers nationwide, located in the states of Baja California, Baja California Sur, Sonora and Veracruz de Ignacio de la Llave.

The country's rivers and streams represent a network of 633,000 km. This includes 50 rivers representing 87% of the runoff of the country whose basins cover 65% of the continental land area at the national level. Two-thirds of runoff belongs to seven rivers:

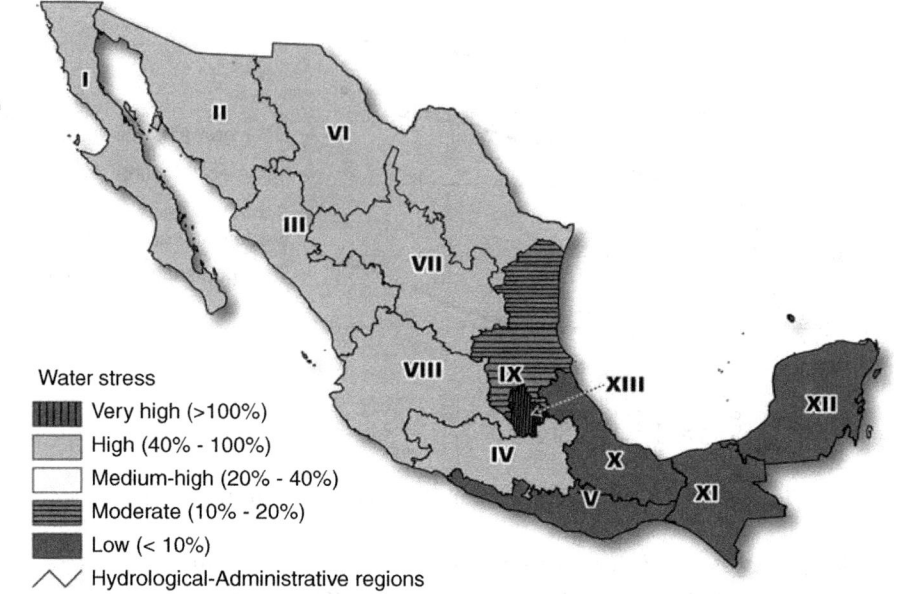

Figure 7. Water stress by Hydrological-Administrative Region.

Grijalva-Usumacinta, Papaloapan, Coatzacoalcos, Balsas, Pánuco, Santiago and Tonalá. The surface of their basins accounts for 22% of the country. The Balsas and Santiago rivers drain to the Pacific and the other five rivers drain to the Gulf of Mexico. The largest basins in terms of surface are the Rio Grande and Balsas, and the longest ones are Rio Grande and Grijalva-Usumacinta. Finally, the Lerma, Aguanaval and Nazas rivers drain to the inside of the country. At present, 100% of surface water availability in the country has now been published in the Official Government Gazette.

The NWC recognizes that prevention and control of pollution is one of the most important challenges in the management of water. It has therefore established a National Network for Measuring Water Quality which includes a monitoring network and a network of laboratories as well as an assessment system of water quality which identifies the bodies of water that are polluted according to the Ecological Criteria of Water Quality which are defined in accordance with the water use. This enables the development of effective strategies to control the pollution of bodies of water that are considered a priority for the country (Figure 8).

The evaluation of water quality is carried out using three indicators: five-day biochemical oxygen demand (BOD_5), chemical oxygen demand (COD) and total suspended solids (TSS). BOD_5 and COD are used to determine the quantity of organic matter present in water bodies, mainly from municipal and non-municipal wastewater discharges. The sites with water quality monitoring are situated in areas with a high influence of human activities.

In Mexico, drinking water services, together with sewage treatment and sewage disposal, are the responsibility of municipalities. For public water supply, groundwater is the predominant source, with 62% of the volume; however, in the last 10 years, surface water concessions for this use grew by 28%.

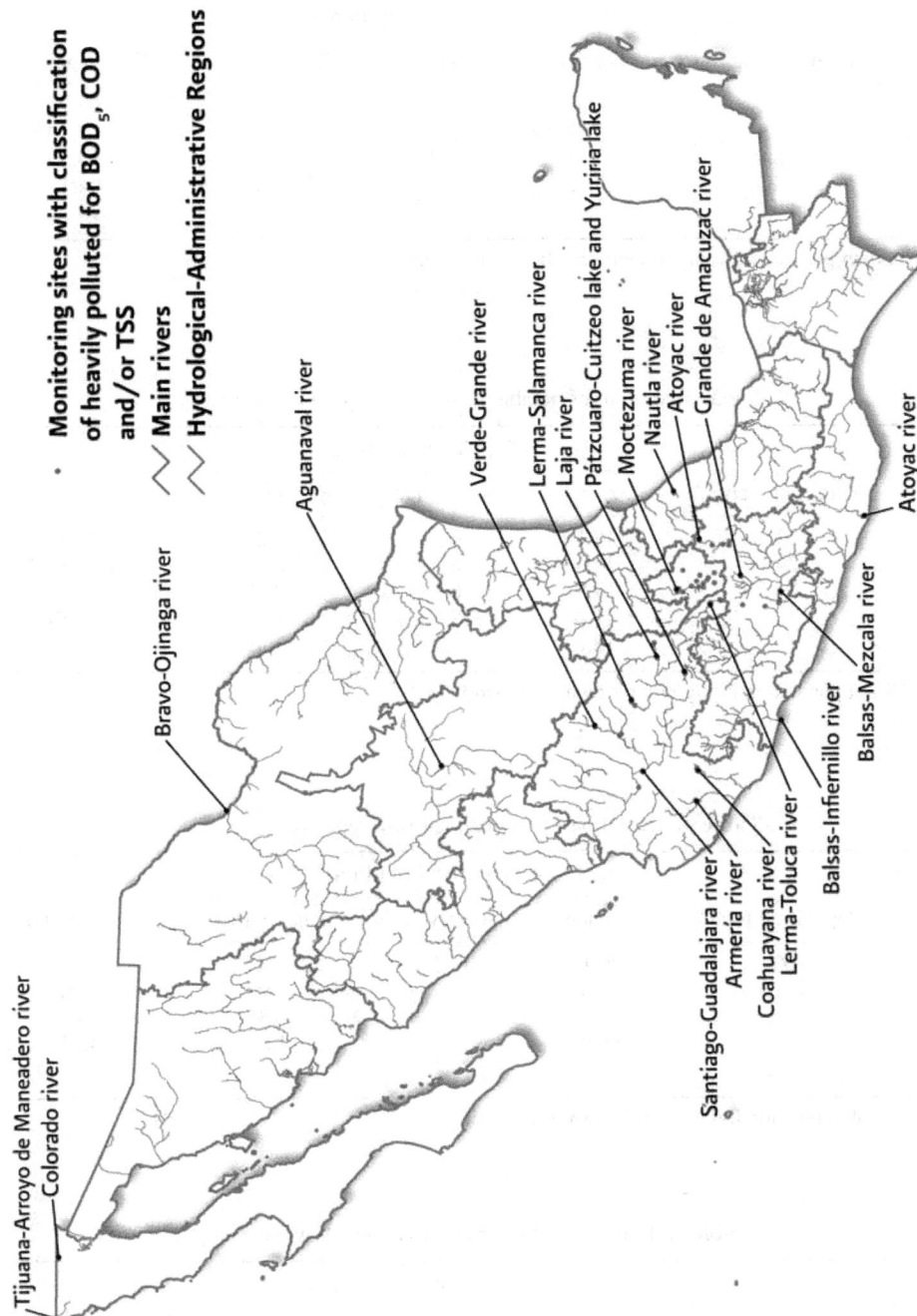

Figure 8. Monitoring sites with a classification of heavy pollution for BOD5, COD and/or TSS.

Table 2. Evolution of potable water coverage in urban areas.

Year	Urban population[a]	Population (millions)			Coverage (%)
		Service	No service	Beneficiaries	
2005	76.3	72.6	3.7	0.3	95.2
2006	77.1	73.3	3.8	0.7	95.1
2007	78.0	73.7	4.3	0.4	94.5
2008	78.8	74.2	4.6	0.9	94.1
2009	79.7	75.1	4.6	1.8	94.3

Note: [a] Urban population living in their own house (millions).

Table 3. Evolution of potable water coverage in rural areas.

Year	Rural population[a]	Population (million)			Coverage (%)
		Service	No service	Beneficiaries	
2005	23.9	17.1	6.8	0.2	71.5
2006	24.1	17.4	6.7	0.3	72.3
2007	24.0	17.9	6.1	0.5	74.7
2008	23.9	18.4	5.6	1.0	76.8
2009	23.9	18.8	5.1	1.4	78.6

Note: [a] Rural population living in their own house (millions).

Table 4. Evolution of the national sewerage coverage.

Year	Population total[a]	Population (million)			Coverage (%)
		Service	No service	Beneficiaries	
2005	100.2	86.1	14.1	0.5	85.9
2006	101.1	87.0	14.1	0.9	86.0
2007	101.9	87.8	14.1	0.8	86.1
2008	102.8	88.8	14.0	1.0	86.4
2009	103.6	89.9	13.7	1.1	86.8

Note: [a] Total population living in their own house (millions).

Table 5. Evolution of the water treatment coverage.

Year	2005	2006	2007	2008	2009
Increased flow (m^3/s)	7.3	2.6	4.9	4.3	4.5
Accumulative (m^3/s)	71.8	74.4	79.3	83.6	88.1
Collected wastewater (m^3/s)	205.0	206.0	207.0	208.0	209.0
Percentage	35.0	36.1	38.3	40.2	42.1

The NWC considers that potable water coverage includes populations who have piped water inside their house, outside their house but within their premises, public taps or another home. Those people with coverage do not necessarily have quality water for human consumption.

Taking the above into account, it is estimated that by the end of 2009 water coverage was 94.3% in urban areas, 76.8% in rural areas and 90.7% nationwide. Tables 2 and 3 show the changes in coverage of potable water to the country's population between 2005 and 2009. The populations in cities with more than 100,000 people have better coverage compared with smaller towns. On the other hand, the NWC considers that sewerage coverage includes people who are connected to the sewerage network or have a septic tank or drain, or dispose the sewage in a canyon, chasm, lake or sea. It is estimated that by the end of 2009 sewerage coverage was 86.8%, comprising 93.9% coverage in urban areas and 61.8% in rural areas. Table 4 shows the sewerage coverage nationwide.

The National Water Program 2007–2012 set a target of 60% treatment for municipal wastewater collected in sewerage systems. It is currently 42.1% (Table 5). To achieve the goals in this area it is necessary to improve the participation of federal, state and municipal institutions and the private sector users, particularly in sanitation projects in the metropolitan areas of Guadalajara and Mexico Valley (National Water Commission of Mexico, 2009b).

Water Policy

Based on the sector's problems, the President of the country has made a commitment to deal with all present and future water challenges.

In support of the objectives of the National Development Plan 2007–2012, regarding the strengthening of national sovereignty, social development, and comprehensive and sustainable economic growth, general policy objectives were established to provide a framework for the formulation and implementation of programmes and actions in the water sector in the short-, medium- and long-terms (National Water Commission of Mexico, 2008):

- Reduce the delays in water supply services to vulnerable social groups.
- Improve integrated watershed sanitation management, starting where pollution causes most negative effects on health, the economy and the environment.
- Grant legal certainty to the rights to use national waters and related properties.
- Contribute to the process of transition to sustainable development through the rationalization of water pricing using economic and environmental criteria.
- Expand the channels for the participation of society in the planning and use of water.
- Manage the resource more efficiently through constant and gradual decentralization of programmes and functions to users and local authorities within the framework of the new federalism.
- Induce patterns of more efficient water use in irrigation, domestic and industrial sectors in order to preserve the future availability and quality of the resource.

From the late 1970s the international community noted with concern the trends towards continuous depletion and degradation of water resources, and recommended adopting a holistic approach to its administration and management as well as the implementation of economic mechanisms to influence their rational utilization.

The Rio de Janeiro Conference on Environment and Development picked up on the same concerns, only now the problems had worsened and the solutions had to be developed within a different framework because the water problems were closely linked to economic efficiency, social development and environmental sustainability. These are the vertices of a triangle with interactions and complementary objectives, which are not exclusive but represent a new form of development. This is how the concept of sustainable development is defined in terms of natural resource management and technological and institutional change in such a way that ensures the continued satisfaction of human needs for present and future generations. This development does not degrade the environment, is economically viable, technically feasible and socially acceptable.

In this order of ideas, Mexico has implemented substantial changes to its laws and institutions, in its schemes to manage the natural resources, and in its economic and regulatory instruments to achieve sustainable development (United Nations, 2006).

Conclusions

Provision of drinking water, sewerage and sanitation services is the major challenge for the coming years, especially in rural and urban parts of the country where the lag in the provision of these services is higher.

It is extremely important to increase coverage of sanitation, inspection and monitoring of discharges as a mechanism for the prevention and control of pollution in national waters. In addition one must consider the risk assessment of environmental damage. According to data from the Ministry of Health, one dollar invested in sanitation is equivalent to 10 dollars saved in health.

Because of the degree of deterioration of water quality in the most polluted rivers, the diversity of pollutants and the effects of climate change, it is necessary to increase the number of monitoring sites and rethink their overall objectives.

For the next few years, the selection of alternative sources of water supply, the reduction of groundwater depletion and the protection of aquifers will necessarily have to be faced in an integrated manner.

By 2030 it is estimated that the population will reach 125 million inhabitants and that the need for water will increase by 25%. In addition, the increased demand for water will be in areas of low availability.

Policies concerning the rational use of water resources should point to the predominant use: agriculture.

It is of utmost importance to implement, in practice, the environmental flow reserve to protect the aquatic ecosystem functions considering that Mexico is one of the five most biodiverse countries worldwide.

Adaptation and mitigation measures must be implemented to address the phenomenon of climate change that will cause rising sea levels, prolonged droughts or heavy rainfall.

For proper and timely decision making by the authorities involved in the management of water resources, it is vital to modernize and expand networks measuring water quantity and quality.

Finally, although significant progress has been achieved, the development of Mexico and the global trends in economic, social and environmental issues have created new challenges for water management.

References

Ministry of Agriculture and Water Resources (1976) *Water Resources Report 1970–1976* (Mexico City: Ministry of Agriculture and Water Resources).

National Water Commission of Mexico (1989) *Guidelines for the National Water and Sewerage. Ministry of Environment, Natural Resources and Fishing* (Mexico City: National Water Commission of Mexico).

National Water Commission of Mexico (2008) *National Water Program 2007–2012. Ministry of Environment and Natural Resources* (Mexico City: National Water Commission of Mexico).

National Water Commission of Mexico (2009a) *Historical sketch of Water in Mexico. Ministry of Environment and Natural Resources* (Mexico City: National Water Commission of Mexico).

National Water Commission of Mexico (2009b) *Status Subsector Water and Wastewater, 2009 Edition* (Mexico City: Ministry of Environment and Natural Resources).

National Water Commission of Mexico (2010) *Statistics on Water in Mexico, 2010 Edition* (Mexico City: Ministry of Environment and Natural Resources).

United Nations (1977) *Report of the United Nations Conference on Water, Mar del Plata, 14 to 25 March 1977* (New York, NY: United Nations Organization).

United Nations (2006) *Water a Shared Responsibility*. UNESCO (New York, NY: United Nations Organization).

United Nations (2009) *World Water Development Report 3: Water in a Changing World*. UNESCO (New York, NY: United Nations Organization).

Agriculture and Water Pollution: Farmers' Perceptions in Central Mexico

ROSARIO PEREZ-ESPEJO*, ALONSO AGUILAR IBARRA* & JOSE LUIS ESCOBEDO-SAGAZ**

*Instituto de Investigaciones Economicas, UNAM, Circuito Mario de la Cueva, Ciudad Universitaria, Mexico City, Mexico; **Facultad de Economia, Unidad Campo Redondo, Universidad Autonoma de Coahuila, Saltillo, Coahuila, Mexico

ABSTRACT *Agricultural nonpoint discharges represent a major problem in Mexico. However, the perception of farmers toward water-quality issues is critical for the potential acceptance of environmental measures. In order to assess farmers' perceptions on water quality and agricultural practices, questionnaires were given to 145 farmers in an irrigation district in Central Mexico. It was found that farmers do not reckon water quality in the Lerma River to be a serious environmental problem and the stated willingness to diminish the use of pesticides and fertilizers depended on farm size. Smaller farmers were more reluctant to adopt sustainable practices than bigger ones. Therefore, differentiated agro-environmental policies might be more effective for dealing with non-point source water pollution.*

Introduction

Agriculture and livestock operations are a major source of water nonpoint source (NPS) pollution. Indeed, agricultural activities have an important contribution to nonpoint pollutants as more intensive uses of pesticides and fertilizers have led to a rise in both phosphates and nitrates in both soils and water. In spite of ample evidence on this matter, dealing with agriculture-originated NPS pollution is a comparatively new concern (Horan & Shortle, 2001).

Optimal instruments (i.e. first-best policy) for NPS control imply high information, transaction and enforcement costs which in practice are rarely applied. Hence, given the number or real-world restrictions, only second-best instruments which are not cost-effective have been developed (Shortle *et al.*, 2001). Thus, given the difficulties in implementing even second-best instruments to agriculture, governments have focused on voluntary compliance programmes, including public persuasion and technical assistance in order to encourage green technologies or sustainable agricultural practices (Horan *et al.*, 2001). Examples of such programmes include the Best Management Practices (BMPs) programme in the United States and the Agro-Environmental Measures (AEMs) programme in Europe. In contrast, most developing countries lack agro-environmental policies to cope with NPS problems (e.g. Ibarra & Pérez Espejo, 2008); as a consequence, voluntary programmes for water pollution control are not as widespread as in industrialized economies.

It is well documented that designing effective agro-environmental measures, such as voluntary programmes, implies knowing the perception and attitudes of farmers before and after their implementation (e.g. Davies & Hodge, 2006; Ajayi, 2007; Wei *et al.*, 2007; Defrancesco *et al.*, 2008; Barnes *et al.*, 2009). In fact, Wossink (2005) reckons that a deeper understanding of farmer perceptions is one of several major emerging research issues in agro-environmental policy analysis. Accordingly, the analysis of perception of farmers towards water-quality issues and potential implementation of agro-environmental policies, such as voluntary programmes, is paramount in order to deal with NPS pollution in developing countries. Hence, the objectives of this paper are: to assess the perception of farmers on water quality and voluntary programmes, before any agro-environmental measure is implemented; and to determine whether there are differences on theses perceptions among farmers according to their plot size with a case study in Central Mexico.

Acceptance of Agro-environmental Programmes

Defrancesco *et al.* (2008) point out that although there is a general agreement on the fact that participation in voluntary programmes depends on farmers' behaviour, gaps remain in the knowledge of which factors influence their decision-making. In general, factors inducing acceptance of agro-environmental programmes include:

- An awareness of or concerns about environmental issues that directly affect farms' productivity (Knowler & Bradshaw, 2007; He *et al.*, 2008), or farmers having a sense of land or environmental stewardship (Wossink, 2005; Davies & Hodge, 2006).
- Access to information sources (Defrancesco *et al.*, 2008), credit and training (He *et al.*, 2008), and non-farm extra income (Wei *et al.*, 2007).
- Both public and private transaction costs are an important issue for successful agro-environmental programmes (Falconer, 2000; Ducos *et al.*, 2009). For example, the acceptance among farmers of such programmes increases as costs and technical efforts diminish, and benefits are also clearly perceived (Vanslembrouck *et al.*, 2002; He *et al.*, 2008).

In contrast, reticence to participate in agro-environmental programmes is, among other factors, due to the following:

- A perception of the potential risk, higher costs or technical difficulties for implementing new agro-environmental practices (Wossink, 2005; Davies & Hodge, 2006; Keraita *et al.*, 2008).
- A lack of trust in government officials and scepticism towards water quality indicators (Kraft *et al.*, 1996; Barnes *et al.*, 2009).
- Kraft *et al.* (1996) pointed out that farmers who owned their land were less willing to adopt new practices than farmers who leased it.

Some studies point out that there is a difference in farmers' plot size with regard to acceptance or participation in agro-environmental measures. On the one hand, there is evidence pointing out that small farmers would adopt sustainable practices (Vanslembrouck et al., 2002; He et al., 2008). On the other hand, several studies have found that smaller farmers would be more reluctant than larger farmers to enter agro-environmental programmes (Ducos et al., 2009; Damianos & Giannakopoulos, 2002; Perret & Stevens, 2006). Such differences have important policy implications. For instance, pollution control complexity (and therefore transaction costs) increases with both the spatial scale and the scale of polluters' nature. If a command-and-control approach is taken, the enforcement cost will be higher for controlling many small farmers than a few big farmers. But if a voluntary approach is adopted, the amount of payments for many farmers could be huge for authorities.

This study tests the hypothesis that perceptions among farmers toward water-quality issues and, consequently, their potential acceptance of environmental measures depend on farm or plot size. The following sections present methods and results of such an analysis.

Methods

The study was carried out at Irrigation District 011 (ID-11) in Guanajuato state, located in the Lerma River basin, Central Mexico, which is one of the most important agricultural regions of the country. ID-11 comprises about 100,000 ha (48% of the cultivated surface of Guanajuato state) and over 22,000 farmers working in 11 irrigation modules over two crop cycles: autumn–winter (A-W) and spring–summer (S-S). Agricultural activities use 78% of water extractions and, along with industrial and urban sources, discharge towards the Lerma and its tributaries, resulting in one of the most polluted rivers in Mexico. The 55% of the water comes from storage dams, 35% from wells and the rest is pumped directly from the river. As in the rest of the country, the land is highly fragmented; official figures show that the general average is about 4 ha/user (Figure 1), as 80% of farmers possess units from 0.1 to 5.0 ha, 13% between 5.0 and 11.0 ha, and 7% over 11 ha (information from the Irrigation Modules in 2008). Most irrigation districts in Mexico comprise less than 10 ha/user, so that ID-11 might be reckoned as a representative Mexican irrigation district.

Prior to fieldwork, the Mexican Water Commission provided the official registers of farmers of ID-11 in an Arc View GIS 3.2 format. Using this database, seven irrigation modules located in the banks of Lerma were chosen; representing about 72% of the total area of ID-11 and comprising 69% of total farmers.

Preliminary fieldwork was carried out in November 2007 in order to determine a representative sample size under the constraints of available human and financial resources, for improving the questionnaire, and to become acquainted with local authorities, modules' supervisors and several farmers' leaders, which later facilitated the major field survey. Users of the selected modules were classified into three categories: small, medium and large producers. Based on the size of the unit, the crop and the type of irrigation system (either by gravity irrigation or well-pumping) a representative sample of farmers was selected for both agricultural cycles. The questionnaire gathered information on agricultural practices during the crop cycles of A-W 2007–2008 and S-S 2008. Fieldwork was carried out in June–December 2008, obtaining responses for 145 questionnaires. Furthermore, semi-structured interviews were conducted with modules' staff and officials from the Guanajuato state Water Commission, the state Ministry of Agricultural Development and the Fund for Agriculture (FIRA).

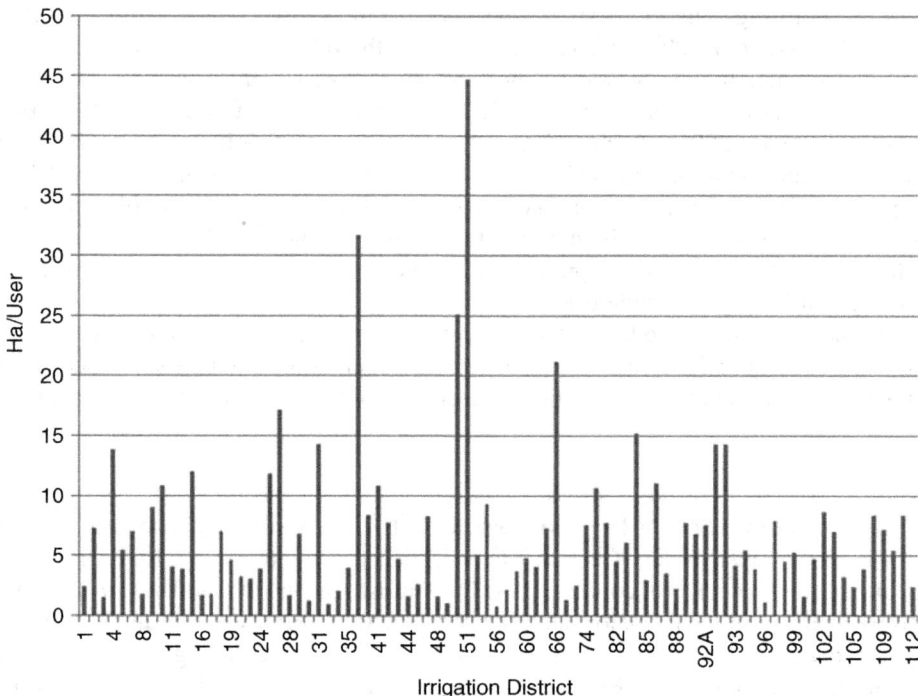

Figure 1. Average size of irrigation districts surface per user in Mexico. The study was carried out for Irrigation District 11.

Farmers' perceptions on water quality and agricultural practices were assessed with a 1–5 Likert's scale using six items. The Likert scale or Likert-type methods are frequently used for estimating perceptions and attitudes of farmers with respect to agro-environmental programmes (e.g. Wei et al., 2007; Keraita et al., 2008; Barnes et al., 2009). Although being a semi-qualitative method, scores obtained from respondents can be used for hypothesis testing (Bryman & Cramer, 2004).

We included two items related to water quality perception, two on pollution sources perception and two on agro-environmental policy perception under the following format: How much do you agree with the following sentences (1 = nothing; 5 = very much):

- N1. Water quality I use for irrigation is optimal or adequate for production.
- N2. Other water users of Lerma River have problems due to water quality.
- N3. Agriculture is a major source of water pollution in Lerma River.
- N4. There are other activities that generate more water pollution than agriculture.
- N5. I am willing to diminish the use of pesticides and fertilizers in a voluntary way.
- N6. Changing agricultural practices should be supported with government funds.

To test the hypothesis that perception towards water quality and environmental policies depends, among other factors, on the size of farms, the sample of surveyed farmers was divided in three groups according to plot area.

The surfaces of farms were classified into three categories, using the standard deviation of the whole sample (19.69 ha). Questionnaires with incomplete responses on perception were deleted, resulting in a total of 121 observations (Table 1).

Differences among groups (small = S, medium = M, large = L) were estimated with analysis of variance (ANOVA), where the null hypothesis was:

H_0: $\mu S = \mu M = \mu L$
H_A: otherwise

where the decision criterion was:

H_0 is rejected if F-value (tables) $>$ F-value (observed)

In other words, if $p \geq 0.05$, then significant differences of means among groups exist.

Results

From the questionnaire responses, semi-structured interviews and field observations, we were able to characterize farmers as follows:

- Male population: 97% of respondents were men.
- High fragmentation of land: the average area per user ranged between 2.9 and 6.6 ha.
- Land tenure: estimated land leasing reaches about 60% of the total ID-11 surface.
- Low level of education: 10% of farmers declared they had no schooling, 38% had an incomplete primary education and 17% had an incomplete high school education.
- Aged population: the average age was 53 years, and half the farmers with no education were more than 59 years of age.
- Overuse of water and pesticides are common practices, especially among small producers and wheat producers.
- 35% of large farmers are educated to a professional level while 59% of small farmers have only basic or incomplete elementary education.

About 25% of farmers, especially large and medium-sized, declared that they carried out either minimum tillage, zero tillage, conservation tillage or no tillage. Among small farmers these practices are hardly performed because of a lack of adequate machinery; besides, they are not sufficiently organized to acquire it.

Table 1. Groups of farm size in the sample at ID-11, June–December 2008.

Farm size	Lower limit	Upper limit	Number of observations
Small	0.0	19.7	90
Medium	19.8	39.5	17
Large	39.6		14
Total			121

Table 2. Mean Likert scores for each farm size group.

	Item					
	Water quality		Pollution source		Agro-environ- mental policy	
Farm size	N1	N2	N3	N4	N5	N6
Small	3.99	3.38	2.06	4.14	3.03	4.47
Medium	4.35	3.53	2.06	4.24	3.41	4.35
Large	4.07	3.86	2.43	4.36	3.57	4.21
Total sample	4.05	3.45	2.10	4.18	3.15	4.42

Table 3. ANOVA results for Likert's scale items.

Item	Title	F-value	Significance	Result
N1	Water quality I use for irrigation is optimal or adequate for production	0.18	0.162	H0 is rejected
N2	Other water users of Lerma River have problems due to water quality	0.59	0.442	H0 is rejected
N3	Agriculture is a major source of water pollution in Lerma River	0.45	0.364	H0 is rejected
N4	There are other activities that generate more water pollution than agriculture	0.18	0.162	H0 is rejected
N5	I am willing to diminish the use of pesticides and fertilizers in a voluntary way	0.90	0.589	H0 is rejected
N6	Changing agricultural practices should be supported with government fun	0.31	0.265	H0 is rejected

Note: F-value (tables) = 0.051.

Agrochemicals' empty packaging was thrown away without any treatment by 25% of the farmers, 38% of them burned it and only 9% took it to a special container.

Health problems (e.g. hospitalization, nausea, dizziness, vomiting, headache) related to frequent agrochemicals use was reported by 25% of respondents.

The technology used is the "Green Revolution" type. Although these practices were implemented more than 40 years ago, farmers still do not use it properly. In fact, only 6% of surveyed farmers declared having taken any training on the use of pesticides, weed control, application of agrochemicals and pest control.

With respect to Likert scores, Table 2 shows the results for each item according to farm size groups.

Table 3 presents ANOVA results for each of the items presented to farmers. The results show that perceptions of water quality and environmental measures are different among farmers according to their plantation size. In most answers, the larger the farm, the higher the score, except for question N1, where medium-size farmers obtained the highest score, and for question N6, where the larger the farm, the lower the score.

Discussion

Perceptions on Water Quality

In Guanajuato, water availability has diminished from 2,800 to 800 l/capita/day from 1950 to the end of the 1990s; the number of exploited wells increased from 2,000 to 16,000 in that period; and ground level fell by 2 m/year (Sandoval-Minero & Almeida-Jara, 2006). In spite of this, this study found that farmers were not concerned about environmental problems and water quality for irrigation. This is normal perception in both developing and industrialized nations (Keraita et al., 2008). In fact, about three-quarters of the total irrigated area in the world is located in developing countries and from this total about 10% is irrigated using wastewater (Jiménez, 2006).

In spite of being aware that water for irrigation does not need to be of high quality, farmers do not reckon that the water quality of the Lerma is a serious environmental problem, neither for them as an input nor for other users. As in many cases, they are not even aware of potential health hazards, mostly when dealing with pesticides or waste water-irrigated produce. It has been frequently recorded elsewhere that farmers express either indifference or a lack of awareness of environmental problems that agriculture generates, like for example pesticides (Lichtenberg & Zimmerman, 1999), fertilizers (Izcara Palacios, 2002), nitrate leaching (Wei et al., 2007) or water-quality degradation (Barnes et al., 2009).

In this study, when comparing with recommendations of diverse research and support institutions (INIFAP, CESAVEG, SAGARPA), 37% of small wheat producers and 17% of small corn producers were overusing pesticides. The numbers for big producers are 16% for wheat and 18% for corn.

Incomplete information is a major factor for farmers to be unaware of environmental consequences. Hence, more complex problems, which are not easily observable, such as leaching, soil erosion, nutrients runoff or biodiversity loss, make farmers not completely aware of environmental problems unless impairments become exacerbated.

The priorities for people in developing countries are more directed at economic activities for survival rather than at conservation measures, even when they are aware of a problem or because they do not understand the consequences, or because their poor social organization does not help to correct any water impairment.

With respect to the perception of pollution sources, farmers in DR-11 reckoned that agriculture was not a major source of NPS pollution, whereas other activities generated more pollution. This is also a frequent result among farmers as they seem frequently reluctant to accept responsibility, frequently blaming other activities on environmental degradation (Lichtenberg & Zimmerman, 1999; Barnes et al., 2009).

Accepting responsibilities or being aware of environmental damage are closely related to a sound perception of such problems. In fact, when farmers are aware of environmental problems, they are more prone to adopt agro-environmental programmes (Vanslembrouck et al., 2002; He et al., 2008; Knowler & Bradshaw, 2007). In our survey, more educated farmers stated that they would adopt or, in fact, had adopted certain sustainable practices, but they still refused any responsibility for water pollution. A related issue is the institutional setting, which facilitates enterprising activities among larger farmers. For example, financial aids from the Mexican government (Irrigation Modernization Programs) are suited for large exploitations, and have technical capabilities for the implementation of new practices such as drip irrigation. In contrast, smaller farmers in Mexico lack capital, are ill-educated, and struggle with both administrative and technical operations. The burning of crop by-products, although banned by the Technical Environmental Standard 005 of Guanajuato, is widely practised in the region. In our study, 6% of the producers recognized that they burnt crops. Most of them were small farmers who lacked the necessary machinery for practices such as zero tillage. Furthermore, reaching government officers in urban areas is more burdensome than for larger farmers; therefore, the likelihood of adopting voluntary environmental practices among small Mexican farmers is negligible.

In fact, a drawback of voluntary agreements is that they do not directly induce changes on polluters' behaviour in the short-term, as in the case of France (Lacroix *et al.*, 2005) or California (Young & Karkoski, 2000). In contrast, when combined with economic incentives, voluntary instruments have promising results. Indeed, as farmers seek profitability in their activities, environmentally friendly practices are seldom taken into account unless a surplus income is implied. In the Mexican case, agricultural policy is dispersed in 59 programmes and only five of them (mostly in forest) correspond to the environmental area; two compulsory standards on wastewater discharges applied only to point discharges. To date, none of the voluntary programmes for environmental management in Mexico aims to improve water quality impairments directly. The impact of such programmes has been limited because they operate on very small portions of agricultural land and benefit only a few producers.

Implications of Farm Size to Agro-environmental Policies

Although farm size has been found to be one significant factor in empirical analyses of farmers' attitudes toward agro-environmental practices, there are other studies where this variable is negatively correlated or is not statistically significant (for reviews on these issues, see Wossink & van Wenum, 2003; Knowler & Bradshaw, 2007; and Defrancesco *et al.*, 2008). However, analysing the size of farmers' exploitation is of relevance to controlling NPS water pollution because an important question here is how to use financial resources effectively to alleviate water impairments.

One approach to deal with water pollution in developing and transition economies is to focus on big polluters in order to implement alleviation programmes. This is a cost-effective way to abate a large share of pollution which is very often produced by a few big farms (Kathuria, 2006). For example, Wei *et al.* (2007) recommend first targeting larger farmers for improving environmental perceptions, as they found that farmers had a better perception on environmental issues in Northern China, a similar result found in this study at ID-11 in Central Mexico. With respect to smaller farmers, Damianos & Giannakopoulos (2002) suggest that they should be somehow encouraged by governments to participate in agro-environmental programmes. As stated above, if acceptability of voluntary programmes depends on a farmer's size, then the design of such programmes should distinguish between small and reticent farmers, and larger and more willing-to-participate farmers, for an efficient agro-environmental policy. Thus, substantial transaction costs would be curtailed due to a lower adverse selection problem. Indeed, costs are a critical factor for the success of the implementation of agricultural practices devoted to reducing water pollution since agro-environment transaction costs are borne not only by the farmer, but also by the authorities (Ducos *et al.*, 2009).

Analysing the attitudes of farmers toward environmental problems according to their size is thus of paramount importance in transition economies, where fewer larger farmers are more willing to adopt voluntary programmes than many small farmers who are more risk averse and more reluctant to change agricultural practices. In the Mexican case, big farmers producing vegetables and fruits for the export market must comply with US quality regulations, which is a strong incentive to be more sustainable. Important management implications arise since a differentiated agro-environmental policy would be warranted in such diversified contexts. Hence, large producers might be more like their European or North American counterparts, while smaller producers resemble small rural farmers elsewhere. Thus, differentiating producers more willing to accept voluntary practices and searching for instruments specially designed for reticent farmers would reduce adverse selection problems, resulting in a more cost-effective agro-environmental policy. Although Shortle *et al.* (2001) state that economic incentives and market-based instruments may be more cost-effective than command-and-control instruments and might be more effective than voluntary agreements in order to improve water quality among small farmers, caution has to be granted because the heterogeneity of agricultural practices, types of producers, and crops in developing and emerging economies make analyses and agro-environmental policies difficult to compare with those set up in developed nations. In fact, as in the case of point-source water pollution (Kathuria, 2006), policy instruments should be according to the specific local conditions.

Finally, in emerging economies further research questions would be to assess the decision-making behaviour, the willingness to accept agro-environmental programmes, and the marginal environmental damage of farmers according to their size in different economic settings. In our case, the stated willingness to diminish the use of pesticides, fertilizers and water in a voluntary basis depended on farm size. Yet, the ambiguity of farm size is not solved. However, as in the present study, there is evidence elsewhere that smaller farmers (who outnumber medium and big producers) are more reluctant to adopt sustainable practices than larger ones. We argue, therefore, that differentiated agro-environmental policies might be more effective for dealing with NPS pollution in transition and developing economies, and further research should be directed to assess the environmental effects of many small farmers, in contrast to few big producers, on water quality.

Conclusions

Agricultural pollution of water is a major issue in Mexico because agriculture uses 78% of water extractions. Mexican agriculture is quite heterogeneous and provides income for more than 20% of the population.

Agricultural water pollution is due, among other factors, to a poor perception of producers on water quality impairments. Their regular agricultural practices include: overuse of water, fertilizers and agrochemicals for crops with a low economic value; no fallow of agricultural land; and the burning of crop by-products.

Water pollution by agriculture is also the result of misguided government intervention. Public policies do not differentiate among producers' size. Subsidies foster overuse of both water and pesticides, leading to a zero-cost pollution situation in the whole irrigation district.

Agricultural practices and producers' perception are related to the size of plot: larger producers, more capitalized, educated and export-oriented, carry on more sustainable practices than smaller ones. Small and medium farmers, with a low educational level, lack the skills for the proper use of "Green Revolution" inputs. Therefore, differentiated agro-environmental policies might be more effective for dealing with NPS pollution.

Acknowledgements

This study was partially funded by the PAPIIT-UNAM Program (Contract No. IN305107). Alethya Jara Duran and Andrea Santos Baca helped with field surveys and data base management. Part of this investigation was presented at the World Water Week in Stockholm, September 2010, where it benefited from useful comments from several participants. A former draft was improved thanks to the comments of anonymous reviewers.

References

Ajayi, O. C. (2007) User acceptability of sustainable soil fertility technologies: lessons from farmers' knowledge, attitude and practice in Southern Africa, *Journal of Sustainable Agriculture*, 30, pp. 21–40.

Barnes, A. P., Toma, L., Hall, C. & Willock, J. (2009) Farmer perspectives and practices regarding water pollution control programmes in Scotland, *Agricultural Water Management*, 96, pp. 1715–1722.

Bryman, A. & Cramer, D. (2004) Constructing variables, in: M. Hardy & A. Bryman (Eds) *Handbook of Data Analysis*, pp. 17–34 (London: Sage).

Damianos, D. & Giannakopoulos, N. (2002) Farmers' participation in agri-environmental schemes in Greece, *British Food Journal*, 104, pp. 261–273.

Davies, B. B. & Hodge, I. D. (2006) Farmers' preferences for new environmental policy instruments: determining the acceptability of cross compliance for biodiversity benefits, *Journal of Agricultural Economics*, 57, pp. 393–414.

Defrancesco, E., Gatto, P., Runge, F. & Trestini, S. (2008) Factors affecting farmers' participation in agri-environmental measures: a northern Italian perspective, *Journal of Agricultural Economics*, 59, pp. 114–131.

Ducos, G., Dupraz, P. & Bonnieux, F. (2009) Agri-environment contract adoption under fixed and variable compliance costs, *Journal of Environmental Planning and Management*, 52, pp. 669–687.

Falconer, K. (2000) Farm-level constraints on agri-environmental scheme participation: a transactional perspective, *Journal of Rural Studies*, 16, pp. 379–394.

He, X., Li, F. & Cao, H. (2008) Factors influencing the adoption of pasture crop rotation in the semiarid area of China's Loess Plateau, *Journal of Sustainable Agriculture*, 32, pp. 161–180.

Horan, R. D., Ribaudo, M. & Abler, D. G. (2001) Voluntary and indirect approaches for reducing externalities and satisfying multiple objectives, in: J. S. Shortle & D. G. Abler (Eds) *Environmental Policies for Agricultural Pollution Control*, pp. 67–84 (New York, NY: CABI Publ.).

Horan, R. D. & Shortle, J. S. (2001) Environmental instruments for agriculture, in: J. S. Shortle & D. G. Abler (Eds) *Environmental Policies for Agricultural Pollution Control*, pp. 19–65 (New York, NY: CABI Publ.).

Ibarra, A. A. & Pérez Espejo, R. (2008) La contaminación agrícola del agua en México: retos y perspectivas, *Problemas del Desarrollo*, 39, pp. 205–215.

Izcara Palacios, S. P. (2002) Farmers and the implementation of the EU Nitrates Directive in Spain, *Sociología Rural*, 38, pp. 144–162.

Jiménez, B. (2006) Irrigation in developing countries using wastewater, *International Review of Environment and Strategy*, 6, pp. 229–250.

Kathuria, V. (2006) Controlling water pollution in developing and transition countries—lessons from three successful cases, *Journal of Environmental Management*, 78, pp. 405–426.

Keraita, B., Drechsela, P. & Konradsen, F. (2008) Perceptions of farmers on health risks and risk reduction measures in wastewater-irrigated urban vegetable farming in Ghana, *Journal of Risk Research*, 11, pp. 1047–1061.

Knowler, D. & Bradshaw, B. (2007) Farmers' adoption of conservation agriculture: a review and synthesis of recent research, *Food Policy*, 32, pp. 25–48.

Kraft, S. E., Lant, C. & Gillman, K. (1996) WQIP: an assessment of its chances for acceptance by farmers, *Journal of Soil and Water Conservation*, 51, pp. 494–498.

Lacroix, A., Beaudoin, N. & Makowski, D. (2005) Agricultural water nonpoint pollution control under uncertainty and climate variability, *Ecological Economics*, 53, pp. 115–127.

Lichtenberg, E. & Zimmerman, R. (1999) Information and farmers' attitudes about pesticides, water quality, and related environmental effects, *Agriculture, Ecosystems and Environment*, 73, pp. 227–236.

Perret, S. R. & Stevens, J. B. (2006) Socio-economic reasons for the low adoption of water conservation technologies by smallholder farmers in southern Africa: a review of the literature, *Development Southern Africa*, 23, pp. 461–476.

Sandoval-Minero, R. & Almeida-Jara, R. (2006) Public policies for urban wastewater treatment in Guanajuato, in: A. K. Biswas, C. Tortajada, B. Braga & D. J. Rodriguez (Eds) *Water Quality Management in the Americas* (Berlin & Heidelberg: Springer & ANA).

Shortle, J. S., Abler, D. G. & Ribaudo, M. (2001) Agriculture and water quality: the issues, in: J. S. Shortle & D. G. Abler (Eds) *Environmental Policies for Agricultural Pollution Control*, pp. 1–18 (New York, NY: CABI Publ.).

Vanslembrouck, I., van Huylenbroeck, G. & Verbeke, W. (2002) Determinants of the willingness of Belgian farmers to participate in agri-environmental measures, *Journal of Agricultural Economics*, 53, pp. 489–511.

Wei, Y. P., White, R. E., Chen, D., Davidson, B. A. & Zhang, J. B. (2007) Farmers' perception of sustainability for crop production on the north China plain, *Journal of Sustainable Agriculture*, 30, pp. 129–147.

Wossink, A. (2005) Farm policy developments and emerging research issues: an international perspective, in: APRN Workshop on Farm Level Policy (Edmonton, AB: University of Alberta).

Wossink, A. & van Wenum, J. H. (2003) Biodiversity conservation by farmers: analysis of actual and contingent participation, *European Review of Agricultural Economics*, 30, pp. 461–485.

Young, T. F. & Karkoski, J. (2000) Green evolution: are economic incentives the next step in nonpoint source pollution control?, *Water Policy*, 2, pp. 151–173.

Index

Page numbers in *Italics* represent tables.
Page numbers in **Bold** represent figures.

Abdel-Dayem, S. 181–225
accountability 113
acidity 20
aggregates extraction 143
agri-environmental payments *47*
Agricultural Outlook projections 50
Agricultural Policy Reform Program (APRP) **187**
agricultural water pollution 34–49; payments 46–8; taxes 48–9
agriculture 22–3; activities 44; land area **41**; permits 49; research 49; runoff 56; total emissions **39**
Agriculture & Agri-Food Canada: national monitoring network **192**
agrifood uses 131–3
Agro-Environmental Measures (AEMs) 263–4
agro-environmental policy 270–1; perception 266
agro-environmental programmes 264–5
Almozara: flow volumes *155*; water quality *154*
Anahuac: Valley of 247
animal waste treatment plan 109
aquaculture 143
Aragon: authorities **120**; demographic figures *102*; Regional Government of 103; wastewater treatment plans management model **112**; water treatment plans 110–11
Aragon Imperial Canal 150, 155; water quality *155*
Aragon Ministry of Environment of the Government 1
Aragon Wastewater Treatment Plans **107**
Aragon Water Commission (CAA) 113
Aragon water governance 101–16
Aragon Water Institute 105
Aragon Water Plan 105
Argentina 220–1
Asian countries 84
Australia 57–60
Australian Constitution 57

Australian Great Barrier Reef 39–40
Automatic Hydrological Information System (SAIH) 144
Autonomous Communities 121, 131, 240

Baltic Sea 79
Besen, R.: and Jacobi, P. 224
Best Management Practice (BMPs) 263
biochemical oxygen demand 256
biological community 31
biomass production 15
Biswas, A.K. 1; and Tortajada, C. 5–11
Bolivia 221
Botswana 95
Boyd, L.: and Tompkins, R. 203–17
Brazil 221
Breede Water Management Area (WMA) 213
Bronze Age 25
built-operate-transfer (BOT) 89
business process form *214*

Cairo urban population 182
California Legislature 56
Camp de Tarragona 130
capacity building 116
Catchment Management Agency (CMA) 61, 63
Catchment Management Strategy (CMS) 63
Celma, J. 149–65
CEMAS network 124
Central Pollution Control Board (CPCB) 64, 93
Checklist for Public Action 93
chemicals: bombs 15; contamination evolution **28**; oxygen demand 256; time bombs 24–5; waste sites 22
Chile 96
China: development 169; government 93; water quality management 167–79; water-related environmental issues 168
Chinese Academy of Sciences (CAS) 172
cholera explosion 29–30
Clark, M.: *et al* 92

INDEX

Clean Water Act (CWA) 54
climate change 76–7, 104
combined division responsibility (CDR) 170
Committee of Competent Authorities 240
community engagement 51
concentration variation 25
confidence scales 104
consumption management 152
contaminated consumption 223
Contaminated Land Management Act (CLMA) 60
contaminations 6
COP index *231*; carbonate water bodies **238**
Cortés, F.I.A.: and Maravilla, E.M. 245–61
Critical Control Points (CCPs) 212; definition *215*
Critical Risk Factors (CRFs) 211–12; definition *215*
cycling water system **14**

Dairying for Tomorrow (Dft) 58–9
Damianos, D.: and Giannakopoulos, N. 270
decentralization process 253
Decision-Making System (SAD) 144
Defrancesco, E.: et al 264
degradation 26
Denmark 23; nutrient pollution 42; water prices 42
Department of Agriculture, Forestry and Fisheries (DAFF) 62
Department of Environment 10–11
Department of Environmental Affairs and Tourism (DEAT) 62
Department of Water Affairs & Forestry 61
developing countries 6
Dhabi, A. 91
Díaz, R.J.: and Rosenberg, R. 71–9
diffuse sources 31
discharge zone processes 22
dissolved oxygen (DO) 71
distribution network 152
domestic consumption 158–9
Draft Indian Standard Requirements for Good Agricultural Prices 2008 (IndiaGAP) 65
Drainage Research Institute (DRI) 188
drainage water 184
DRASTIC index **237**
Drinking Water Protected Areas (DWPAs) 241
drought management 144

Earth system controls 26
Ebro Delta areas **127**
Ebro Depression 149
Ebro River Basin 119–47; industrial effluents **142**; institutions *146*; international sections **147**; natural contributions **128**, *129*; power plants **140**; recreational activities **143**; recreational use *143*; water bodies 123
Ebro River Basin Confederation 121, 146
Ebro River Hydrographic Confederation **150**, 153
Ebro Valley: food-farming complex **134**
Ebro Valley Foundation 133
ecohydrological phenomena 13
economy 158
ecosystem 74
effluent management 84
Egypt 94; agricultural drains 198; industrial enterprises 185; map **182**; water quality management 181–225
Egyptian Environmental Affairs Agency (EEAA) 188
energy uses 139
environment 17–18; characteristics 127; deterioration 207; measures 268; objectives 145; problems 56; recovery 26; state evaluation *125*; vulnerability 221
Environment (Protection) Act (EPA) 64
Environment Protection and Heritage Council (EPHC) 58
Environmental Action Plan of Egypt (1992) 191
Environmental Management of Groundwater Resources (EMGR) 190
Environmental Protection Agency 54
Environmental Protection Ministry (EPM) 176
Environmental Quality Standard 168
European Commission 48, 234
European Community Directive 144
European Directive 107
European Parliament 227
European regulation 164
European System of Accounts (ESA) 110
European Union 88, 106, 228; INTERREG Research Projects 115; water policy 122; water protection legislation 239
European Union Environment Programme 151
European Water Framework Directive 48, 107, 111, 122–3, 149, 157

Falkenmark, M. 13–31
farm pollutants 34
farm size groups *267*; Likert scores *268*
federal funding 55
Federal Public Works and Improvement of Water Infrastructure 250
fisheries 76
flood management 143–4
food-farming complex 131
Franceys, R.: and Gerlach, E. 87

INDEX

fresh water availability 181

García, L.: and Riofrío, G. 224
General Law of Paths of Communication 248
Geographical Information System (GIS) 232–3, 241
Geological Survey of Spain (IGME) 236
Gerlach, E.: and Franceys, R. 87
Giannakopoulos, N.: and Damianos, D. 270
global food production needs 50
Global Water Intelligence 92
globalization 115
Graham, S.: et al 53–67
gravel bed hydroponics (GBH) 200
Great Hungarian Plain 23
Great Lakes of North America 39
Greater Cairo area 91
Green Revolution 271
gross added value (VAB) 101
gross nitrogen balance **35**
groundwater 126; abstractions 229; bodies *126*, 230–1; bodies status **135**; nitrate contents 198; overexploitation 22; pollution 24; quality 227–41; types **19**
Guardia Civil barracks 112
Gulf of Mexico 56

health problems 268
heterogeneity 271
High Aswan Dam (HAD) 183; water quality parameters **196–7**
Hochstrat, R.: et al 92
households: problem perception *222*
Howard, G.: et al 89
Huai River Basin 170–2, **176**; location **171**; pollutant discharge amount **173**; river quality classes **175**; sources distribution **174**
Huai River Commission 173, 176
human activity 30, 228; water flow *16*
human consumption 241
human health 29, 186; risk 215
human waste 30
hydrocide 30
hydroelectric reservoirs 139, 140
hydroelectricity 139
hydrogeological criteria 231
Hydrological Basin Plan 136
Hydrological Plan (1996) 128; irrigation **137**
Hydrological Plan (1998): economic impacts *138*; irrigation systems *138*
Hydrological-Administrative Region 246, 255; water stress **256**
hypoxia 71–9; climate drivers *78*; global distribution **72**; population response *75*

Iberian Mountain System 119
immigration 105
implementation management tool 104
income structure 160–1
Indah Water Konsortium (IWK) 90
India 64–6
Indiana Department of Environmental Management 56
industrial activity 141
industrial uses 141
industrial wastewater 195
information sources *235*
infrastructure 130
Infrastructure Improvement Plan 162
inherited water pollution 26
input regulations 45–6
institutions: accountability 212; structures 186
Instruction of Hydrological Planning 128
integrated monitoring 17708
Integrated Irrigation Improvement Irrigation Project (IIIMP) 200
Integrated Plan for the Protection of the Ebro Delta (PIPDE) 142
integrated wastewater treatment plan for Aragon (PIDA) 109
integrated water quality management (IWQM) 204; business process 210–12, **211**; model 212–15; network management chain **210**
international activity 115
International Expo (2008) 164
International Centre for Water and Environment (CIAMA) 1
international community 259–60
International Decade for Action 164
International Decade of Water Supply and Sanitation (IDWSSD) 251
international literature review 203
International Meeting of Experts 2
International Year of Freshwater 249
intrinsic vulnerability 237
investment needs 88–9
irrigated land *136*
irrigation 125
Irrigation District 265
irrigation modernization 138
irrigation projects *137*
Izquierdo, R. 101–16

Jacobi, P.: and Besen, R. 224
Japan: sewered population **7**
Japan Sewage Works Association 7
Johannesburg Summit 7

Kauffmann, C. 83–96

Lake Nasser 183; water quality parameters *195*
land development 23

land fill pollutant **23**
land management 77
land use 13–31; water pollution 22–4
Latin America 8, 87, 94, 219–25; MDG goal 9; poor neighbourhoods 224–5; poor urban areas 225
legal framework 247–8
lethal conditions 75
low-density network 191
lump-sum accounts 159

major urban centres 10
management: approaches 204; framework 208–10; processes 203; programme 55
management units 213; CCPs *216*; CRFs *216*
Mar del Plata **251**; conference 252; meeting **254**
Mar del Plata Plan of Action 252
Marañón-Pimentel, B. 84
Maravilla, E.M.: and Cortés, F.I.A. 245–61
marine waters 42
Martínez-Navarrete, C. 227–41
Massachusetts Department of Environmental Protection 57
Matarraña Irrigation Reservoirs 114
methodology phases 232–4
Mexican agriculture 271
Mexican Gulf 30
Mexican Water Commission 265
Mexico: farmers' perceptions 263–71; irrigation districts **266**; urban and rural population **247**; water management 245–61
Mexico City Metropolitan Area 8
Meybeck, M. 26; river system changes 27
Millenium Development Goals (MDGs) 7
mining 23
Ministry of Agricultural Development and the Fund for Agriculture (FIRA) 265
Ministry of Agriculture and Development (MAD) 248
Ministry of the Environment Countryside and Marine Habitat (MARM) 114
Ministry of Environment and Forests (MoEF) 65
Ministry of Environment Protection (MEP) 168
Ministry of the Environment and Rural and Marine Affairs 236
Ministry of Health and Consumption 236
Ministry of Water Resources and Irrigation (MWRI) 186
Ministry of Water Resources (MWR) 167, 249
Mississippi River Basin 77
Mobile Bay 75
Monitoring and Analyses of Drainage Water Quality (MADWQ) 190

monitoring sites **37**, **38**; pollution **257**
mortality 73
Municipal Institute of Public Health 156
municipal management 212
Municipal Regulations on Efficiency and Integral Management of Water 151
municipal solid waste 185
municipal wastewater 184–5
municipality 209

National Development Plan 259
National Environment Policy 65
National Environment Protection Authority (NEPA) 65–6
National Environment Protection Council (NEPC) 58
National Irrigation Commission (NIC) 248
National Key Water Project 174
National Plan of Sanitation and Wastewater Treatment 131
National Pollution Discharge Elimination System (NPDES) 55
National Water Act (NWA) 61
National Water Commission in Mexico (NWC) 248, 249
National Water Program (2007–2012) 259
National Water Quality Management Strategy (NWQMS) 58
National Water Quality Monitoring network 192–3
National Water Quality Plan 110, 125
National Water Research Center (NWRC) national monitoring network **192**
National Water Resource Strategy (NWRS) 61
National Water Resources Plan (NWRP) 186; water demands *183*; water supplies *184*
National Waters Law (NWL) 248
Native Vegetation Act (2003) 59
natural resources 260
Netherlands 42
New Cairo wastewater-treatment 91
New Jersey: commercial marine fisheries *76*
New South Wales (NSW) 59, *see also* NSW...
New York Bight hypoxia (1976) 73
New York City 89
NEWater Factories 92
Nile Research Institute (NRI) 188
Nile River water quality parameters *195*
nitrate concentrations 169
nitrate contribution **37**
Nitrates Directive 48
nonpoint pollution 133
nonpoint source regulation 57–60, 64–6
nonpoint source water pollution 53–67
Norway 71

INDEX

NSW Department of Environment and Climate Change 60
NSW Department of Environment, Climate Change and Water (DECCW) 59
NSW Diffuse Source Water Pollution Strategy 60
nutrient concentrations 36
nutrient management plans and farms **40**
nutrient management practices 40
nutrient pollution 39
nutrient regulations 46
non-OECD countries 90

OECD countries: sanitation services *87*
Official Government Gazette 255
Official Journal of the Federation 253
Omedas-Margelí, M. 119–47
Onesti, M. 219–25
organic waste 185
Organisation for Economic Co-operation and Development (OECD) 2, 33; commodity projections **50**; Food and Agriculture Organization (FAO) 50; policy objectives 45; water pollution 50–1

Pacific coast 77
Pan American Health Organization 252
Parris, K. 33–52
pathologies 220
Pepper Creek 74
Perez-Espejo, R.: *et al* 263–71
pest management practices 41
pesticides: concentrations 36; regulations 46; use **36**
phosphate concentration **25**
pipeline leak reduction *152*
policy coordination 224
policy implications 223–4
policy instruments 45–6, 49
policy linkages **35**
policy-makers 51
polluted water 169
pollution-abatement 29–30
political commitment 94
population and private sector *91*
portable water coverage: rural areas *258*; urban areas *258*
Porter-Cologne Water Quality Control Act 56
power plants *140*
Priority Action Plan 67
private company 110
private facilities modification 153
private sector participation 90–3
production scale 46
property rights 44
provincial boundary cross-sections 173
Public Utilities Board (PUB) 92
public-private participation (PPP) 110
Puerto Vallarta **246**
Pyrenees Water Treatment Plan 108–9, 114

quality management 168

recharge zone processes 20, **21**
recreational fishers 75
Regulation and Supervision Bureau 91
regulatory framework 95
remote sense (RS) monitoring 177
Research Institute for Groundwater (RIGW) 188
Resource Quality Objectives (RQOs) 61
Resource Water Quality Objectives (RWQOs) 208
Rio de Janeiro Conference 260
Riofrío, G.: and García, L. 224
river basin management plan 238
River Delta surface water monitoring locations **193**
river flow 175
river management mechanisms **176**
River Nile **194**
river quality: changes 26–8; outcome 25–6
river symptoms main causes **27**
river system changes: evolution **27**
Rosenberg, R.: and Díaz, R.J. 71–9
rural sanitation 199–200

safeguard zones **236**, **239**; boundaries 234; conceptual framework **232**; initial delineation 238; territory coverage *239*; working phases **233**
sanitation infrastructure 199
Sanitation tax 103
Sanitation and Waste Water Treatment (PASD) 107
Sanitation and Wastewater plans 125
São Paulo water supply **222**
Second World War 26
service provision 94
sewerage coverage *258*
sewered population **7**
Shortle, J.S.: *et al* 271
Sites of Special Scientific Interest 43
social participation 116
social transformations 102
socio-economic activity 228
Soil Conservation Act (1938) 59
solid waste 184
South Africa: institutional boundary map 213; integrated water quality management 203–17; nonpoint source regulation 61–4; Water Management Areas 209; water resources 215; Western Cape Province 205

South African Department of Water Affairs Drinking Water Quality Framework 205
South African National Water Act 63–4
Southern Border Region XI 246
Spain 229; river basin confederations 121–2; safeguard zones *234*
State Pollution Control Boards (SPCBs) 64
state regulation 66
Statue of Autonomy for Aragon 114
Stockholm Water Prize 1
stream restoration 17
stream tube: categories 19; perception 19–20; pollution **24**; water quality changes **20**; water quality related processes 20
stream water pollution 15–16
streamflow depletion 17
supply systems **132**
support institutions 269
sustainable water management **44**, 51–2
Syndicated River Basin Confederation 121
syndrome development 28
Szesztay, K. 23

Technical Advisory Department 248
technology 153
Texcoco Lake 247
Third World Centre for Water Management 1, 6–7
Tompkins, R.: and Boyd, L. 203–17
Tortajada, C.: and Biswas, A.K. 5–11
total maximum daily loads (TMDLs) 54, 174
Toulouse Treaty 146
tourism 182
transfer-own-transfer (TOT) contracts 92
transferring competences 114
transparency 158
transport phase processes 21–2
trophic transfer 74

Union Cabinet (2006) 65
United Nations Conference on Water 249, 251
United Nations Framework Convention on Climate Change (UNFCCC) 89
United Nations Water Conference (1977) 7
United States of America (USA): nonpoint source regulation 54–7
United States Environmental Protection Agency (USEPA) 88
urban development 220
urban network 164

Veolia Water Solutions & Technologies 91
Virginia Department of Conservation and Recreation (VDCR) 57

Washington Department of Ecology (DOE) 56
waste gas 15
wastewater 83–9; agricultural 184
wastewater treatment infrastructures **199**
wastewater treatment plants (WWTPs) 108
water: abstractions **230**; biological characteristics 192; body and forecast *124*; community 31; consumption 128–30; crisis 5; fractions 18; nonpoint source 263; parallel functions 16; regional comparison **247**; resources 62, 83; sales 206
Water Conservatory Bureaus 176
Water Framework Directive (WFD) 88, 123, 127, 227, 241
Water Institute of Aragon (IAA) 103; interrelationships **116**
Water Law (2002) 93, 170
water management 115; regional governments 102–3
water management areas (WMAs) 63
water management institutions 207
water policy 253; consensus 112–13
water pollution 33–52; abatement **18**; economic costs 42–3; management practice 40–2; mitigation 17; recent trends 34–40; risks 28–30
Water Pollution Control Act 56
water quality 9; changes 13–14; degradation 269; disturbances 16; improvement 152; issues and policy responses 43; management framework 5–11; monitoring 55; parameters **10**, 192; perceptions 269–70; problems 54
Water Quality Management (WQM) 183; improvement **200**; laws 190; national policies 186
Water Resources Ministry (WRM) 169, 176
Water Services Authority (WSA) 209
water supply quality *85–6*
water treatment: coverage **255**, *258*; plants 111
Water Use Cycle 205–7; precipitation **206**
Wei, Y.P.: *et al* 270
wellhead protection areas 229, *236*
Windhoek Goreangab Operating Company (WINGOC) 92
World Health Organization (WHO) 89, 252; Water Safety Plan 211
World Water Forums 102, 249

Xia, J.: *et al* 167–79

Yangtze Rivers 167
Yesa Reservoir 153–5; water quality *155*; water quality improvement *156*

INDEX

Zaragoza: consumption and accounts *159*; domestic accounts *160*; domestic consumption *159*, **163**; domestic consumption income *161*; domestic consumption prices *160*; gross domestic product **162**; income forecast *161*; income structure *161*; industrial accounts *160*; non-domestic consumption *160*; water quality 149–65; water supply **154**, *156*, **163**; water supply improvements *157*

Zaragoza City Council 151

9781138798076